D0571126

A Friendly Guide to Wavelets

To Yadwiga–Wanda and Stanisław Włodek,
to Teofila and Franciszek Kowalik,
and to their children Janusz, Krystyn, Mirek, and Renia.
Risking their collective lives, they took a strangers' infant
into their homes and hearts. Who among us would have
the heart and the courage of these simple country people?

To my mother Cesia, who survived the Nazi horrors,
and to my father Bernard, who did not.
And to the uncounted multitude of others, less fortunate than I,
who did not live to sing their song.

GERALD KAISER

A Friendly Guide to Wavelets

Birkhäuser
Boston · Basel · Berlin
1994

Gerald Kaiser
Department of Mathematics
University of Massachusetts at Lowell
Lowell, MA 01854

Library of Congress Cataloging-in-Publication Data

Kaiser, Gerald.
A friendly guide to wavelets / Gerald Kaiser.
p. cm.
Includes bibliographical references and index.
ISBN 0-8176-3711-7 (acid-free). -- ISBN 3-7643-3711-7 (acid-free)
1. Wavelets. I. Title.
QA403.3.K35 1994 94-29118
515'.2433--dc20 CIP

Printed on acid-free paper

ISBN 0-8176-3711-7
ISBN 3-7643-3711-7
Typeset by Marty Stock, Cambridge, MA.
Printed and bound by Quinn-Woodbine, Woodbine, NJ.
Printed in the U.S.A.

9 8 7 6 5 4

Contents

Part I: Basic Wavelet Analysis

1. Preliminaries: Background and Notation

2. Windowed Fourier Transforms

3. Continuous Wavelet Transforms

4. Generalized Frames: Key to Analysis and Synthesis

5. Discrete Time-Frequency Analysis and Sampling

6. Discrete Time-Scale Analysis

7. Multiresolution Analysis

8. Daubechies' Orthonormal Wavelet Bases

Part II: Physical Wavelets

9. Introduction to Wavelet Electromagnetics

10. Applications to Radar and Scattering

11. Wavelet Acoustics

Preface

This book consists of two parts. Part I (Chapters 1–8) gives a basic introduction to wavelets and the related mathematics. It is designed for a one-semester course on wavelet analysis aimed at graduate students or advanced undergraduates in science, engineering, and mathematics. This part is an outgrowth of lecture notes to just such a course that I have now taught, with variations, for several years at the University of Massachusetts at Lowell. It can also be used as a self-study or reference book by practicing researchers in signal analysis and related areas or by anyone else interested in wavelet theory. Since the expected audience is not presumed to have a high level of mathematical background, much of the needed machinery is developed from the beginning, with emphasis on motivation and explanation rather than mathematical rigor. The underlying mathematical ideas are explained in a conversational, common-sense way rather than the standard definition-theorem-proof fashion. A notation is introduced that makes it possible to present signal analysis in a clean, general, modern mathematical language having its roots in linear algebra. Each chapter ends with a set of straightforward problems designed to drive home the concepts just covered. The only prerequisites for the first part are elementary matrix theory, Fourier series, and Fourier integral transforms.

Part II (Chapters 9–11) consists of original research and is written in a more advanced style. It can be used for a higher-level second-semester course or, when combined with Chapters 1 and 3, as a reference for a research seminar. Its theme may be described as an attempt to unify signal analysis and physics. This quest is based on the observation that the electromagnetic and acoustic waves used as carriers of information lend themselves naturally to a particular space-time form of wavelet analysis. The existence, and virtual uniqueness, of this analysis is guaranteed by the very structure of the differential equations governing these physical waves (Maxwell's equations for electromagnetic waves and the wave equation for acoustic waves). Namely, those equations are invariant under the group of *conformal transformations of space-time*, which includes dilations and translations – the basic operations of wavelet analysis – as a subgroup.

Although wavelet analysis is a young field, several excellent books have already appeared on the subject. *Ten Lectures on Wavelets* by Ingrid Daubechies (SIAM, 1992) is a comprehensive account containing (but not restricted to) the wonderful lectures she gave at the NSF–CBMS wavelet conference held at the University of Lowell in June 1990. I have relied heavily on Ingrid Daubechies' monograph while writing Chapters 5–8 of this book. *An Introduction to Wavelets* by Charles Chui (Academic Press, 1992) is readable and concise, emphasizing especially the important connection with splines. *Wavelets and Operators* by Yves Meyer (Cambridge University Press, 1993) is a translated and updated version of the earlier *Ondelettes et Opérateurs*, by a master of classical analysis who is also one of the founders of wavelet theory. In addition, there exist by now several collections of essays and research papers on wavelets, including *Wavelets and Their Applications*, edited by Mary-Beth Ruskai et al. (Jones and Bartlett, 1992); *Wavelets: A Tutorial in Theory and Applications*, edited by Charles Chui (Academic Press, 1992); *Wavelets: Mathematics and Applications*, edited by John Benedetto and Michael Frazier (CRC Press, 1993); and *Progress in Wavelet Analysis and Applications*, edited by Yves Meyer and Sylvie Roques (Editions Frontières, Gif-sur-Yvette, France, 1993).

Why write yet another book? Having taught several courses in wavelet analysis to a variety of audiences with backgrounds in science, engineering and signal analysis, I discovered that most of my students found the existing books quite difficult – mainly because the presupposed level of mathematical sophistication seemed to be quite high. I believe that some of this difficulty is a language problem. Most scientists, engineers, and applied researchers are unfamiliar with modern mathematical notation and concepts. Thus, a casual reference to "measurable functions" or "$L^2(\mathbf{R})$" can be enough to make an aspiring pilgrim weary. Also, none of the monographs with which I am familiar contain exercises that can be assigned as homework problems, and this makes them less suitable as textbooks. I hope that this volume will fill the need for a book truly aimed at the level of a competent and motivated student or researcher, even if he or she may not be highly trained in modern mathematics.

Numerous graphics illustrate the concepts as they are introduced. The notation and basic concepts needed for a clean and fairly general treatment of wavelet analysis are developed from the beginning without, however, going into such technical detail that the book becomes a mathematics course in itself and loses its stated purpose. There is no attempt to be comprehensive in the coverage; one of the goals of the book is to enable readers to extract more specialized information from the literature for themselves. Nor is there an attempt to be comprehensive in the literature citations. The number of research papers on wavelets and related topics has grown astronomically in the past few years, and it is very difficult for anyone to keep up with all the aspects of this literature. An extensive survey of wavelet literature, including abstracts, was prepared by Stefan Pittner, Josef Scheid, and Christoph Ueberhuber. It is available from the Institute for Applied and Numerical Mathematics, Technical University of Vienna. Another large bibliography on wavelets and digital signal processing was compiled by Reiner Creutzburg and appears in *Progress in Wavelet Analysis and Applications*, edited by Yves Meyer and Sylvie Roques (cited earlier). A survey of the literature on time-frequency methods and related topics is available upon request from Hans Feichtinger at the electronic mail address fei@tyche.mat.univie.at.

Although I have done my very best to make this book accessible, I cannot claim that it is easy reading for everyone. The level of presentation becomes more advanced as the language is established and the concepts internalized. Certain aspects of signal analysis that are usually glossed over are explained carefully here, such as the fundamental algebraic distinction between the space-time domain $\mathbf{R}^n$ and the wave-number-frequency domain $\mathbf{R}_n$ (the latter is the *dual space* of the former; see Section 1.4). What *is* claimed is that, apart from the stated prerequisites of matrix theory, Fourier series, and Fourier integrals, the book is largely self-contained, enabling a diligent reader without much previous mathematical background to understand the concepts sufficiently well to apply them, or to tackle some of the more mathematically oriented books or delve into the research literature.

Every chapter begins with a brief summary of its contents and a list of prerequisites. In all but the first chapter, the prerequisites are an understanding of some of the previous chapters. This should make the book more useful as a reference for working researchers, since they will be able to study their topics of interest without reading the entire volume.

Part I, called *Basic Wavelet Analysis*, consists of Chapters 1–8. A brief description of the contents is as follows: Chapter 1 contains a review of linear algebra, especially the relation between matrices and linear operators. A streamlined formalism, called *star notation*, is developed, which makes the construction of dual bases and resolutions of unity appealing and intuitive. By reinterpreting finite-dimensional vectors as functions on finite sets, the formalism is seen to generalize seamlessly to an infinite number of dimensions. When the functions depend on a discrete variable, it usually suffices to restrict the analysis to spaces of square-summable sequences, giving the so-called ℓ^2 spaces. In order to accommodate functions of a continuous variable such as time signals, the concepts of measure and integration are explained, leading to L^2 spaces. Next, the analysis of periodic functions by Fourier series and the associated resolutions of unity for ℓ^2 spaces are developed. As the periods become infinite, the Fourier series go over to the Fourier integral transforms on L^2 spaces. In Chapters 2 and 3, continuous time-frequency and time-scale (wavelet) analyses are motivated and developed, along with the associated continuous resolutions of unity. In Chapter 4 we introduce the concept of *generalized frames*, which combines the idea of resolutions of unity (both continuous and discrete) with the usual (discrete) notion of *frames.* The continuous resolutions of unity constructed in Chapters 2 and 3 are special cases. In Chapters 5 and 6 we discretize the resolutions of unity of Chapters 2 and 3 to obtain various time-frequency and time-scale *sampling theorems.* In the context of Chapter 4, this amounts to constructing *discrete subframes* of the continuous frames found in Chapters 2 and 3. Chapters 7 and 8 give an algebraic presentation of multiresolution analysis and orthonormal wavelet bases. Section 8.4 introduces a new algorithm for the construction of scaling functions and wavelets from a given filter sequence. This method is similar to diadic interpolation but is somewhat more efficient since no eigen-equation needs to be solved to obtain the initial values. It also inspires a new approach to the construction of multiresolution analyses, in particular orthonormal filter sequences, based on the statistical concept of *cumulants.*

Part II, called *Physical Wavelets*, consists of Chapters 9–11. It attempts to bridge the gap between signal analysis and physics, motivated by the observation that signals are often communicated by electromagnetic or acoustic waves. These waves satisfy Maxwell's equations and the wave equation, respectively, and we show that the structure of those equations implies the existence of *electromagnetic and acoustic wavelets* that can be used as building blocks to compose arbitrary electromagnetic and acoustic waves. The construction of these "physical wavelets" is implemented in a simple and elegant way by means of the *analytic-signal transform*, which extends the physical waves to a complex space-time domain $\mathcal{T}$, the *causal tube.* These extensions, which generalize Gabor's idea of "analytic signals," tend to *unfold* the physical waves, displaying their informational contents. All the physical wavelets in any representation can be obtained from a single "reference wavelet" by conformal space-time transformations, just as one-dimensional wavelets can all be obtained as dilations and translations of a single "mother wavelet." In Chapter 9 this construction and some of its consequences are explored in detail for electromagnetic waves. In Chapter 10 the electromagnetic wavelets are applied to radar and other electromagnetic imaging. My interest in radar was sparked by an invitation to teach a short course on wavelet electrodynamics at the tenth annual ACES (Applied Computational Electromagnetics Society) conference at Monterey, California in March 1994. The naive model I presented there has since undergone considerable evolution. The analytic signal of a given electromagnetic wave plays the role of an extended space-time *cross ambiguity function.* A novel geometric model is proposed for electromagnetic scattering, based on the simple transformation properties of the wavelets under conformal transformations. In Chapter 11 we construct the acoustic wavelets. A major difference between electromagnetic and acoustic waves is that whereas the former can travel in vacuum, the latter need a medium in which to propagate. Since all reference frames in uniform relative motion are equivalent in vacuum, relativity theory requires that Lorentz transformations be represented by unitary operators on the Hilbert space of solutions of Maxwell's equations. No such requirement applies to acoustic waves, since for them the medium determines a unique reference frame. We construct a one-parameter family of nonunitary wavelet representations of the conformal group on spaces of solutions of the wave equation.

I believe the single most exciting result in Part II is the discovery that the physical wavelets Ψ_z, which are solutions of the *homogeneous* Maxwell and wave equations, naturally split into two parts: one is an incoming wavelet Ψ_z^- that gets *absorbed* (or detected) and the other is an outgoing wavelet Ψ_z^+ that gets *emitted* just when the incoming wavelet is absorbed. This splitting is far from trivial, because of the analyticity constraints. Neither Ψ_z^- nor Ψ_z^+ are global solutions of the homogeneous equation. Rather, they each solve the corresponding *inhomogeneous* equation, and the source terms are "currents" given by one-dimensional wavelets, as in Equation (11.49). That establishes a deep connection between the physical wavelets and a certain special class of one-dimensional wavelets, and therefore between the wavelet analysis of physical waves and that of communication signals.

I want to take this opportunity to thank Hans Feichtinger, Chris Heil, Mark Kon, Gilbert Strang, and John Weaver, who read parts of the manuscript and made many excellent suggestions for improvements or pointed out some errors. (Of course, I take full responsibility for any remaining mistakes.) Thanks also to my colleagues Ron Brent, Charlie Byrne, James Graham-Eagle, Yuli Makovoz, Raj Prasad, and Alex Samarov for helpful discussions, and to Ingrid Daubechies for informative e-mail exchanges. Thanks to Waterloo Maple Software for providing me with their product, which was used for some of the computations in Chapter 8, and to Math Works, the makers of MATLAB®, for their generous help with the cover graphic. Several of the graphics in Chapters 3 and 5 were produced by Hans Feichtinger with MATLAB using his "irregular sampling toolbox" (IRSATOL), which will be made public via anonymous ftp on 131.130.22.36. I thank Jerry Doty and Alon Schatzberg for their help in producing some of the other MATLAB graphics. Thanks also to Martin Stock for drawing the mathematical illustrations. I am grateful to Arje Nachman of the Air Force Office of Scientific Research, who gave me valuable and well-timed encouragement in my early, naive attempts to apply electromagnetic wavelets to radar and scattering, and to Marvin Bernfeld for many informative discussions on this subject and for reading parts of Chapter 10. I am especially grateful to Ann Kostant and the staff at Birkhäuser-Boston for their enthusiasm, editorial advice, and great patience throughout this enterprise. Warm thanks to Marty Stock, my TEX maven and friend, for being always available with expert advice and for his speedy and excellent job of cleaning up and correcting the typesetting of this manuscript. Most of all, I thank my wife Lynn for her understanding and support during this arduous project.

Gerald Kaiser

Lowell, Massachusetts

June, 1994

Suggestions to the Reader

The reader unfamiliar with modern mathematics, especially the extensive use of mappings between vector spaces and other compound sets, is advised to read the first chapter several times, since he or she will find the ideas and notation somewhat foreign on first reading. Even mathematical readers may need some repeated exposure to get used to the "star notation," which is introduced in Sections 1.2 and 1.3 and used throughout the book. The exercises at the end of each chapter are generally straightforward. Their main purpose is to familiarize the reader with the ideas and notation introduced in that chapter. Proofs, which tend to be informal, end with the symbol ■. This makes it easy to skip the proof on first reading, if desired.

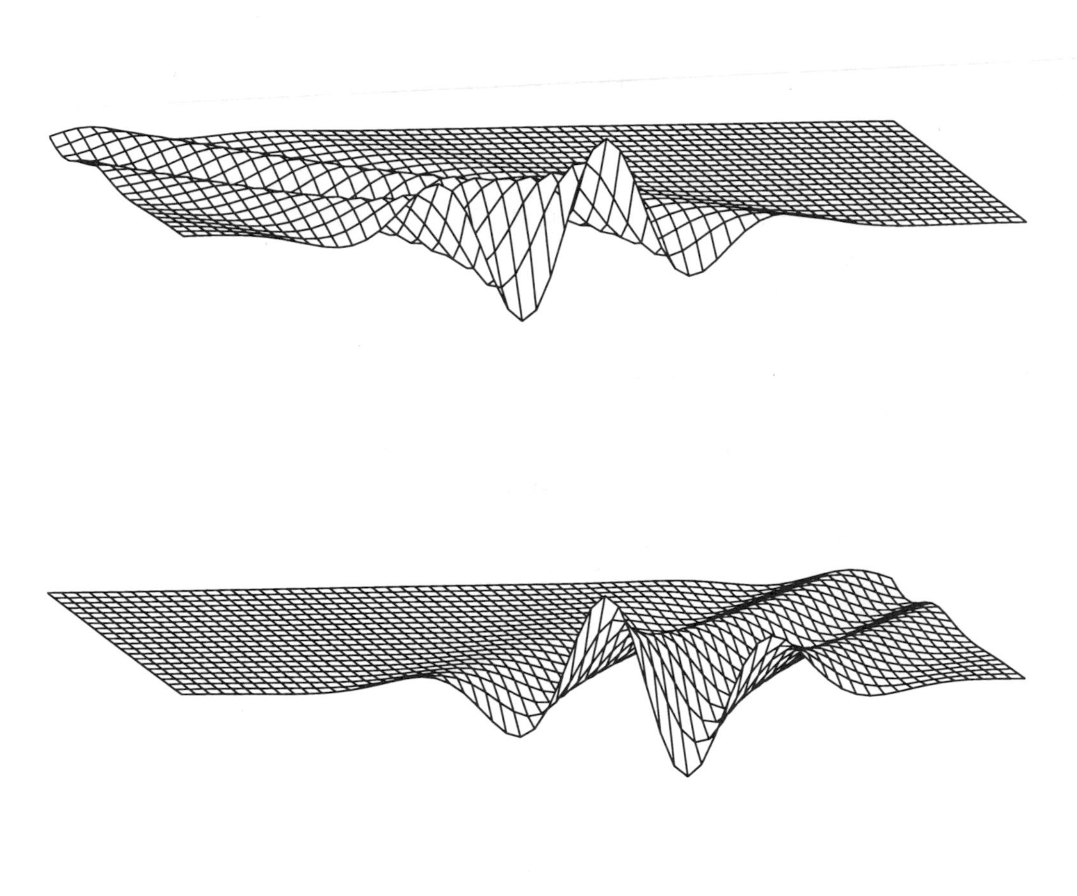

Symbols, Conventions, and Transforms

Unit imaginary: i (not j).
Complex conjugation: $\bar{c}$ (not c^*) for $c \in \mathbf{C}$, $\bar{f}(t)$ for $f(t) \in \mathbf{C}$.
Inner products in $L^2(\mathbf{R})$: $\langle g, f \rangle = \int dt\, \bar{g}(t) f(t)$, $\|f\|^2 = \langle f, f \rangle$.
Hermitian adjoints: A^* for matrices or operators.
Star notation:

For vectors $\mathbf{v} \in \mathbf{C}^N$: $\mathbf{v}^*$ is the Hermitian adjoint, a row vector.
$\mathbf{v}^*\mathbf{u} \equiv \langle \mathbf{v}, \mathbf{u} \rangle$, and $\mathbf{u}\mathbf{v}^*$ is the operator $\mathbf{u}\mathbf{v}^*\mathbf{w} \equiv \mathbf{u}\,(\mathbf{v}^*\mathbf{w})$.
For functions: $g^*f \equiv \langle g, f \rangle$, and fg^* is the operator $fg^*h \equiv f(g^*h)$.

Fourier coefficients and Fourier series for T-periodic functions:

$$c_n = \int_0^T dt\; e^{-2\pi\, int/T} f(t), \quad f(t) = \frac{1}{T} \sum_n c_n\, e^{2\pi\, int/T}.$$

Fourier transform and inverse transform (Chapters 1–8):

$$\text{Analysis:} \quad \hat{f}(\omega) = \int dt\; e^{-2\pi\, i\omega t} f(t)$$

$$\text{Synthesis:} \quad f(t) = \int d\omega\; e^{2\pi\, i\omega t} \hat{f}(\omega).$$

Continuous windowed Fourier transform (WFT) with window g:

$$g_{\omega,t}(u) = e^{2\pi\, i\omega u} g(u-t), \quad C \equiv \|g\|^2 < \infty$$

$$\text{Analysis:} \quad \tilde{f}(\omega, t) = g_{\omega,t}^* f = \int du\; e^{-2\pi\, i\omega u}\, \bar{g}(u-t)\, f(u)$$

$$\text{Synthesis:} \quad f(u) = C^{-1} \iint d\omega\, dt\, g_{\omega,t}(u)\, \tilde{f}(\omega, t)$$

$$\text{Resolution of unity:} \quad C^{-1} \iint d\omega\, dt\, g_{\omega,t}\, g_{\omega,t}^* = I.$$

Continuous wavelet transform (CWT) with wavelet ψ (all scales $s \neq 0$):

$$\psi_{s,t}(u) = |s|^{-1/2} \psi\left(\frac{u-t}{s}\right), \quad C \equiv \int \frac{d\omega}{|\omega|} |\hat{\psi}(\omega)|^2 < \infty$$

$$\text{Analysis:} \quad \tilde{f}(s, t) = \psi_{s,t}^* f = \int du\, \bar{\psi}_{s,t}(u)\, f(u)$$

$$\text{Synthesis:} \quad f(u) = C^{-1} \iint \frac{ds\, dt}{s^2}\, \psi_{s,t}(u)\, \tilde{f}(s, t)$$

$$\text{Resolution of unity:} \quad C^{-1} \iint \frac{ds\, dt}{s^2}\, \psi_{s,t}\, \psi_{s,t}^* = I.$$

PART I

Basic Wavelet Analysis

Chapter 1

Preliminaries: Background and Notation

Summary: In this chapter we review some elements of linear algebra, function spaces, Fourier series, and Fourier transforms that are essential to a proper understanding of wavelet analysis. In the process, we introduce some notation that will be used throughout the book. All readers are advised to read this chapter carefully before proceeding further.

Prerequisites: Elementary matrix theory, Fourier series, and Fourier integral transforms.

1.1 Linear Algebra and Dual Bases

Linear algebra forms the mathematical basis for vector and matrix analysis. Furthermore, when viewed from a certain perspective to be explained in Section 1.3, it has a natural generalization to *functional analysis*, which is the basic tool in signal analysis and processing, including wavelet theory. We use the following notation:

$\mathbf{N} = \{1, 2, 3, \ldots\}$ is the set of all positive integers,

$M \in \mathbf{N}$ means that M belongs to $\mathbf{N}$, i.e., M is a positive integer,

$\mathbf{Z} = \{0, \pm 1, \pm 2, \ldots\}$ is the set of all integers,

$\mathbf{N} \subset \mathbf{Z}$ means that $\mathbf{N}$ is a subset of $\mathbf{Z}$,

$\mathbf{R}$ is the set of all real numbers, and

$\mathbf{C}$ is the set of all complex numbers.

The expression

$$\mathbf{C} = \{x + iy : x, y \in \mathbf{R}\}$$

means that $\mathbf{C}$ is the set of all possible combinations $x + iy$ with x and y in $\mathbf{R}$. Given a positive integer M, $\mathbf{R}^M$ denotes the set of all ordered M-tuples $(u^1, u^2, \ldots, u^M)$ of real numbers, which will be written as column vectors

$$\mathbf{u} = \begin{bmatrix} u^1 \\ u^2 \\ \vdots \\ u^M \end{bmatrix}. \tag{1.1}$$

We often use superscripts rather than subscripts, and these must not be confused with exponents. The difference should be clear from the context. Equation (1.1) shows that the superscript m in u^m denotes a *row index*. $\mathbf{R}^M$ is a real vector space, with vector addition defined by $(\mathbf{u}+\mathbf{v})^k = u^k + v^k$ and scalar multiplication by $(c\,\mathbf{u})^k = cu^k$, $c \in \mathbf{R}$. We will sometimes write the scalar to the *right* of the vector: $\mathbf{u}c = c\,\mathbf{u}$. This will facilitate our "star" notation for resolutions of unity. In fact, we can view the expression $\mathbf{u}c$ as the *matrix product* of the column vector $\mathbf{u}$ ($M \times 1$ matrix) and the scalar c (1×1 matrix). This merely amounts to a change of point of view: Rather than thinking of c as multiplying $\mathbf{u}$ to obtain $c\,\mathbf{u}$, we can think of $\mathbf{u}$ as multiplying c to get $\mathbf{u}\,c$. The result is, of course, the same.

Similarly, the set of all column vectors with M complex entries is denoted by $\mathbf{C}^M$. $\mathbf{C}^M$ is a *complex vector space*, with vector addition and scalar multiplication defined exactly as for $\mathbf{R}^M$, the only difference being that now the scalar c may be complex. A *subspace* of a vector space $\mathbf{V}$ is a subset $\mathbf{S} \subset \mathbf{V}$ that is "closed" under vector addition and scalar multiplication. That is, the sum of any two vectors in $\mathbf{S}$ also belongs $\mathbf{S}$, as does any scalar multiple of any vector in $\mathbf{S}$. For example, any plane through the origin is a subspace of $\mathbf{R}^3$, whereas the unit sphere is not. $\mathbf{R}^M$ is a subspace of $\mathbf{C}^M$, provided we use only real scalars in the scalar multiplication. We therefore say that $\mathbf{R}^M$ is a *real* subspace of $\mathbf{C}^M$. It often turns out that even when analyzing real objects, complex methods are simpler than real methods. (For example, the complex exponential form of Fourier series is formally simpler than the real form using sines and cosines.) For $M = 1$, $\mathbf{R}^M$ and $\mathbf{C}^M$ reduce to $\mathbf{R}$ and $\mathbf{C}$, respectively:

$$\mathbf{C}^1 \equiv \mathbf{C}, \qquad \mathbf{R}^1 \equiv \mathbf{R}.$$

A *basis* for $\mathbf{C}^M$ is a collection of vectors $\{\mathbf{b}_1, \mathbf{b}_2, \ldots, \mathbf{b}_N\}$ such that (a) *any* vector $\mathbf{u} \in \mathbf{C}^M$ can be written as a linear combination of the $\mathbf{b}_n$'s, i.e., $\mathbf{u} = \sum_n c^n\,\mathbf{b}_n$, and (b) there is just *one* set of coefficients $\{c^n\}$ for which this can be done. The scalars c^n are called *the components of* $\mathbf{u}$ *with respect to the basis* $\{\mathbf{b}_n\}$, and they necessarily depend on the choice of basis. It can be shown that any basis for $\mathbf{C}^M$ has exactly M vectors, i.e., we must have $N = M$ above. If $N < M$, then some vectors in $\mathbf{C}^M$ cannot be expressed as linear combinations of the $\mathbf{b}_n$'s and we say that the collection $\{\mathbf{b}_n\}$ is *incomplete*. If $N > M$, then every vector can be expressed in an *infinite* number of different ways, hence the c^m's are not unique. We then say that the collection $\{\mathbf{b}_n\}$ is *overcomplete*. Of course, not *every* collection of M vectors in $\mathbf{C}^M$ is a basis! In order to form a basis, a set of M vectors must be *linearly independent*, which means that no vector in it can be expressed as a linear combination of all the other vectors.

The *standard basis* $\{\mathbf{e}_1, \mathbf{e}_2, \ldots, \mathbf{e}_M\}$ in $\mathbf{C}^M$ (as well as in $\mathbf{R}^M$) is defined as follows: For each fixed m, $\mathbf{e}_m$ is the column vector whose only nonvanishing component is $(\mathbf{e}_m)^m = 1$. Thus

$$(\mathbf{e}_m)^k = \delta^k_m \equiv \begin{cases} 1, & k = m \\ 0, & k \neq m \end{cases}. \tag{1.2}$$

For example, the standard basis for $\mathbf{C}^2$ is $\mathbf{e}_1 = \begin{bmatrix} 1 \\ 0 \end{bmatrix}$, $\mathbf{e}_2 = \begin{bmatrix} 0 \\ 1 \end{bmatrix}$. The symbol $\equiv$ in (1.2) means "equal by definition." Hence, the expression on the right *defines* the symbol δ^k_m, which is called the *Kronecker delta*.† The components of the vector $\mathbf{u}$ in (1.1) with respect to the standard basis of $\mathbf{C}^M$ are just the numbers $u^1, u^2, \ldots, u^M$, as can be easily verified. The standard basis for $\mathbf{C}^M$ is also a standard basis for the subspace $\mathbf{R}^M$.

Example 1.1: General Bases. Let us show that the vectors

$$\mathbf{b}_1 = \begin{bmatrix} 2 \\ 0 \end{bmatrix}, \quad \mathbf{b}_2 = \begin{bmatrix} 1 \\ -1 \end{bmatrix} \tag{1.3}$$

form a basis in $\mathbf{C}^2$, and find the components of $\mathbf{v} = \begin{bmatrix} 3i \\ -1 \end{bmatrix}$ with respect to this basis. To express any vector $\mathbf{u} = \begin{bmatrix} u^1 \\ u^2 \end{bmatrix} \in \mathbf{C}^2$ as a linear combination of $\mathbf{b}_1, \mathbf{b}_2$, we must solve a set of two linear equations:

$$\begin{bmatrix} u^1 \\ u^2 \end{bmatrix} = c^1 \begin{bmatrix} 2 \\ 0 \end{bmatrix} + c^2 \begin{bmatrix} 1 \\ -1 \end{bmatrix}. \tag{1.4}$$

The solution exists and is unique: $c^1 = .5(u^1 + u^2)$, $c^2 = -u^2$. Since $\mathbf{u}$ is an *arbitrary* vector in $\mathbf{C}^2$, this proves that $\mathbf{b}_1, \mathbf{b}_2$ is a basis. The components of the given vector $\mathbf{v}$ are $c^1 = .5(3i - 1) = 1.5i - .5$ and $c^2 = 1$.

Given $M, N \in \mathbf{N}$, a *function* F from $\mathbf{C}^M$ to $\mathbf{C}^N$ is defined by specifying a rule that assigns a vector $F(\mathbf{u})$ in $\mathbf{C}^N$ to each vector $\mathbf{u}$ in $\mathbf{C}^M$. For example,

$$F\left(\begin{bmatrix} u^1 \\ u^2 \end{bmatrix}\right) = \begin{bmatrix} u^1 + 2u^2 \\ u^1 - u^2 \\ 3u^1 + u^2 \end{bmatrix} \tag{1.5}$$

defines a function from $\mathbf{C}^2$ to $\mathbf{C}^3$. The fact that F takes elements of $\mathbf{C}^M$ to elements of $\mathbf{C}^N$ is denoted by the shorthands

$$F : \mathbf{C}^M \to \mathbf{C}^N \quad \text{or} \quad \mathbf{C}^M \xrightarrow{\;F\;} \mathbf{C}^N, \tag{1.6}$$

and the fact that F takes the vector $\mathbf{u} \in \mathbf{C}^M$ to the vector $F(\mathbf{u}) \in \mathbf{C}^N$ is denoted by

$$F : \mathbf{u} \mapsto F(\mathbf{u}).$$

† Later, when dealing with functions, we will also use $\equiv$ to denote "identically equal to." For example, $f(x) \equiv 1$ means that f is the constant function taking the value 1 for all x.

We say that F *maps* $\mathbf{u}$ to $F(\mathbf{u})$, or that $F(\mathbf{u})$ is the *image* of $\mathbf{u}$ under F. F is said to be *linear* if

$$\begin{aligned} &\text{(a)} \quad F(c\,\mathbf{u}) = c\,F(\mathbf{u}) \quad \text{for all } \mathbf{u} \in \mathbf{C}^M \text{ and all } c \in \mathbf{C} \\ &\text{(b)} \quad F(\mathbf{u}+\mathbf{v}) = F(\mathbf{u}) + F(\mathbf{v}) \quad \text{for all } \mathbf{u}, \mathbf{v} \in \mathbf{C}^M. \end{aligned}$$

This means simply that F "respects" the vector space structures of both $\mathbf{C}^M$ and $\mathbf{C}^N$. These two equations together are equivalent to the single equation

$$F(c\,\mathbf{u}+\mathbf{v}) = c\,F(\mathbf{u}) + F(\mathbf{v}) \quad \text{for all } \mathbf{u}, \mathbf{v} \in \mathbf{C}^M \text{ and all } c \in \mathbf{C}.$$

Clearly, the function $F : \mathbf{C}^2 \to \mathbf{C}^3$ in the above example is linear, whereas the function $F : \mathbf{C}^2 \to \mathbf{C}$ given by $F(\mathbf{u}) = u^1 u^2$ is not. If F is linear, it is called an *operator* (we usually drop the word "linear"), and we then write $F\mathbf{u}$ instead of $F(\mathbf{u})$, since operators act in a way similar to multiplication. The *set* of all operators $F : \mathbf{C}^M \to \mathbf{C}^N$ is commonly denoted by $L(\mathbf{C}^M, \mathbf{C}^N)$. We will denote it by $\mathbf{C}_M^N$ for brevity.

Now let $\{\mathbf{a}_1, \mathbf{a}_2, \ldots, \mathbf{a}_M\}$ be any basis for $\mathbf{C}^M$ and $\{\mathbf{b}_1, \mathbf{b}_2, \ldots, \mathbf{b}_N\}$ be any basis for $\mathbf{C}^N$. If $F : \mathbf{C}^M \to \mathbf{C}^N$ is an operator (i.e., $F \in \mathbf{C}_M^N$), then linearity implies that

$$F\mathbf{u} = F\left(\sum_{m=1}^{M} \mathbf{a}_m\, u^m\right) = \left(\sum_{m=1}^{M} F\mathbf{a}_m\right) u^m. \tag{1.7}$$

(Note that we have placed the scalars to the right of the vectors; the reason will become clear later.) Since $F\mathbf{a}_m$ belongs to $\mathbf{C}^N$, it can be written as a linear combination of the basis vectors $\mathbf{b}_n$:

$$F\mathbf{a}_m = \sum_{n=1}^{N} \mathbf{b}_n F_m^n, \tag{1.8}$$

where F_m^n are some complex numbers. Thus

$$F\mathbf{u} = \sum_{m=1}^{M} \sum_{n=1}^{N} \mathbf{b}_n F_m^n\, u^m = \sum_{n=1}^{N} \sum_{m=1}^{M} \mathbf{b}_n F_m^n\, u^m, \tag{1.9}$$

and therefore the components of $F\mathbf{u}$ with respect to the basis $\{\mathbf{b}_n\}$ are

$$(F\mathbf{u})^n = \sum_{m=1}^{M} F_m^n\, u^m. \tag{1.10}$$

Since m is a row index in u^m and n is a row index in $(F\mathbf{u})^n$, the set of coefficients $[F_m^n]$ is an $N \times M$ matrix with row index n and column index m. Conversely, any such matrix, together with a choice of bases, can be used to define a unique operator $F \in \mathbf{C}_M^N$ by the above formula. This gives a *one-to-one correspondence* between operators and matrices. As with vector components, the matrix coefficients F_m^n depend on the choice of bases. We say that the matrix $[F_m^n]$

represents the operator F with respect to the choice of bases $\{\mathbf{a}_m\}$ and $\{\mathbf{b}_n\}$. To complete the identification of operators with matrices, it only remains to show how operations such as scalar multiplication of matrices, matrix addition, and matrix multiplication can be expressed in terms of operators. To do this, we must first define the corresponding operations for operators. This turns out to be extremely natural.

Given two operators $F, G \in \mathbf{C}_M^N$ and $c \in \mathbf{C}$, we define the functions

$$cF : \mathbf{C}^M \to \mathbf{C}^N \quad \text{and} \quad F + G : \mathbf{C}^M \to \mathbf{C}^N$$

by

$$(cF)(\mathbf{u}) \equiv c(F\mathbf{u}), \qquad (F+G)(\mathbf{u}) \equiv F\mathbf{u} + G\mathbf{u}.$$

Then cF and $F+G$ are both linear (Exercise 1.1a,b), hence they also belong to $\mathbf{C}_M^N$. These operations make $\mathbf{C}_M^N$ itself into a complex vector space, whose dimension turns out to be MN. The operators cF and $F+G$ are called the *scalar multiple* of F by c and the *operator sum* of F with G, respectively. Now let $H \in \mathbf{C}_N^K$ be a third operator. The *composition* of F with H is the function $HF : \mathbf{C}^M \to \mathbf{C}^K$ defined by

$$\mathbf{C}^M \xrightarrow{\ F\ } \mathbf{C}^N \xrightarrow{\ H\ } \mathbf{C}^K. \tag{1.11}$$

Then HF is linear (Exercise 1.1c), hence it belongs to $\mathbf{C}_M^K$.

We can now give the correspondence between matrix operations and the corresponding operations on operators: cF is represented by the scalar multiple of the matrix representing F, i.e., $(cF)_m^n = cF_m^n$; $F+G$ is represented by the sum of the matrices representing F and G, i.e., $(F+G)_m^n = F_m^n + G_m^n$; and finally, HF is represented by the *product* of the matrices representing H and F, i.e., $(HF)_m^k = \sum_{n=1}^N H_n^k F_m^n$. Hence the scalar multiplication, addition, and composition of operators corresponds to the scalar multiplication, addition, and multiplication of their representative matrices. For example, the matrix representing the operator (1.5) with respect to the standard bases in $\mathbf{C}^2$ and $\mathbf{C}^3$ is

$$F = \begin{bmatrix} 1 & 2 \\ 1 & -1 \\ 3 & 1 \end{bmatrix}.$$

Since $\mathbf{C} = \mathbf{C}^1$, an operator $F : \mathbf{C} \to \mathbf{C}^N$ is represented by an $N \times 1$ matrix, i.e. a column vector. This shows that any $F \in \mathbf{C}_1^N$ may also be viewed simply as a vector in $\mathbf{C}^N$, i.e., that we may *identify* $\mathbf{C}_1^N$ with $\mathbf{C}^N$. In fact, we have already made use of this identification when we regarded the scalar product $\mathbf{u}\,c$ as the operation of the column vector $\mathbf{u} \in \mathbf{C}^N$ on the scalar $c \in \mathbf{C}$. The *formal* correspondence between $\mathbf{C}^N$ and $\mathbf{C}_1^N$ is as follows: The vector $\mathbf{u} \in \mathbf{C}^N$ corresponds to the operator $F_{\mathbf{u}} \in \mathbf{C}_1^N$ defined by $F_{\mathbf{u}}(c) \equiv \mathbf{u}\,c$. Conversely, the operator $F \in \mathbf{C}_1^N$ corresponds to the vector $\mathbf{u}_F \in \mathbf{C}^N$ given by $\mathbf{u}_F \equiv F(1)$. *We will not distinguish between* $\mathbf{C}_1^N$ *and* $\mathbf{C}^N$*, variously using* $\mathbf{u}$ *to denote* $F_{\mathbf{u}}$ *or* $\mathbf{u}_F$ *to denote* F. This is really another way of saying that we do not distinguish between $\mathbf{u}$ as a column vector and $\mathbf{u}$ as an $N \times 1$ matrix.

As already stated, the matrix $[F_m^n]$ representing a given operator $F : \mathbf{C}^M \to \mathbf{C}^N$ depends on the choice of bases. This shows one of the advantages of operators over matrices: *Operators are basis-independent, whereas matrices are basis-dependent.* A similar situation exists for vectors: They may be regarded in a geometric, basis-independent way as arrows, or they may be represented in a basis-dependent way as columns of numbers. In fact, the above correspondence between operators $F \in \mathbf{C}_1^N$ and vectors $\mathbf{u}_F \in \mathbf{C}^N$ shows that the basis-independent vector picture is just a *special case* of the basis-independent operator picture. If we think of F as dynamically mapping $\mathbf{C}$ into $\mathbf{C}^N$, then the *image* of the interval $[0,1] \subset \mathbf{C}$ under this mapping, i.e., the set

$$F([0,1]) \equiv \{Fu : 0 \le u \le 1\} \subset \mathbf{C}^N,$$

is an arrow from the origin $\mathbf{0} = F(0) \in \mathbf{C}^N$ to $\mathbf{u}_F \equiv F(1) \in \mathbf{C}^N$. Another advantage of operators over matrices is that operators can be readily generalized to an infinite number of dimensions, a fact of crucial importance for signal analysis.

Next, consider an operator $F : \mathbf{C}^N \to \mathbf{C}$, i.e. $F \in \mathbf{C}_N^1$. Because such operators play an important role, they are given a special name. They are called *linear functionals* on $\mathbf{C}^N$. F corresponds to a $1 \times N$ matrix, or row vector: $F = [F_1 \, F_2, \ldots F_N]$. The *set* of all linear functionals on $\mathbf{C}^N$ will be written as $\mathbf{C}_N$. Similarly, the set of linear functionals on $\mathbf{R}^N$ will be written as $\mathbf{R}_N$:

$$\mathbf{C}_N^1 \equiv \mathbf{C}_N, \qquad \mathbf{R}_N^1 \equiv \mathbf{R}_N.$$

Like $\mathbf{C}^N$, $\mathbf{C}_N$ forms a complex vector space. This is just a special case of the fact that $\mathbf{C}_N^K$ forms a vector space, as discussed earlier. Given two linear functionals $F, G : \mathbf{C}^N \to \mathbf{C}$ and a scalar $c \in \mathbf{C}$, we have

$$(c\,F)\mathbf{u} \equiv c\,F\mathbf{u}, \qquad (F+G)\,\mathbf{u} \equiv F\mathbf{u} + G\mathbf{u}.$$

Note that although F and G are (row) vectors, we do not write them in boldface for the same reason we have not been writing operators in boldface: to avoid a proliferation of boldface characters. We will soon have to wean ourselves of boldface characters anyway, since vectors will be replaced by functions of a continuous variable.

Since $\mathbf{C}_N$ consists of row vectors with N complex entries, we expect that, like $\mathbf{C}^N$, it is N-dimensional. This will now be shown directly. Let $\{\mathbf{b}_1, \mathbf{b}_2, \ldots, \mathbf{b}_N\}$ be any basis for $\mathbf{C}^N$ (not necessarily orthogonal). We will construct a corresponding basis $\{B^1, B^2, \ldots, B^N\}$ for $\mathbf{C}_N$ as follows: Any vector $\mathbf{u} \in \mathbf{C}^N$ can be written uniquely as $\mathbf{u} = \sum_{n=1}^N \mathbf{b}_n \, u^n$, where $\{u^n\}$ are now the components of $\mathbf{u}$ with respect to $\{\mathbf{b}_n\}$. For each n, define the function $B^n : \mathbf{C}^N \to \mathbf{C}$ by $B^n(\mathbf{u}) \equiv u^n$. That is, when applied to $\mathbf{u} \in \mathbf{C}^N$, B^n simply gives the (unique!) n-th component of $\mathbf{u}$ with respect to the basis $\{\mathbf{b}_n\}$. It is easy to see that B^n is linear, hence it belongs to $\mathbf{C}_N$. Since $\mathbf{b}_k = \sum_{n=1}^N \mathbf{b}_n \delta_k^n$, the components of the basis vectors $\mathbf{b}_k$ themselves are given by

$$B^n\, \mathbf{b}_k \equiv (\mathbf{b}_k)^n = \delta_k^n\,, \qquad k, n = 1, 2, \ldots, N. \tag{1.12}$$

We now prove that $\{B^1, B^2, \ldots, B^n\}$ is a basis for $\mathbf{C}_N$. To do this, we must show that (a) the set $\{B^1, B^2, \ldots, B^N\}$ *spans* $\mathbf{C}_N$ (i.e., that every linear functional $F \in \mathbf{C}_N$ can be expressed as a linear combination $\sum_{n=1}^N F_n B^n$ of the B^n's with appropriate scalar coefficients F_n) and (b) they are linearly independent. Given any $F \in \mathbf{C}_N$, let $F_n \equiv F(\mathbf{b}_n) \in \mathbf{C}$. Then by the linearity of F and B^n,

$$\begin{aligned} F\mathbf{u} = F\left(\sum_{n=1}^N \mathbf{b}_n u^n\right) &= \sum_{n=1}^N F(\mathbf{b}_n)\, u^n = \sum_{n=1}^N F_n\, u^n \\ &= \sum_{n=1}^N F_n\, B^n\, \mathbf{u} = \left(\sum_{n=1}^N F_n\, B^n\right) \mathbf{u}\,, \end{aligned} \tag{1.13}$$

hence $F = \sum_{n=1}^N F_n\, B^n$, i.e., F is a linear combination of the B^n's with the scalar coefficients F_n. This shows that $\{B^1, B^2, \ldots, B^N\}$ spans $\mathbf{C}_N$. To show that the B^n's are linearly independent, suppose that some linear combination of them is the zero linear functional: $\sum_{n=1}^N c_n B^n = 0$. This means that for *every* $\mathbf{u} \in \mathbf{C}^N$, $\sum_{n=1}^N c_n B^n \mathbf{u} = 0$. In particular, choosing $\mathbf{u} = \mathbf{b}_k$ for any fixed $k = 1, 2, \ldots, N$, (1.12) gives $B^n \mathbf{u} = \delta_k^n$, and it follows that $c_k = 0$, for all k. Thus $\{B^n\}$ are linearly independent as claimed, which concludes the proof that they form a basis for $\mathbf{C}_N$.

Note the complete symmetry between $\mathbf{C}^N$ and $\mathbf{C}_N$: Elements of $\mathbf{C}_N$ are operators $F : \mathbf{C}^N \to \mathbf{C}$, and elements of $\mathbf{C}^N$ may be viewed as operators $\mathbf{u} : \mathbf{C} \to \mathbf{C}^N$. $\mathbf{C}_N$ is called the *dual space* of $\mathbf{C}^N$. We call $\{B^1, B^2, \ldots, B^N\}$ the *dual basis*† of $\{\mathbf{b}_1, \mathbf{b}_2, \ldots, \mathbf{b}_N\}$. The relation (1.12), which can be used to determine the dual basis, is called the *duality relation.* It can also be shown that the dual of $\mathbf{C}_N$ can be identified with $\mathbf{C}^N$; the operator $G_{\mathbf{u}} : \mathbf{C}_N \to \mathbf{C}$ corresponding to a vector $\mathbf{u} \in \mathbf{C}^N$ is defined by $G_{\mathbf{u}}(F) \equiv F\mathbf{u}$. This reversal of directions is the mathematical basis for the notion of duality, a deep and powerful concept.

---

† The term "dual basis" is often used to refer to a basis in $\mathbf{C}^N$ (rather than $\mathbf{C}_N$) that is "biorthogonal" to $\{\mathbf{b}_n\}$. We will call such a basis the *reciprocal basis* of $\{\mathbf{b}_n\}$. As will be seen in Section 1.2, the dual basis is more general since it determines a reciprocal basis once an inner product is chosen.

Example 1.2: Dual Bases. Find the basis of $\mathbf{C}_2$ that is dual to the basis of $\mathbf{C}^2$ given in Example 1.1. Setting $B^1 = [a \quad b]$ and $B^2 = [c \quad d]$, the duality relation implies

$$[a \quad b]\begin{bmatrix}2\\0\end{bmatrix} = 1, \quad [a \quad b]\begin{bmatrix}1\\-1\end{bmatrix} = 0$$
$$[c \quad d]\begin{bmatrix}2\\0\end{bmatrix} = 0, \quad [c \quad d]\begin{bmatrix}1\\-1\end{bmatrix} = 1.$$

This gives $B^1 = [.5 \quad .5]$ and $B^2 = [0 \quad -1]$.

In addition to the duality relation, the pair of dual bases $\{B^n\}$ and $\{\mathbf{b}_n\}$ have another important property. The *identity operator* $I : \mathbf{C}^N \to \mathbf{C}^N$ is defined by $I\mathbf{u} \equiv \mathbf{u}$ for all $\mathbf{u} \in \mathbf{C}^N$. It is represented by the $N \times N$ unit matrix with respect to *any* basis, i.e., $I_k^n = \delta_k^n$. Since every vector in $\mathbf{C}^N$ can be written uniquely as $\mathbf{u} = \sum_{n=1}^N \mathbf{b}_n u^n$ and its components are given by $u^n = B^n \mathbf{u}$, we have

$$\mathbf{u} = \sum_{n=1}^{N} \mathbf{b}_n (B^n \mathbf{u}). \tag{1.14}$$

Regarding $\mathbf{b}_n \in \mathbf{C}^N$ as an operator from $\mathbf{C}$ to $\mathbf{C}^N$, we can write

$$\mathbf{b}_n (B^n \mathbf{u}) = (\mathbf{b}_n B^n)\mathbf{u}, \tag{1.15}$$

where $\mathbf{b}_n B^n : \mathbf{C}^N \to \mathbf{C}^N$ is the composition

$$\mathbf{C}^N \xrightarrow{\;B^n\;} \mathbf{C} \xrightarrow{\;\mathbf{b}_n\;} \mathbf{C}^N. \tag{1.16}$$

Thus

$$\mathbf{u} = \sum_{n=1}^{N} (\mathbf{b}_n B^n)\, \mathbf{u} = \left[\sum_{n=1}^{N} \mathbf{b}_n B^n\right] \mathbf{u}$$

for all $\mathbf{u} \in \mathbf{C}^N$, which proves that the operator in brackets equals the identity operator in $\mathbf{C}^N$:

$$\sum_{n=1}^{N} \mathbf{b}_n B^n = I. \tag{1.17}$$

Equation (1.17) is an example of a *resolution of unity*, sometimes also called a *resolution of the identity*, in terms of the pair of dual bases. The operator $\mathbf{b}_n B^n$ *projects* any vector to its vector component parallel to $\mathbf{b}_n$, and the sum of all these projections gives the complete vector. Let us verify this important relation by choosing for $\{\mathbf{b}_n\}$ the standard basis for $\mathbf{C}^N$:

$$\mathbf{b}_1 = \begin{bmatrix}1\\0\\ \vdots\\0\end{bmatrix}, \ldots, \mathbf{b}_N = \begin{bmatrix}0\\0\\ \vdots\\1\end{bmatrix}. \tag{1.18}$$

Then the duality relation (1.12) gives

$$B^1 = [1 \quad 0 \quad \cdots \quad 0], \quad \ldots, \quad B^N = [0 \quad 0 \quad \cdots \quad 1]. \tag{1.19}$$

Hence the matrix products $\mathbf{b}_n B^n$ are

$$\mathbf{b}_1 B^1 = \begin{bmatrix} 1 & 0 & \cdots & 0 \\ 0 & 0 & \cdots & 0 \\ \vdots & \vdots & \cdots & \vdots \\ 0 & 0 & \cdots & 0 \end{bmatrix}, \quad \ldots, \quad \mathbf{b}_N B^N = \begin{bmatrix} 0 & 0 & \cdots & 0 \\ 0 & 0 & \cdots & 0 \\ \vdots & \vdots & \cdots & \vdots \\ 0 & 0 & \cdots & 1 \end{bmatrix} \tag{1.20}$$

and $\sum_{n=1}^N \mathbf{b}_n B^n = I$. To show that this works equally well for nonstandard bases, we revisit Example 1.2.

Example 1.3: Resolutions of Unity in Dual Bases. For the dual pair of bases in Examples 1.1 and 1.2, we have

$$\begin{aligned} \sum_n \mathbf{b}_n B^n &= \begin{bmatrix} 2 \\ 0 \end{bmatrix} [.5 \quad .5] + \begin{bmatrix} 1 \\ -1 \end{bmatrix} [0 \quad -1] \\ &= \begin{bmatrix} 1 & 1 \\ 0 & 0 \end{bmatrix} + \begin{bmatrix} 0 & -1 \\ 0 & 1 \end{bmatrix} = \begin{bmatrix} 1 & 0 \\ 0 & 1 \end{bmatrix}. \end{aligned}$$

As a demonstration of the conciseness and power of resolutions of unity in organizing vector identities, let $\{\mathbf{d}_n\}$ be another basis for $\mathbf{C}^N$ and let $\{D^n\}$ be its dual basis, so that $\sum_{k=1}^N \mathbf{d}_k D^k = I$. Given any vector $\mathbf{u} \in \mathbf{C}^N$, let $\{u^n\}$ be its components with respect to $\{\mathbf{b}_n\}$ and $\{w^k\}$ be its components with respect to $\{\mathbf{d}_k\}$. Then

$$\begin{aligned} u^n &= B^n \mathbf{u} = B^n I \mathbf{u} = B^n \sum_k \mathbf{d}_k D^k \mathbf{u} = B^n \sum_k \mathbf{d}_k \, w^k \\ &= \sum_k (B^n \mathbf{d}_k) w^k \equiv \sum_k T_k^n \, w^k, \end{aligned} \tag{1.21}$$

which shows that the two sets of components are related by the $N \times N$ transformation matrix $T_k^n = B^n \mathbf{d}_k$.

Resolutions of unity and their generalizations (frames and generalized frames) play an important role in the analysis and synthesis of signals, and this role extends also to wavelet theory. But in the above form, our resolution of unity is not yet a practical tool for computations. Recall that the dual basis vectors were defined by $B^n \mathbf{u} \equiv u^n$. This assumes that we already know how to compute the components u^n of any vector $u \in \mathbf{C}^N$ with respect to $\{\mathbf{b}_n\}$. That is in general not easy: In the N-dimensional case it amounts to solving N simultaneous linear equations for the components. Our resolution of unity (1.17), therefore, does not really solve the problem of expanding vectors in the basis $\{\mathbf{b}_n\}$. To make it practical for computations, we now introduce the notion of *inner products.*

1.2 Inner Products, Star Notation, and Reciprocal Bases

The *standard inner product* in $\mathbf{C}^N$ generalizes the dot product in $\mathbf{R}^N$ and is defined as

$$\langle\, \mathbf{u}, \mathbf{v}\,\rangle \equiv \sum_{n=1}^{N} \bar{u}^n\, v^n\,, \qquad \mathbf{u}, \mathbf{v} \in \mathbf{C}^N, \tag{1.22}$$

where u^n , v^n are the components of $\mathbf{u}$ and $\mathbf{v}$ with respect to the standard basis and $\bar{u}^n$ denotes the complex conjugate of u^n. Note that when $\mathbf{u}$ and $\mathbf{v}$ are real vectors, this reduces to the usual ("Euclidean") inner product in $\mathbf{R}^N$, since $\bar{u}^n = u^n$. The need for complex conjugation arises because it will be essential that the *norm* $\|\mathbf{u}\|$ of $\mathbf{u}$, defined by

$$\|\mathbf{u}\|^2 \equiv \langle\, \mathbf{u}, \mathbf{u}\,\rangle = \sum_{n=1}^{N} |u^n|^2, \tag{1.23}$$

be nonnegative. $\|\mathbf{u}\|$ measures the "length" of the vector $\mathbf{u}$. Note that when $N = 1$, $\|\mathbf{u}\|$ reduces to the absolute value of $\mathbf{u} \in \mathbf{C}$. The standard inner product and its associated norm satisfy the following fundamental properties:

(P) *Positivity*: $\|\mathbf{u}\| > 0$ for all $\mathbf{u} \neq \mathbf{0}$ in $\mathbf{C}^N$, and $\|\mathbf{0}\| = 0$.

(H) *Hermiticity*: $\overline{\langle\, \mathbf{u}, \mathbf{v}\,\rangle} = \langle\, \mathbf{v}, \mathbf{u}\,\rangle$ for all $\mathbf{u}, \mathbf{v} \in \mathbf{C}^N$.

(L) *Linearity*: $\langle\, \mathbf{u}, c\mathbf{v} + \mathbf{w}\,\rangle = c\langle\, \mathbf{u}, \mathbf{v}\,\rangle + \langle\, \mathbf{u}, \mathbf{w}\,\rangle$, $\mathbf{u}, \mathbf{v}, \mathbf{w} \in \mathbf{C}^N$ and $c \in \mathbf{C}$.

Note: $\langle\, \mathbf{u}, \mathbf{v}\,\rangle$ is linear only in its *second* factor $\mathbf{v}$. From (L) and (H) it follows that it is *antilinear* in its first factor:

$$\langle\, c\mathbf{u} + \mathbf{w}, \mathbf{v}\,\rangle = \bar{c}\langle\, \mathbf{u}, \mathbf{v}\,\rangle + \langle\, \mathbf{w}, \mathbf{v}\,\rangle. \tag{1.24}$$

(This is the convention used in the physics literature; in the mathematics literature, $\langle\, \mathbf{u}, \mathbf{v}\,\rangle$ is linear in the first factor and antilinear in the second factor. We prefer linearity in the second factor, since it makes the map $\mathbf{u}^*$, to be defined below, linear.)

Inner products other than the standard one will also be needed, but they all have these three properties. Hence (P), (H), and (L) are *axioms* that must be satisfied by all inner products. They contain everything that is essential, and only that, for any inner product. Note that (L) with $c = 1$ and $\mathbf{w} = 0$ implies that $\langle\, \mathbf{u}, \mathbf{0}\,\rangle = 0$ for all $\mathbf{u} \in \mathbf{C}^N$, hence in particular $\|\mathbf{0}\| = 0$. This is stated explicitly in (P) for convenience. An example of a more general inner product in $\mathbf{C}^N$ is

$$\langle\, \mathbf{u}, \mathbf{v}\,\rangle = \sum_{n=1}^{N} \mu_n \bar{u}^n\, v^n\,, \tag{1.25}$$

where $\mu_1, \mu_2, \ldots, \mu_N$ are fixed positive "weights" and u^n, v^n are the components of $\mathbf{u}, \mathbf{v}$ with respect to the standard basis. This inner product is just as good as the one defined in (1.22) since it satisfies (P), (H), and (L). In fact, an inner product of just this type occurs naturally in $\mathbf{R}^3$ when different dimensions are measured in different units, e.g., width in feet, length in inches, and height in meters. A similar notion of "weights" in the case of (possibly) infinite dimensions leads to the idea of *measure* (see Section 1.5). Given any inner product in $\mathbf{C}^N$, the vectors $\mathbf{u}$ and $\mathbf{v}$ are said to be *orthogonal with respect to this inner product* if $\langle \mathbf{u}, \mathbf{v} \rangle = 0$. A basis $\{\mathbf{b}_n\}$ of $\mathbf{C}^N$ is said to be *orthonormal* with respect to the inner product if $\langle \mathbf{b}_k, \mathbf{b}_n \rangle = \delta_n^k$. Hence, *the concept of orthogonality is relative to a choice of inner product, as is the concept of length.* It can be shown that given *any* inner product in $\mathbf{C}^N$, there exists a basis that is orthonormal with respect to it. (In fact, there is an infinite number of such bases, since any orthonormal basis can be "rotated" to give another orthonormal basis.) The standard inner product is distinguished by the fact that with respect to it, the standard basis is orthonormal.

One of the most powerful concepts in linear algebra, whose generalization to function spaces will play a fundamental role throughout this book, is the concept of adjoints. Suppose we fix any inner products in $\mathbf{C}^M$ and $\mathbf{C}^N$. The *adjoint* of an operator $F : \mathbf{C}^M \to \mathbf{C}^N$ is an operator

$$F^* : \mathbf{C}^N \to \mathbf{C}^M$$

satisfying the relation

$$\langle \mathbf{u}, F^*\mathbf{v} \rangle_{\mathbf{C}^M} = \langle F\mathbf{u}, \mathbf{v} \rangle_{\mathbf{C}^N} \quad \text{for all} \quad \mathbf{v} \in \mathbf{C}^N, \mathbf{u} \in \mathbf{C}^M. \tag{1.26}$$

The subscripts are a reminder that the inner products are the particular ones chosen in the respective spaces. Note that (1.26) is *not a definition* of F^*, since it does not tell us directly how to find $F^*\mathbf{v}$. Rather, it is a nontrivial consequence of the properties of inner products that such an operator exists and is unique. We state the theorem without proof. See Akhiezer and Glazman (1961).

Theorem 1.4. *Choose any inner products in $\mathbf{C}^M$ and $\mathbf{C}^N$, and let $F : \mathbf{C}^M \to \mathbf{C}^N$ be any operator. Then there exists a unique operator F^* satisfying (1.26).*

To get a concrete picture of adjoints, choose bases $\{\mathbf{a}_1, \mathbf{a}_2, \ldots, \mathbf{a}_M\}$ and $\{\mathbf{b}_1, \mathbf{b}_2, \ldots, \mathbf{b}_N\}$, which are orthonormal with respect to the given inner products. Relative to these two bases, F is represented by an $N \times M$ matrix $[F_m^n]$ and F^* is represented by an $M \times N$ matrix $[(F^*)_n^m]$. We now compute both of these matrices in order to see how they are related. The definitions of the two matrices are given by (1.8):

$$F\,\mathbf{a}_m = \sum_n \mathbf{b}_n F_m^n, \qquad F^*\,\mathbf{b}_n = \sum_m \mathbf{a}_m (F^*)_n^m. \tag{1.27}$$

$\ldots, \mathbf{b}^N\}$ the *reciprocal basis*† of $\{\mathbf{b}_1, \mathbf{b}_2, \ldots, \mathbf{b}_N\}$. In fact, the relation between $\{\mathbf{b}_n\}$ and $\{\mathbf{b}^n\}$ is completely symmetrical; $\{\mathbf{b}_n\}$ is also the reciprocal basis of $\{\mathbf{b}^n\}$, since (1.38), which *characterizes* the reciprocal basis, implies $\langle\, \mathbf{b}_n\,, \mathbf{b}^k \,\rangle = \delta_n^k$. The biorthogonality relation generalizes the idea of an orthonormal basis: $\{\mathbf{b}_n\}$ is orthonormal if and only if $\langle\, \mathbf{b}_k, \mathbf{b}_n \,\rangle = \delta_n^k$, which means that $\langle\, \mathbf{b}^k - \mathbf{b}_k\,, \mathbf{b}_n \,\rangle = 0$ for $k, n = 1, 2, \ldots, N$, hence $\mathbf{b}^k = \mathbf{b}_k$. Thus, *a basis in* $\mathbf{C}^N$ *is orthonormal if and only if it is self-reciprocal.*

Given an inner product and a basis $\{\mathbf{b}_n\}$ in $\mathbf{C}^N$, the relation $B^n = (\mathbf{b}^n)^*$ between the dual and the reciprocal bases can be substituted into the resolution of unity (1.17) to yield

$$\sum_{n=1}^{N} \mathbf{b}_n\,(\mathbf{b}^n)^* = I. \tag{1.39}$$

This gives an expansion of any vector in terms of $\mathbf{b}_n$:

$$\mathbf{u} = \sum_{n=1}^{N} \mathbf{b}_n\, u^n, \qquad \text{where} \quad u^n = (\mathbf{b}^n)^*\mathbf{u} = \langle\, \mathbf{b}^n, \mathbf{u} \,\rangle.$$

On the other hand, we can take the adjoint of (1.39) to obtain

$$\sum_{n=1}^{N} \mathbf{b}^n\, \mathbf{b}_n^* = I, \tag{1.40}$$

where we used $(\mathbf{b}_n(\mathbf{b}^n)^*)^* = (\mathbf{b}^n)^{**}\mathbf{b}_n^* = \mathbf{b}^n\mathbf{b}_n^*$, which holds by (1.36) and (1.37). That gives an expansion of any vector in terms of $\mathbf{b}^n$:

$$\mathbf{u} = \sum_{n=1}^{N} \mathbf{b}^n\, u_n, \qquad \text{where} \quad u_n = \mathbf{b}_n^*\mathbf{u} = \langle\, \mathbf{b}_n, \mathbf{u} \,\rangle. \tag{1.41}$$

Both expansions are useful in practice, since the two bases often have quite distinct properties. When the given basis happens to be orthonormal with respect to the inner product, then $\mathbf{b}^n = \mathbf{b}_n$ and the resolutions in (1.39) and (1.40) reduce to the single resolution

$$\sum_{n=1}^{N} \mathbf{b}_n\mathbf{b}_n^* = I, \tag{1.42}$$

with the the usual expansion for vectors, $u^n = \langle\, \mathbf{b}_n\,, \mathbf{u} \,\rangle$.

Example 1.7: Reciprocal (Biorthogonal) Bases. Returning to Example 1.2, choose the standard inner product in $\mathbf{R}^2$. Then the reciprocal basis of

† The usual term in the wavelet literature is *dual basis*, but we reserve that term for the more general basis $\{B^n\}$ of $\mathbf{C}_N$, which determines $\{\mathbf{b}^n\}$ once an inner product is chosen.

$\{\mathbf{b}_1, \mathbf{b}_2\}$ is obtained by taking the ordinary Hermitian conjugates of the row vectors B^1 and B^2, i.e.,

$$\mathbf{b}^1 = (B^1)^* = \begin{bmatrix} .5 \\ .5 \end{bmatrix}, \qquad \mathbf{b}^2 = (B^2)^* = \begin{bmatrix} 0 \\ -1 \end{bmatrix}.$$

Both bases are displayed in Figure 1.1, where the meaning of biorthogonality is also visible. To see how the bases work, we compute the components of $\mathbf{v} = \begin{bmatrix} 3i \\ -1 \end{bmatrix}$ with respect to both bases:

$$\begin{aligned}
\mathbf{v} = \sum_n \mathbf{b}_n (\mathbf{b}^n)^* \mathbf{v} &= \begin{bmatrix} 2 \\ 0 \end{bmatrix} [.5 \quad .5] \begin{bmatrix} 3i \\ -1 \end{bmatrix} + \begin{bmatrix} 1 \\ -1 \end{bmatrix} [0 \quad -1] \begin{bmatrix} 3i \\ -1 \end{bmatrix} \\
&= \begin{bmatrix} 2 \\ 0 \end{bmatrix} (1.5i - .5) + \begin{bmatrix} 1 \\ -1 \end{bmatrix} (1) \\
\mathbf{v} = \sum_n \mathbf{b}^n \mathbf{b}_n^* \mathbf{v} &= \begin{bmatrix} .5 \\ .5 \end{bmatrix} [2 \quad 0] \begin{bmatrix} 3i \\ -1 \end{bmatrix} + \begin{bmatrix} 0 \\ -1 \end{bmatrix} [1 \quad -1] \begin{bmatrix} 3i \\ -1 \end{bmatrix} \\
&= \begin{bmatrix} .5 \\ .5 \end{bmatrix} (6i) + \begin{bmatrix} 0 \\ -1 \end{bmatrix} (3i + 1).
\end{aligned}$$

Note that in this case, neither basis is (self-) orthogonal, but that does not make the computation of the components any more difficult, since that computation was already performed, once and for all, in finding the biorthogonal basis!

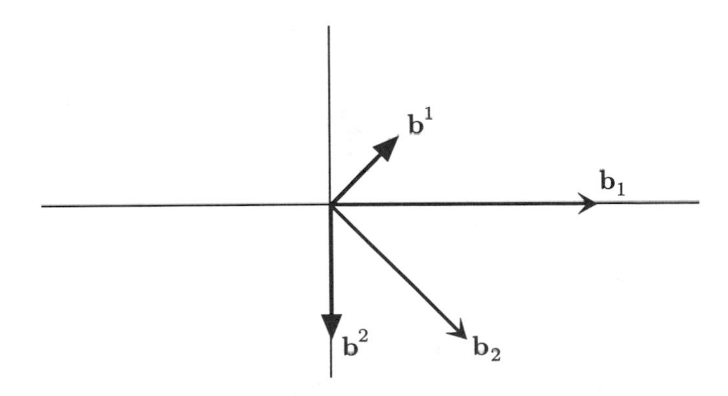

Figure 1.1. Reciprocal Bases. Biorthogonality means that $\mathbf{b}^1$ is orthogonal to $\mathbf{b}_2$, $\mathbf{b}^2$ is orthogonal to $\mathbf{b}_1$, and the inner products of $\mathbf{b}^1$ with $\mathbf{b}_1$ and $\mathbf{b}^2$ with $\mathbf{b}_2$ are both 1 (that gives the normalization).

We refer to the notation in (1.39) and (1.40) and (1.42) as *star notation.* It is particularly useful in the setting of generalized frames (Chapter 4), where the idea of resolutions of unity is extended far beyond bases to give a widely applicable method for expanding functions and operators. Its main advantage is its simplicity. In an expression otherwise loaded with symbols and difficult concepts, it is very important to keep the clutter to a minimum so that the mathematical message can get through.

The significance of the reciprocal basis is that it allows us to compute the coefficients of $\mathbf{u}$ *with respect to a (not necessarily orthogonal) basis by taking inner products – just as in the orthogonal case, but using the reciprocal vectors.* For this to be of practical value, we must be able to actually *find* the reciprocal basis. One way to do so is to solve the biorthogonality relation, i.e., find $\{\mathbf{b}^1, \mathbf{b}^2, \ldots, \mathbf{b}^N\}$ such that (1.38) is satisfied. (The properties (P), (H), and (L) of the inner product imply that there is one and only one such set of vectors!) A more systematic approach is as follows: Define the *metric operator*† $G : \mathbf{C}^N \to \mathbf{C}^N$ by

$$G \equiv \sum_{n=1}^{N} \mathbf{b}_n \mathbf{b}_n^*. \tag{1.43}$$

The positivity condition (P) implies (using $u_n \equiv \langle \mathbf{b}_n, \mathbf{u} \rangle$, which does not involve $\mathbf{b}^n$) that

$$\langle \mathbf{u}, G\mathbf{u} \rangle = \sum_{n=1}^{N} |u_n|^2 > 0 \quad \text{for all} \quad \mathbf{u} \neq \mathbf{0}. \tag{1.44}$$

An operator with this property is said to be *positive-definite.* It can be shown that any positive-definite operator on $\mathbf{C}^N$ has an inverse. This inverse can be computed by matrix methods if we represent G as an $N \times N$ matrix with respect to any basis, such as the standard basis. To construct $\mathbf{b}^n$, note first that

$$G\mathbf{b}^k = \sum_{n=1}^{N} \mathbf{b}_n \mathbf{b}_n^* \mathbf{b}^k = \sum_{n=1}^{N} \mathbf{b}_n \, \delta_n^k = \mathbf{b}_k. \tag{1.45}$$

Hence the reciprocal basis is given by

$$\mathbf{b}^n = G^{-1}\mathbf{b}_n, \qquad n = 1, 2, \ldots, N. \tag{1.46}$$

Note that (1.46) actually represents $\{\mathbf{b}^n\}$ as a generalized "reciprocal" of $\{\mathbf{b}_n\}$. (The individual basis vectors have no reciprocals, of course, but the basis as a whole does!) A generalization of the operator G will be used in Chapter 4 to compute the *reciprocal frame* of a given (generalized) frame in a Hilbert space.

We now state, without proof, two important properties shared by all inner products and their associated norms. Both follow directly from the general properties (P), (H), and (L) and have obvious geometric interpretations.

Triangle inequality: $\|\mathbf{u} + \mathbf{v}\| \leq \|\mathbf{u}\| + \|\mathbf{v}\|$ for all $\mathbf{u}, \mathbf{v} \in \mathbf{C}^N$.

Schwarz inequality: $|\langle \mathbf{u}, \mathbf{v} \rangle| \leq \|\mathbf{u}\| \, \|\mathbf{v}\|$ for all $\mathbf{u}, \mathbf{v} \in \mathbf{C}^N$.

---

† The name "metric operator" is based on the similarity of $\mathbf{b}_n$ and $\mathbf{b}^n$ to "covariant" and "contravariant" vectors in differential geometry, and the fact that G mediates between them.

Equality is attained in the first relation if and only if $\mathbf{u}$ and $\mathbf{v}$ are proportional by a nonnegative scalar and in the second relation if and only if they are proportional by an arbitrary scalar.

1.3 Function Spaces and Hilbert Spaces

We have seen two interpretations for vectors in $\mathbf{C}^N$: as N-tuples of complex numbers or as linear mappings from $\mathbf{C}$ to $\mathbf{C}^N$. We now give yet another interpretation, which will immediately lead us to a far-reaching generalization of the vector concept. Let $\mathcal{N} = \{1, 2, \ldots, N\}$ be the set of integers from 1 to N, and consider a function $f : \mathcal{N} \to \mathbf{C}$. This means that we have some rule that assigns a complex number $f(n)$ to each $n \in \mathcal{N}$. The *set* of all such functions will be denoted[†] by $\mathbf{C}^{\mathcal{N}}$. Obviously, specifying a function $f \in \mathbf{C}^{\mathcal{N}}$ amounts to giving N consecutive complex numbers $f(1), f(2), \ldots, f(N)$, which may be written as a column vector $\mathbf{u}_f \in \mathbf{C}^N$ with components $(\mathbf{u}_f)^n \equiv f(n)$. This gives a one-to-one correspondence between $\mathbf{C}^{\mathcal{N}}$ and $\mathbf{C}^N$. Furthermore, it is natural to define scalar multiplication and vector addition in $\mathbf{C}^{\mathcal{N}}$ by $(cf)(n) = c\,f(n)$ and $(f+g)(n) \equiv f(n) + g(n)$ for $c \in \mathbf{C}$ and $f, g \in \mathbf{C}^{\mathcal{N}}$. These operations make $\mathbf{C}^{\mathcal{N}}$ into a complex vector space. Finally, the vector operations in $\mathbf{C}^{\mathcal{N}}$ and $\mathbf{C}^N$ correspond under $f \leftrightarrow \mathbf{u}_f$. That is, $cf \leftrightarrow c\,\mathbf{u}_f$ and $f+g \leftrightarrow \mathbf{u}_f + \mathbf{u}_g$. This shows that there is no essential difference between $\mathbf{C}^{\mathcal{N}}$ and $\mathbf{C}^N$, and we may *identify* f with $\mathbf{u}_f$ just as we identified vectors in $\mathbf{C}^N$ with operators from $\mathbf{C}$ to $\mathbf{C}^N$. We write $\mathbf{C}^{\mathcal{N}} \approx \mathbf{C}^N$. Similarly, had we considered only the set of real-valued functions $f : \mathcal{N} \to \mathbf{R}$, we would have obtained a real vector space $\mathbf{R}^{\mathcal{N}} \approx \mathbf{R}^N$. The following point needs to be stressed because it can cause great confusion:

$$f(n) \leftrightarrow u^n \text{ are numbers, not functions or vectors}$$
$$f \leftrightarrow \mathbf{u} \text{ are } \textit{vectors.}$$

It is not customary to use boldface characters to denote functions, even though they are vectors. The same remarks apply when f becomes a function of a continuous variable.

A little thought shows that (a) the reason $\mathbf{C}^{\mathcal{N}}$ could be made into a complex vector space was the fact that the functions $f \in \mathbf{C}^{\mathcal{N}}$ were complex-*valued*, since we used ordinary multiplication and addition in $\mathbf{C}$ to *induce* scalar multiplication and vector addition in $\mathbf{C}^{\mathcal{N}}$, and (b) what made $\mathbf{C}^{\mathcal{N}}$ N-dimensional was the fact

---

[†] This notation actually makes sense for the following reason: Let A and B be *finite* sets with $|A|$ and $|B|$ elements, respectively, and denote the set of all functions from A to B by B^A. Since there are $|A|$ choices for $a \in A$, and for each such choice there are $|B|$ choices for $f(a) \in B$, there are exactly $|B|^{|A|}$ such functions. That is, $|B^A| = |B|^{|A|}$.

that the independent variable n could assume N distinct values. This suggests a generalization of the idea of vectors: Let S be an *arbitrary* set, and let $\mathbf{C}^S$ denote the set of all functions $f : S \to \mathbf{C}$. Define vector operations in $\mathbf{C}^S$ by $(cf)(s) \equiv c\,f(s)$, $(f+g)(s) \equiv f(s)+g(s)$, for $s \in S$. Then $\mathbf{C}^S$ is a complex vector space under these operations. If S has an infinite number of elements, $\mathbf{C}^S$ is an infinite-dimensional vector space. For example, $\mathbf{C}^{\mathbf{R}}$ is the vector space of *all possible* (not necessarily linear!) functions $f : \mathbf{R} \to \mathbf{C}$, which includes all real and complex *time signals.* Thus, $f(t)$ could be the voltage entering a speaker at time $t \in \mathbf{R}$. What makes this idea so powerful is that most of the tools of linear algebra can be extended to spaces such as $\mathbf{C}^S$ once suitable restrictions on the functions f are made. We consider two types of infinite-dimensional spaces: those with a *discrete* set S, such as $\mathbf{N}$ or $\mathbf{Z}$, and those with a *continuous* set S, such as $\mathbf{R}$ or $\mathbf{C}$. The most serious problem to be solved in going to infinite dimensions is that matrix theory involves sums over the index n that now become infinite series or integrals whose convergence is not assured. For example, the inner product and its associated norm in $\mathbf{C}^{\mathbf{Z}}$ might be defined formally as

$$\langle f, g \rangle \equiv \sum_{n=-\infty}^{\infty} \overline{f(n)}\, g(n), \qquad \|f\|^2 \equiv \langle f, f \rangle = \sum_{n=-\infty}^{\infty} |f(n)|^2, \tag{1.47}$$

but these sums diverge for general $f, g \in \mathbf{C}^{\mathbf{Z}}$. The typical solution is to consider only the subset of functions with finite norm:

$$\mathcal{H} \equiv \{f \in \mathbf{C}^{\mathbf{Z}} : \|f\|^2 < \infty\}. \tag{1.48}$$

This immediately raises another question: Is $\mathcal{H}$ still a vector space? That is, is it "closed" under scalar multiplication and vector addition? Clearly, if $f \in \mathcal{H}$, then $fc \in \mathcal{H}$ for any $c \in \mathbf{C}$, since $\|cf\| = |c|\|f\| < \infty$. Hence the question reduces to finding whether $f+g \in \mathcal{H}$ whenever $f, g \in \mathcal{H}$. This is where the triangle inequality comes in. It can be shown that the above infinite series for $\langle f, g \rangle$ and $\|f\|^2$ converge absolutely when $f, g \in \mathcal{H}$ and that $\langle f, g \rangle$ satisfies the axioms (P), (H), and (L). Hence $\|f+g\| \le \|f\| + \|g\| < \infty$ if $f, g \in \mathcal{H}$, which means that $f+g \in \mathcal{H}$. Thus $\mathcal{H}$ is an infinite-dimensional vector space and $\langle f, g \rangle$ defines an inner product on $\mathcal{H}$. When equipped with this inner product, $\mathcal{H}$ is called the *space of square-summable complex sequences* and denoted by $\ell^2(\mathbf{Z})$.

Let us now consider the case where the index set S is continuous, say $S = \mathbf{R}$. A natural definition of the inner product and associated norm is

$$\langle f, g \rangle \equiv \int_{-\infty}^{\infty} dt\ \bar{f}(t)\, g(t), \qquad \|f\|^2 \equiv \int_{-\infty}^{\infty} dt\ |f(t)|^2, \tag{1.49}$$

where we have written $\bar{f}(t) \equiv \overline{f(t)}$ for notational convenience. There are two problems associated with these definitions:

(a) The first difficulty is that the usual Riemannian integrals defining $\langle f, g \rangle$ and $\|f\|^2$ *do not exist* for many (in fact, most) functions $f, g : \mathbf{R} \to \mathbf{C}$. The reason is not merely that the (improper) integrals diverge. Even when the integrals are restricted to a bounded interval $[a, b] \subset \mathbf{R}$, the lower and upper Riemann sums defining them do not converge to the same limit in general! For example, let $q(t) = 1$ if t is rational and $q(t) = 0$ if t is irrational. If we partition $[a, b]$ into n subintervals, then each subinterval contains both rational and irrational numbers, thus all the lower Riemann sums for $\int_a^b dt\, q(t)$ are $R_N^{\text{lower}} = 0$ and all the upper Riemann sums are $R_N^{\text{upper}} = b - a$. Hence $\int_a^b dt\, q(t)$ cannot be defined in the sense of Riemann, and therefore neither can the improper integral $\int_{-\infty}^{\infty} dt\ q(t)$. This difficulty is resolved by generalizing the integral concept along lines originally developed by Lebesgue. Lebesgue's theory of integration is highly technical, and we refer the interested reader to Rudin (1966). *It is not necessary to understand Lebesgue's theory in order to proceed.* For the reader's convenience, we give a brief and intuitive summary of the theory in the appendix to this chapter (Section 1.5). Here, we merely summarize the relevant results. The theory is based on the concept of *measure*, which generalizes the idea of *length*. The Lebesgue measure of a set $A \subset \mathbf{R}$ is, roughly, its total length. Only certain types of functions, called *measurable*, can be integrated. The class of measurable functions is much larger than that for which Riemann's integral is defined. In particular, it includes the function q defined above. The scalar multiples, sums, products, quotients, and complex conjugates of measurable functions are also measurable. Thus if $f, g : \mathbf{R} \to \mathbf{C}$ are measurable, then so are the functions $\bar{f}(t)g(t)$ and $|f(t)|^2$ in (1.49). In particular, the integral defining $\|f\|^2$ makes sense (i.e., the lower and upper Lebesgue sums converge to the same limit), although the result may still be infinite. Therefore, the set of functions

$$\mathcal{H} \equiv \{f : \mathbf{R} \to \mathbf{C} : \ f \text{ is measurable and } \|f\|^2 \equiv \int_{-\infty}^{\infty} dt\ |f(t)|^2 < \infty\} \qquad (1.50)$$

is a well-defined subset of $\mathbf{C}^{\mathbf{R}}$. This will be our basic space of signals of a real variable. (Engineers refer to such signals as having a *finite energy*, although $\|f\|^2$ is not necessarily the *physical* energy of the signal.) It can be shown that for $f, g \in \mathcal{H}$, the integral in (1.49) defining $\langle f, g \rangle$ is absolutely convergent; hence $\langle f, g \rangle$ is also defined. This is a *possible* inner product in $\mathcal{H}$. However, that brings us to the second difficulty.

(b) Even if we restrict our attention to $f, g \in \mathcal{H}$, such that the integrals in (1.49) for $\|f\|^2$ and $\langle f, g \rangle$ exist and are absolutely convergent, $\langle f, g \rangle$ does not define a true inner product in $\mathcal{H}$ because the positivity condition (P) fails. To see this, consider a function $f(t)$ that vanishes at all but a finite number of points $\{t_1, t_2, \ldots, t_N\}$. Then f is measurable, and the set of points where $f(t) \neq 0$ has a total "length" of zero, i.e., it has *measure zero.* We say that $f(t) = 0$ "almost everywhere" and write "$f(t) = 0$ a.e." For such a function, Lebesgue's theory gives $\int_{-\infty}^{\infty} dt\ f(t) = 0$. More generally, if f vanishes at all but a *countable* (discrete) set of points, then $f(t) = 0$ a.e. and $\int_{-\infty}^{\infty} dt\ f(t) = 0$. For example, the function $q(t)$ defined in difficulty (a) above vanishes at all but the rational numbers, and these form a countable set. Hence $q(t) = 0$ a.e., and it follows that $\|q\|^2 = 0$, even though q is not the zero-function. (The situation is even worse: There also exist sets with zero measure and an *uncountably infinite* number of elements!) This proves that the proposed "inner product" (1.49) violates the positivity condition. Why worry? Because the positivity axiom will play an essential role in proving of many necessary properties. The solution is to regard any two measurable functions f and g as *identical* if the set of points on which they differ, i.e.,

$$D \equiv \{t \in \mathbf{R} : f(t) \neq g(t)\}, \tag{1.51}$$

has measure zero. Generalizing the above notation, we then say that $f(t) = g(t)$ *almost everywhere*, written "$f(t) = g(t)$ a.e." If f and g belong to $\mathcal{H}$, we then regard them as the *same vector* and write $f = g$. (For this to work, certain consistency relations must be verified, such as: If $f(t) = g(t)$ a.e. and $g(t) = h(t)$ a.e., then $f(t) = h(t)$ a.e. It must also be shown that identifying functions in this way is compatible with vector addition, etc. All the results come up positive.) Thus $\mathcal{H}$ no longer consists of *single* functions but of sets, or *classes*, of functions that are regarded as identical because they are equal to one another a.e. Therefore, the problems surrounding the interpretation of the inner product (1.49) have a positive and very important resolution: They force us to generalize the whole concept of a function, from a pointwise idea to an "almost everywhere" idea.[†] In particular, the zero-element $0 \in \mathcal{H}$ consists of all measurable functions which vanish a.e. With this provision, $\langle \cdot, \cdot \rangle$ becomes an inner product in $\mathcal{H}$. That is, it satisfies

(P) *Positivity*: $\|f\| > 0$ for all $f \in \mathcal{H}$ with $f \neq 0$, and $\|0\| = 0$

(H) *Hermiticity*: $\overline{\langle f, g \rangle} = \langle g, f \rangle$ for all $f, g \in \mathcal{H}$

(L) *Linearity*: $\langle f, cg + h \rangle = c\langle f, g \rangle + \langle f, h \rangle$ for $f, g, h \in \mathcal{H}$, $c \in \mathbf{C}$.

[†] A further generalization was achieved by L. Schwartz around 1950 in the theory of *distributions*, also called *generalized functions*, which admit very singular objects that do not even fit into Lebesgue's theory, such as the "Dirac delta function." Before Schwartz's theory physicists happily used the delta function, but most mathematicians sneered at it. It is now a perfectly respectable member of the distribution community. See Gel'fand and Shilov (1964).

As in the finite-dimensional case, these properties imply the triangle and Schwarz inequalities

$$\|f+g\| \leq \|f\|+\|g\|, \qquad |\langle f,g\rangle| \leq \|f\|\,\|g\| \tag{1.52}$$

for all $f, g \in \mathcal{H}$, with equality in the second relation if and only if f and g are proportional. The triangle inequality ensures that $\mathcal{H}$ is closed under vector addition, which makes it a complex vector space. $\mathcal{H}$ is called the *space of square-integrable complex-valued functions on* $\mathbf{R}$ and is denoted by $L^2(\mathbf{R})$. The space of square-integrable functions $f: \mathbf{R}^n \to \mathbf{C}$ is defined similarly and denoted by $L^2(\mathbf{R}^n)$.

The spaces $\ell^2(\mathbf{Z})$, $\ell^2(\mathbf{N})$, and $L^2(\mathbf{R})$ introduced above are examples of *Hilbert space.* The general definition is as follows: A Hilbert space is any vector space with an inner product satisfying (P), (H), and (L) that is, moreover, *complete* in the sense that any sequence $\{f_1, f_2, \ldots\}$ in $\mathcal{H}$ for which $\|f_n - f_m\| \to 0$ when m and n both go to ∞ converges to some $f \in \mathcal{H}$ in the sense that $\|f - f_n\| \to 0$ as $n \to \infty$. Such a sequence is called a *Cauchy sequence.* (An example of an incomplete vector space with an inner product is the set $\mathcal{C}$ of *continuous* functions in $L^2(\mathbf{R})$, which forms a subspace. It is easy to construct Cauchy sequences in $\mathcal{C}$ that converge to discontinuous functions, so their limit is not in $\mathcal{C}$.) $\mathbf{C}^N$ is a (finite-dimensional) Hilbert space when equipped with any inner product.

A function $f \in L^2(\mathbf{R})$ is said to be *supported* in an interval $[a,b] \subset \mathbf{R}$ if $f(t) = 0$ a.e. outside of $[a,b]$, i.e., if the set of points outside of $[a,b]$ at which $f(t) \neq 0$ has zero measure. We then write $\operatorname{supp} f \subset [a,b]$. The set of square-integrable functions supported in $[a,b]$ is denoted by $L^2([a,b])$. It is a subspace of $L^2(\mathbf{R})$ and is itself also a Hilbert space. Given an arbitrary function $f \in L^2(\mathbf{R})$, we say that f has *compact support* if $\operatorname{supp} f \subset [a,b]$ for *some* bounded interval $[a,b] \subset \mathbf{R}$.

Given any two Hilbert spaces $\mathcal{H}$ and $\mathcal{K}$ and a function $A: \mathcal{H} \to \mathcal{K}$, we say that A is *linear* if

$$A(cf+g) = cA(f) + A(g) \quad \text{for all } c \in \mathbf{C} \text{ and } f, g \in \mathcal{H}.$$

A is then called an *operator*, and we write Af instead of $A(f)$. But (unlike the finite-dimensional case) linearity is no longer sufficient to make A "nice." A must also be *bounded*, meaning that there exists a (positive) constant C such that $\|Af\|_{\mathcal{K}} \leq C\|f\|_{\mathcal{H}}$ for all $f \in \mathcal{H}$. (Since two Hilbert spaces are involved, we have indicated which norm is used with a subscript, although this is superfluous since only $\|Af\|_{\mathcal{K}}$ makes sense for $Af \in \mathcal{K}$ and only $\|f\|_{\mathcal{H}}$ makes sense for $f \in \mathcal{H}$.) C is called a *bound* for A. The idea of boundedness becomes necessary only when $\mathcal{H}$ is infinite-dimensional, since it can be shown that *every* operator on a finite-dimensional Hilbert space is necessarily bounded (regardless of whether the target space $\mathcal{K}$ is finite- or infinite-dimensional).[†] The composition of two bounded operators is another bounded operator.

The concept of adjoint operators can be generalized to the Hilbert space setting, and it will be used as a basis for extending the star notation to the infinite-dimnsional case. First of all, note that as in the finite-dimensional case, every vector $h \in \mathcal{H}$ may also be regarded as a linear map $h : \mathbf{C} \to \mathcal{H}$, defined by $hc \equiv ch$. This operator is bounded, with bound $\|h\|$ (exercise). On the other hand, h also defines a *linear functional* on $\mathcal{H}$,

$$h^* : \mathcal{H} \to \mathbf{C}, \quad \text{by} \quad h^*g \equiv \langle\, h, g \,\rangle, \quad \text{for all } g \in \mathcal{H}. \tag{1.53}$$

h^* is also bounded, since

$$|h^*g| = |\langle\, h, g \,\rangle| \leq \|h\|\, \|g\| \tag{1.54}$$

by the Schwarz inequality. (The norm in the target space $\mathbf{C}$ of h^* is the absolute value.) Equation (1.54) shows that $\|h\|$ is also a bound for h^*. That is, *every vector in $\mathcal{H}$ defines a bounded linear functional h^*.* As in the finite-dimensional case, the reverse is equally true.

Theorem 1.8: (Riesz Representation Theorem). *Let $\mathcal{H}$ be any Hilbert space. Then every bounded linear functional $H : \mathcal{H} \to \mathbf{C}$ can be represented in the form $H = h^*$ for a unique $h \in \mathcal{H}$. That is, Hg can be written as*

$$Hg = h^*g \equiv \langle\, h, g \,\rangle, \quad \textit{for every} \quad g \in \mathcal{H}. \tag{1.55}$$

Theorems 1.4 and 1.6 extend to arbitrary Hilbert spaces, provided we are dealing with bounded operators. The generalizations of those theorems are summarized below.

Theorem 1.9. *Let $\mathcal{H}$ and $\mathcal{K}$ be Hilbert spaces and*

$$A : \mathcal{H} \to \mathcal{K}$$

be a bounded operator. Then the there exists a unique operator

$$A^* : \mathcal{K} \to \mathcal{H}$$

satisfying

$$\langle\, h, A^*k \,\rangle_{\mathcal{H}} = \langle\, Ah, k \,\rangle_{\mathcal{K}} \quad \textit{for all} \quad h \in \mathcal{H}, \ k \in \mathcal{K}. \tag{1.56}$$

[†] Strictly speaking, an unbounded operator from $\mathcal{H}$ to $\mathcal{K}$ can only be defined on a (dense) *subspace* of $\mathcal{H}$, so it is not actually a function from (all of) $\mathcal{H}$ to $\mathcal{K}$. However, such operators must often be dealt with because they occur naturally. For example, the derivative operator $Df(t) \equiv f'(t)$ defines an unbounded operator on $L^2(\mathbf{R})$. The analysis of unbounded operators is much more difficult than that of bounded operators.

A^* *is called the* adjoint *of* A. *The adjoint of* A^* *is the original operator* A. *That is,*

$$A^{**} \equiv (A^*)^* = A. \tag{1.57}$$

Furthermore, if $B : \mathcal{K} \to \mathcal{L}$ *is a bounded operator to a third Hilbert space* $\mathcal{L}$, *then the adjoint of the composition* $BA : \mathcal{H} \to \mathcal{L}$ *is the composition* $A^*B^* : \mathcal{L} \to \mathcal{H}$:

$$(BA)^* = A^*B^*. \tag{1.58}$$

In diagram form,

$$\begin{array}{ccccc} \mathcal{H} & \xrightarrow{A} & \mathcal{K} & \xrightarrow{B} & \mathcal{L} \\ \mathcal{H} & \xleftarrow{A^*} & \mathcal{K} & \xleftarrow{B^*} & \mathcal{L}. \end{array} \tag{1.59}$$

As in the finite-dimensional case, the linear functional h^* defined in (1.53) is a special case of the adjoint operator. (Thus our notation is consistent.)

Example 1.10: Linear Functionals in Hilbert Space. Let ε be a given positive number. For any "nice" function $g : \mathbf{R} \to \mathbf{C}$, let $F_\varepsilon\, g$ be the average of g over the interval $[0, \varepsilon]$:

$$F_\varepsilon\, g \equiv \frac{1}{\varepsilon} \int_0^\varepsilon dt\, g(t). \tag{1.60}$$

First of all, note that F_ε is linear: $F_\varepsilon(cg_1 + g_2) = cF_\varepsilon\, g_1 + F_\varepsilon\, g_2$. $F_\varepsilon\, g$ may be written as the inner product of g with the function

$$\delta_\varepsilon(t) = \begin{cases} 1/\varepsilon & \text{if } 0 \le t \le \varepsilon \\ 0 & \text{otherwise.} \end{cases} \tag{1.61}$$

That is,

$$F_\varepsilon\, f = \int_{-\infty}^{\infty} dt\ \delta_\varepsilon(t) g(t) = \langle\, \delta_\varepsilon\,, g\,\rangle \equiv \delta_\varepsilon^* f. \tag{1.62}$$

(Since $\delta_\varepsilon(t)$ is real, we can omit the complex conjugation.) Now $\delta_\varepsilon \in L^2(\mathbf{R})$, since

$$\|\delta_\varepsilon\|^2 \equiv \int_{-\infty}^{\infty} dt\ \delta_\varepsilon(t)^2 = \frac{1}{\varepsilon} < \infty. \tag{1.63}$$

Therefore, if $g \in L^2(\mathbf{R})$, we have

$$|F_\varepsilon\, g| = |\langle\, \delta_\varepsilon\,, g\,\rangle| \le \|\delta_\varepsilon\|\, \|g\| = \frac{1}{\sqrt{\varepsilon}}\, \|g\|,$$

by the Schwarz inequality. That proves that F_ε is a bounded linear functional on $L^2(\mathbf{R})$ with bound $C = 1/\sqrt{\varepsilon}$. From (1.62) we see that the vector representing F_ε in $L^2(\mathbf{R})$ is δ_ε . Now let us see what happens as $\varepsilon \to 0^+$. If g is a continuous function, then (1.60) implies that

$$F_\varepsilon\, g \to g(0) \quad \text{as} \quad \varepsilon \to 0^+.$$

However, the operator defined by $F_0\, g \equiv g(0)$ cannot be applied to every vector in $L^2(\mathbf{R})$ since the "functions" in $L^2(\mathbf{R})$ do not have well-defined values at individual points. This shows up in the fact that the "limit" of δ_ε is unbounded since $\|\delta_\varepsilon\| = 1/\sqrt{\varepsilon} \to \infty$ as $\varepsilon \to 0^+$. It is possible to make sense of this limit (which is, in fact, the "Dirac delta function" $\delta(t)$) as a *distribution.* The theory of distributions works as follows: Instead of beginning with $L^2(\mathbf{R})$, one considers a space $\mathcal{S}$ of very nice functions: continuous, differentiable, etc.; the details vary from case to case, depending on which particular kinds of distributions one wants. These nice functions are called "test functions." Then distributions are defined as linear functionals on $\mathcal{S}$ that are something like bounded. One no longer has a norm but a "topology," and distributions must be "continuous" linear functionals with respect to this topology. In this case, there is no Riesz representation theorem! Whereas each $f \in \mathcal{S}$ determines a continuous linear functional, there are many more continuous linear functionals than test functions. The set of all such functionals, denoted by $\mathcal{S}'$, is called the distribution space corresponding to $\mathcal{S}$. It consists of very singular "generalized functions" F whose bad behavior is matched by the good behavior of the test functions, such that the "pairing" $\langle F, f \rangle \equiv \int dt\, F(t) f(t)$ still makes sense, even though it can no longer be regarded as an inner product.

1.4 Fourier Series, Fourier Integrals, and Signal Processing

Let $\mathcal{H}_T$ be the set of all measurable functions $f : \mathbf{R} \to \mathbf{C}$ that are *periodic* with period $T > 0$, i.e., $f(t+T) = f(t)$ for (almost) all $t \in \mathbf{R}$, and satisfy

$$\|f\|_T^2 \equiv \int_{t_0}^{t_0+T} dt\, |f(t)|^2 < \infty. \tag{1.64}$$

A periodic function with period T is also called T*-periodic.* The integral on the right-hand side is, of course, independent of t_0 because of the periodicity condition. Nevertheless, we leave t_0 arbitrary to emphasize that the machinery of Fourier series is independent of the location of the interval of integration. $\mathcal{H}_T$ is a Hilbert space under the inner product defined by

$$\langle f, g \rangle_T \equiv \int_{t_0}^{t_0+T} dt\, \bar{f}(t)\, g(t). \tag{1.65}$$

In fact, since every $f \in \mathcal{H}_T$ is determined by its restriction to the interval $[t_0, t_0+T]$, $\mathcal{H}_T$ is essentially the same as $L^2\left([t_0, t_0+T]\right)$. The functions

$$e_n(t) = e^{2\pi\, int/T} = e^{2\pi\, i\omega_n t}, \quad \omega_n \equiv n/T, \quad n \in \mathbf{Z}, \tag{1.66}$$

are T-periodic with $\|e_n\|_T^2 = T$, hence they belong to $\mathcal{H}_T$. In fact, they are orthogonal, with

$$\langle e_n\, , e_m \rangle = T\delta_m^n. \tag{1.67}$$

If a function $f_N \in \mathcal{H}_T$ can be expressed as a finite linear combination of the e_n's, i.e.,

$$f_N = \frac{1}{T} \sum_{n=-N}^{N} c_n \, e_n \,, \tag{1.68}$$

then the orthogonality relation gives

$$c_n = \langle \, e_n \, , f_N \, \rangle_T = \int_{t_0}^{t_0+T} dt \, e^{-2\pi \, i\omega_n t} \, f_N(t). \tag{1.69}$$

Furthermore, the orthogonality relation implies that

$$\|f_N\|_T^2 = \frac{1}{T} \sum_{n=-N}^{N} |c_n|^2. \tag{1.70}$$

Given an *infinite* sequence $\{c_n \,:\, n \in \mathbf{Z}\}$ with $\sum_n |c_n|^2 < \infty$, the sequence of vectors $\{f_1, f_2, f_3, \ldots\}$ in $\mathcal{H}_T$ defined by (1.68) can be shown to be a Cauchy sequence, hence it converges to a vector $f \in \mathcal{H}_T$ (Section 1.3). We therefore write

$$f = \frac{1}{T} \sum_{n=-\infty}^{\infty} c_n \, e_n \,. \tag{1.71}$$

The orthogonality relation (1.67) implies that

$$c_n = \langle \, e_n \, , f \, \rangle_T = \int_{t_0}^{t_0+T} dt \, e^{-2\pi \, i\omega_n t} \, f(t), \tag{1.72}$$

and (1.70) gives

$$\|f\|_T^2 = \frac{1}{T} \sum_{n=-\infty}^{\infty} |c_n|^2 \tag{1.73}$$

in the limit $N \to \infty$. Hence, an infinite linear combination of the e_n's such as (1.71) belongs to $\mathcal{H}_T$ *if and only if* its coefficients satisfy the condition

$$\sum_n |c_n|^2 < \infty, \tag{1.74}$$

i.e., the infinite sequence $\{c_n\}$ belongs to $\ell^2(\mathbf{Z})$. The expansion (1.71) is called the *Fourier series* of $f \in \mathcal{H}_T$, and c_n are called its *Fourier coefficients*.

The Fourier series has a well-known interpretation: If we think of $f(t)$ as representing a *periodic signal*, such as a musical tone, then (1.71) gives a decomposition of f as a linear combination of *harmonic modes* e_n with frequencies ω_n (measured in *cycles per unit time*), including a constant ("DC") term corresponding to $\omega_0 = 0$. All the frequencies are multiples of the *fundamental frequency* $\omega_1 = 1/T$, and this provides a good model for small (linear) vibrations of a guitar string or an air column in a wind instrument, hence the term "harmonic modes." Given $f \in \mathcal{H}_T$, the computation of the Fourier coefficients

c_n is called "analysis" since it amounts to *analyzing* the harmonic content of f. Given $\{c_n\}$, the representation (1.71) is called "synthesis" since it amounts to *synthesizing* the signal f from harmonic modes. Equation (1.73) is interpreted as stating that the "energy" per period of the signal, which is taken to be $\|f\|_T^2$, is distributed among the harmonic modes, the n-th mode containing energy $|c_n|^2/T$ per period.

Fourier series provides the basic model for musical or voice synthesis. Given a signal f, one first analyzes its harmonic content and then synthesizes the signal from harmonic modes using the Fourier coefficients obtained in the analysis stage. We say that the signal has been *reconstructed.* Usually the reconstructed signal is not exactly the same as the original due to errors (round-off or otherwise) or economy. For example, if the original is very complicated in the sense that it contains many harmonics or random fluctuations ("noise"), one may choose to ignore harmonics beyond a certain frequency in the reconstruction stage. An example of *signal processing* is the intentional modification of the original signal by tampering with the Fourier coefficients, e.g., enhancing certain frequencies and suppressing others. An example of *signal compression* is a form of processing in which the objective is to approximate the signal using as few coefficients as possible without sacrificing qualities of the signal that are deemed to be important. All these concepts can be extended in a far-reaching way by replacing $\{e_n\}$ with other sets of vectors, either forming a basis or, more generally, a *frame* (Chapter 4). The choice of vectors to be used in expansions determines the character of the corresponding signal analysis, synthesis, processing, and compression.

Since $\langle\, e_n, f \,\rangle = e_n^* f$, (1.71) and (1.72) can be combined to give

$$f = \frac{1}{T} \sum_{n=-\infty}^{\infty} e_n\, e_n^*\, f \quad \text{for all} \quad f \in \mathcal{H}_T; \tag{1.75}$$

hence we have a resolution of unity in $\mathcal{H}_T$:

$$\frac{1}{T} \sum_{n=-\infty}^{\infty} e_n\, e_n^* = I. \tag{1.76}$$

The functions $T^{-1/2} e_n(t)$ form an orthonormal basis in $\mathcal{H}_T$. However, we will work with the unnormalized set $\{e_n\}$, since that will make the transition to the continuous case $T \to \infty$ more natural.

Choose $t_0 = -T/2$, so that the interval of integration becomes symmetrical about the origin, and consider the limit $T \to \infty$. In this limit the functions are no longer periodic but only satisfy $\int_{-\infty}^{\infty} dt\, |f(t)|^2 < \infty$, hence $\mathcal{H}_T \to L^2(\mathbf{R})$. When T was fixed and finite, we were able to ignore the dependence of the Fourier coefficients c_n on T. This dependence must now be made explicit. Equation (1.72) shows that c_n depends on n only through $\omega_n \equiv n/T$; hence we write it as a function $\hat{f}_T$ of ω_n :

$$c_n = \int_{-T/2}^{T/2} dt\, e^{-2\pi\, i\omega_n t}\, f(t) \equiv \hat{f}_T(\omega_n), \tag{1.77}$$

where the function

$$\hat{f}_T(\omega) \equiv \int_{-T/2}^{T/2} dt\, e^{-2\pi\, i\omega t}\, f(t) \tag{1.78}$$

is defined for *arbitrary* $\omega \in \mathbf{R}$. As $T \to \infty$, $\hat{f}_T(\omega)$ becomes the *Fourier (integral) transform* of $f(t)$:

$$\hat{f}(\omega) \equiv \int_{-\infty}^{\infty} dt\;\, e^{-2\pi\, i\omega t}\, f(t). \tag{1.79}$$

Returning to finite T, note that since $\Delta\omega_n \equiv \omega_{n+1} - \omega_n = 1/T$, (1.71) can be written as

$$f(t) = \sum_{n=-\infty}^{\infty} e^{2\pi\, i\omega_n t}\, \hat{f}_T(\omega_n)\, \Delta\omega_n\,, \tag{1.80}$$

a Riemann sum that in the limit $T \to \infty$ formally becomes the *inverse Fourier transform*

$$f(t) = \int_{-\infty}^{\infty} d\omega\;\, e^{2\pi\, i\omega t}\, \hat{f}(\omega). \tag{1.81}$$

Note the symmetry between the Fourier transform (1.79) and its inverse (1.81). In terms of the interpretation of the Fourier transform in signal analysis, this symmetry does not appear to have a simple intuitive explanation; there seems to be no *a priori* reason why the descriptions of a signal in the time domain and the frequency domain should be symmetrical. Mathematically, the symmetry is based on the fact that $\mathbf{R}$ is self-dual as a locally compact Abelian group; see Katznelson (1976). The "energy relation" (1.73) can be written as

$$\|f\|_T^2 = \sum_{n=-\infty}^{\infty} |\hat{f}_T(\omega_n)|^2 \Delta\omega_n\,, \tag{1.82}$$

again a Riemann sum. In the limit $T \to \infty$, this states that $\hat{f}(\omega)$ belongs to $L^2(\mathbf{R})$ and satisfies

$$\|f\|^2 \equiv \int_{-\infty}^{\infty} dt\; |f(t)|^2 = \int_{-\infty}^{\infty} d\omega\; |\hat{f}(\omega)|^2 = \|\hat{f}\|^2, \tag{1.83}$$

a relation between $f(t)$ and $\hat{f}(\omega)$ known as *Plancherel's theorem.*

The Fourier transform has a straightforward generalization to functions in $L^2(\mathbf{R}^n)$ for any $n \in \mathbf{N}$. Given such a function $f(t)$, where $t = (t^1, t^2, \ldots, t^n) \in \mathbf{R}^n$, we transform it separately in each variable t^k to obtain

$$\hat{f}(\omega_1, \ldots, \omega_n) = \int \cdots \int dt^1 \cdots dt^n \, \exp[-2\pi i \, (\omega_1 t^1 + \cdots + \omega_n t^n)] \, f(t^1, \ldots, t^n). \tag{1.84}$$

The quantity $\omega_1 t^1 + \cdots + \omega_n t^n$ in the exponent is usually interpreted as the *inner product* of $\omega \equiv (\omega_1, \omega_2, \ldots, \omega_n)$ with $t \equiv (t^1, t^2, \ldots, t^n)$, both regarded as vectors in $\mathbf{R}^n$. However, this is not quite correct, since *we have not chosen any inner product in* $\mathbf{R}^n$ *to derive it.* Strictly speaking, we must consider ω as an element of the *dual* $\mathbf{R}_n$ of $\mathbf{R}^n$, i.e., as a *linear functional* $\omega : \mathbf{R}^n \to \mathbf{R}$. Then $\omega_1 t^1 + \cdots + \omega_n t^n = \omega(t) \equiv \omega t$ is the *value* of ω on t.† To make contact with the usual interpretation, we may now choose an arbitrary inner product $\langle \cdot, \cdot \rangle$ in $\mathbf{R}^n$. Then every $\omega \in \mathbf{R}_n$ corresponds to a unique vector $\omega^* \in \mathbf{R}^n$ under this inner product (Section 1.2), with $\omega t = \langle \omega^*, t \rangle$. In particular, when $\langle \cdot, \cdot \rangle$ is the *standard Euclidean* inner product, then ω^* is just the transpose of ω (regarding ω as a row vector and t, ω^* as column vectors), and our interpretation reduces to the usual one.

The above discussion may strike the application-minded reader as unnecessarily esoteric. Why not simply choose the Euclidean inner product in $\mathbf{R}^n$ and avoid all this complication? One reason is that the Euclidean inner product is not always the natural choice. For example, let $n = 4$ and let $t = (t^0, t^1, t^2, t^3) \in \mathbf{R}^4$ represent the coordinates in space-time: t^0 is the time and (t^1, t^2, t^3) is the position vector. Then ω has the following unique interpretation: ω_0 is a *frequency* in cycles per unit time, and $(\omega_1, \omega_2, \omega_3)$ is a *wave vector*, whose components measure the *wave number*, i.e., the number of crests or cycles per unit length, in the x, y, and z directions. Any particular choice of ω corresponds to a unique choice of *plane wave*

$$e_\omega(t) \equiv \exp[2\pi i \, (\omega_0 t^0 + \omega_1 t^1 + \omega_2 t^2 + \omega_3 t^3)] \equiv e^{2\pi i \omega t}. \tag{1.85}$$

The linear map $\omega : \mathbf{R}^4 \to \mathbf{R}$ has the following simple interpretation: $\omega(t) \equiv \omega t$ *measures the number of cycles of* e_ω *passed by an observer when traveling from the origin in* $\mathbf{R}^4$ *to* t. Since that number is independent of the observer's choice of units of length and time, it is also independent of the observer's choice of inner product in $\mathbf{R}^4$. For example, the observer may choose to measure t^0 in seconds, t^1 in meters, t^2 in inches and t^3 in yards. In that case, the inner product will certainly not be the Euclidean one, since conversion factors must be used. But ωt is unaffected, since ω_0 is in cycles per second, ω_1 is in cycles per meter, etc.

---

† This is a special case of *Fourier analysis on groups*, where $\mathbf{R}^n$ is replaced by an arbitrary locally compact Abelian group. In that case, $\mathbf{R}_n$ is replaced by the *dual group*. See Katznelson (1976).

Of course, whatever the observer's choice of inner product, the result must be the same: $\langle\, \omega^*, t\,\rangle = \omega t$. But choosing an inner product and then computing ω^* and $\langle\, \omega^*, t\,\rangle$ is an unnecessary and confusing complication. Furthermore, while the Euclidean inner product is a natural choice in *space* $\mathbf{R}^3$ (corresponding to using the same units to measure x, y, and z), it is certainly *not* a natural choice in space-time, as explained next.

These remarks become especially relevant in the study of electromagnetic or acoustic waves (see Chapters 9–11). Here it turns out that the natural "inner product" in space-time is not the Euclidean one but the *Lorentzian scalar product*

$$\langle\, t, u\,\rangle = c^2 t^0 u^0 - t^1 u^1 - t^2 u^2 - t^3 u^3\,, \qquad t,\, u \in \mathbf{R}^4, \tag{1.86}$$

where c is the propagation velocity of the waves, i.e., the speed of sound or light. (This is not an inner product because it does not satisfy the positivity condition.) This scalar product is "natural" because it respects the space-time symmetry of the wave equation obeyed by the waves, just as the Euclidean inner product respects the rotational symmetry of space (which corresponds to using the same units in all spatial directions). For example, a spherical wave emanating from the origin will have a constant phase along its *wave front* $(t^1)^2 + (t^2)^2 + (t^3)^2 = c^2 (t^0)^2$, which can be written simply as $\langle\, t, t\,\rangle = 0$.

That ω "lives" in a different space from t is also indicated by the fact that it necessarily has different units from t, namely the *reciprocal units*: cycles per unit time for ω_0 and cycles per unit length for ω_1, ω_2, and ω_3, if $\mathbf{R}^n$ is space-time as above. Mathematically, this is reflected in the reciprocal *scaling* properties of ω and t: If we replace t by at for some $a \neq 0$, then we must simultaneously replace ω by $a^{-1}\omega$ in order for $e_\omega(t)$ to be unaffected. Such scaling is also a characteristic feature in wavelet theory.

Using a terminology suggested by the above considerations, we call $\mathbf{R}^n$ the (space-)*time domain* and $\mathbf{R}_n$ the (wave number–)*frequency domain*. Equation (1.84) will be written in the compact form

$$\hat{f}(\omega) \equiv \int_{\mathbf{R}^n} d^n t\, e^{-2\pi\, i\omega t}\; f(t)\,, \tag{1.87}$$

where the integral denotes Lebesgue integration over $\mathbf{R}^n$. Equation (1.87) gives the *analysis* of f into its frequency components. The *synthesis* or *reconstruction* of f can be obtained by applying the inverse transform separately in each variable. That gives

$$f(t) = \int_{\mathbf{R}_n} d^n \omega\, e^{2\pi\, i\omega t}\; \hat{f}(\omega). \tag{1.88}$$

There are three common conventions for writing the Fourier transform and its inverse, differing from one another in the placement of the factor 2π. Since this can be the cause of some confusion, we now give a "dictionary" between them so that the reader can easily compare statements in this book with corresponding statements in other books or the literature.

Convention (1) is the one used here: The factor 2π appears in the exponents of (1.87) and (1.88) and *nowhere else.* This is the simplest convention and, in addition, it makes good sense from the viewpoint of signal analysis: If t represents time or space variables, then ω is just the frequency or wave number, measured in *cycles* per unit time or unit length.

Convention (2): The exponents in (1.87) and (1.88) are $\mp i\omega t$, and there is a factor of $(2\pi)^{-n}$ in front of the integral in (1.88) but no factor in front of (1.87). This notation is a bit more combersome, since some of the symmetry between the Fourier transform and its inverse is now "hidden." However, it still makes perfect sense from the viewpoint of signal analysis: ω now represents frequency or wave number, measured in *radians* per unit time or unit length. We may translate from convention (1) to (2) by substituting $t \to t$ and $\omega \to \omega/2\pi$ in (1.87) and (1.88), which automatically gives the factor $(2\pi)^{-n}$ in front of (1.88). Since the n-dimensional impulse function satisfies $\delta(ax) = |a|^{-n}\delta(x)$ ($0 \neq a \in \mathbf{R}$, $x \in \mathbf{R}^n$), we must also replace $\delta(\omega)$ by $(2\pi)^n\delta(\omega)$.

Convention (3): Again, the exponents in (1.87) and (1.88) are $\mp i\omega t$. But in order to restore the symmetry between the Fourier transform and its inverse, the factor $(2\pi)^{-n/2}$ is placed in front of *both* (1.87) and (1.88). We may translate from convention (1) to (3) by substituting $t \to t/\sqrt{2\pi}$ and $\omega \to \omega/\sqrt{2\pi}$ in (1.87) and (1.88), automatically giving the factor $(2\pi)^{-n/2}$ in front of both integrals. By the above remark, we must also replace $\delta(t)$ and $\delta(\omega)$ by $(2\pi)^{n/2}\delta(t)$ and $(2\pi)^{n/2}\delta(\omega)$, respectively. Although the symmetry is restored, these extra factors (which appear again and again!) are a bother. In the author's opinion, convention (c) also makes the least sense from the signal analysis viewpoint, since now neither t nor ω has a simple interpretation as above.

There is also an n-dimensional version of Plancherel's theorem, which can be derived formally by applying the one-dimensional version to each variable separately.

Theorem 1.11: Plancherel's Theorem. *Let $f \in L^2(\mathbf{R}^n)$. Then $\hat{f} \in L^2(\mathbf{R}_n)$ and*

$$\|f\|^2_{L^2(\mathbf{R}^n)} = \|\hat{f}\|^2_{L^2(\mathbf{R}_n)}\,. \tag{1.89}$$

That is,

$$\int_{\mathbf{R}^n} d^n t\, |f(t)|^2 = \int_{\mathbf{R}_n} d^n\omega\, |\hat{f}(\omega)|^2\,. \tag{1.90}$$

This shows that the Fourier transform gives a one-to-one correspondence between $L^2(\mathbf{R}^n)$ and $L^2(\mathbf{R}_n)$.

The inner product in $L^2(\mathbf{R}^n)$ determines the norm by $\|f\|^2 = \langle f, f \rangle$. It can also be shown that the norm determines the inner product. Using the algebraic identity

$$\bar{a}b = \frac{1}{4}\left(|a+b|^2 - |a-b|^2\right) + \frac{1}{4i}\left(|a+ib|^2 - |a-ib|^2\right), \quad a, b \in \mathbf{C}, \tag{1.91}$$

choosing $a = f(t)$ and $b = g(t)$ and integrating over $\mathbf{R}^n$, we obtain the following result.

Theorem 1.12: Polarization Identity. *For any* $f, g \in L^2(\mathbf{R}^n)$,

$$\langle f, g \rangle = \frac{1}{4}\left(\|f+g\|^2 - \|f-g\|^2\right) + \frac{1}{4i}\left(\|f+ig\|^2 - \|f-ig\|^2\right). \tag{1.92}$$

When combined with Plancherel's theorem, the polarization identity immediately implies the following.

Theorem 1.13: Parseval's Identity. *Let* $f, g \in L^2(\mathbf{R}^n)$. *Then the inner products of* f *and* g *in the time domain and the frequency domain are related by* $\langle f, g \rangle = \langle \hat{f}, \hat{g} \rangle$, *i.e.,*

$$\int_{\mathbf{R}^n} d^n t\ \overline{f(t)}\, g(t) = \int_{\mathbf{R}_n} d^n \omega\ \overline{\hat{f}(\omega)}\, \hat{g}(\omega). \tag{1.93}$$

Although the polarization identity (1.92) was derived here for $L^2(\mathbf{R}^n)$, it also holds for an *arbitrary* Hilbert space (due to the properties (H) and (L) of the inner product, which can be used to replace the identity (1.91)). Hence the norm in any Hilbert space determines the inner product.

Given a function $g \in L^2(\mathbf{R}_n)$ in the *frequency* domain, $\check{g} \in L^2(\mathbf{R}^n)$ will denote its inverse Fourier transform:

$$\check{g}(t) \equiv \int_{\mathbf{R}_n} d^n \omega\ e^{2\pi\, i\omega t}\ g(\omega). \tag{1.94}$$

We will be consistent in using the Fourier transform to go only from the time domain to the frequency domain and the inverse Fourier transform to go the other way. When considering the Fourier transform of a more complicated expression, such as $e^{-t^2} f(t)$, we write $(e^{-t^2} f(t))^\wedge(\omega)$. Similarly, the inverse Fourier transform of $e^{-\omega^2} g(\omega)$ will be written $(e^{-\omega^2} g(\omega))^\vee(t)$. Thus, $(\hat{f})^\vee(t) = f(t)$ and $(\check{g})^\wedge(\omega) = g(\omega)$.

It is tempting to extend the resolution of unity obtained for the Fourier series (1.75) to the case $\mathcal{H} = L^2(\mathbf{R}^n)$. The formal argument goes as follows: Since $\hat{f}(\omega) = \langle e_\omega, f \rangle = e_\omega^* f$, (1.88) gives

$$f(t) = \int_{\mathbf{R}_n} d^n \omega\ e_\omega(t)\, e_\omega^* f, \quad \text{i.e.,} \quad f = \int_{\mathbf{R}_n} d^n \omega\ e_\omega\, e_\omega^* f, \tag{1.95}$$

implying the resolution of unity

$$\int_{\mathbf{R}_n} d^n\omega\; e_\omega\, e_\omega^* = I. \tag{1.96}$$

The trouble with this equation is that e_ω does not belong to $L^2(\mathbf{R}^n)$ since $|e_\omega(t)|^2 = 1$ for all $t \in \mathbf{R}^n$, hence the integral representing the inner product $\langle\, e_\omega, f\,\rangle \equiv e_\omega^* f$ will diverge for general $f \in L^2(\mathbf{R}^n)$. (This is related to the fact that $\hat{f}(\omega)$ is not well-defined for any fixed ω, since $\hat{f}$ can be changed on a set of measure zero, such as the one-point set $\{\omega\}$, without changing $\hat{f}$ as a vector in $L^2(\mathbf{R}_n)$.) It is precisely here that methods such as time-frequency analysis and wavelet (time-scale) analysis have an advantage over Fourier methods. Such methods give rise to genuine resolutions of unity for $L^2(\mathbf{R}^n)$ using the general theory of frames, as explained in the following chapters. Nevertheless, we continue to use the expression $e_\omega^* f$ as a handy shorthand for the Fourier transform.

1.5 Appendix: Measures and Integrals†

The concept of measure necessarily enters when we try to make a Hilbert space of functions $f : \mathbf{R} \to \mathbf{C}$, as in Section 1.3. Because the vector f has a continuous infinity of "components" $f(t)$, we have either to restrict t to a discrete set like $\mathbf{Z}$ (giving a space of type ℓ^2) or find some way of assigning a "measure" (or weight) more democratically among the t's and still have an interesting subspace of $\mathbf{C}^{\mathbf{R}}$ with finite norms. Measure theory is a deep and difficult subject, and we can only give a much-simplified qualitative discussion of it here.

Let us return briefly to the finite-dimensional case, where our vectors $\mathbf{v} \in \mathbf{C}^N$ can also be viewed as functions $f(n)$, $n = 0, 1, \ldots, N$ (Section 1.3). The standard inner product treats all values n of the discrete time variable democratically by assigning each of them the weight 1. More generally, we could decide that some n's are more important than others by assigning a weight $\mu_n > 0$ to n. This gives the inner product (1.25) instead of the standard inner product. The corresponding square norm,

$$\|f\|^2 \equiv \sum_n \mu_n\, |f(n)|^2,$$

can be interpreted as defining the total "cost" (or "energy") expanded in constructing the vector $f = \{f(n)\}$. A situation much like this occurs in probability theory, where n represents a possible outcome of an experiment, such as tossing

† This section is optional.

dice, and μ_n is the *probability* of that outcome. In that case, the weights are required to satisfy the constraint

$$\sum_n \mu_n = 1.$$

One then calls μ_n a *probability distribution* on the set $\{1, 2, \ldots, N\}$, and the norm of $\|f\|^2$ is the *mean* of $|f|^2$. When the domain of the functions becomes countably infinite, say $n \in \mathbf{Z}$, the main modification is that only square-summable sequences $\{f(n)\}$ are considered, leading to the ℓ^2 spaces discussed in Section 1.3, but possibly with variable weights μ_n. The real difficulty comes when the domain becomes continuous, like $\mathbf{R}$. Intervals like $[0, 1]$ cannot be built up by adding points, if "adding" is understood as a sequential, hence *countable*, operation. If we try to construct a weight distribution by assigning each point $\{t\}$ in $\mathbf{R}$ some value value $\mu(t)$, then the total weight of $[0, 1]$ would be infinite unless $\mu(t) = 0$ for "almost all" points. But a new way exists: We could define weights for *subsets* of $\mathbf{R}$ instead of trying to define them for points. For example, we could try to choose the weight democratically by declaring that the weight, now called *measure*, of any set $A \subset \mathbf{R}$ is its total *length* $\lambda(A)$. Thus $\lambda([a, b]) = b - a$ for any interval with $b \geq a$. If $A \subset B$, then the measure of the *complement* $B - A = \{t \in B : t \notin A\}$, is defined as $\lambda(B) - \lambda(A)$. The measure of the union of A and B is defined as $\lambda(A \cup B) = \lambda(A) + \lambda(B) - \lambda(A \cap B)$, since their (possible) intersection must not be counted twice. If A and B do not intersect, then $\lambda(A \cup B) = \lambda(A) + \lambda(B)$, since the measure of the empty set (their intersection) is zero. The measure of a *countable* union $A = \bigcup_k I_k$ of nonintersecting intervals is defined as the sum of their individual measures, i.e., $\lambda(A) = \sum_k \lambda(I_k)$. In particular, any countable set of points $\{a_k\}$ (regarded as intervals $[a_k, a_k]$) has zero measure. (This is the price of democracy in an infinite population.) However, a problem develops: It turns out that not *every* subset of $\mathbf{R}$ can be constructed sequentially from intervals by taking unions and complements! When these operations are pushed to their limit, they yield only a certain class of subsets $A \subset \mathbf{R}$, called *measurable*, and it is possible to conceive of examples of "unmeasurable" sets. (That is where the theory gets messy!) This is the basic idea of *Lebesgue measure.* It can be generalized to $\mathbf{R}^n$ by letting $\lambda(A)$ be the n-dimensional volume of $A \subset \mathbf{R}^n$, provided A can be constructed sequentially from n-dimensional "boxes," i.e., Cartesian products of intervals. Hence the Lebesgue measure is a function λ whose domain (set of possible inputs) consists not of individual points in $\mathbf{R}^n$ but of measurable subsets $A \subset \mathbf{R}^n$. Any point $t \in \mathbf{R}^n$ can also be considered as the subset $\{t\} \subset \mathbf{R}^n$, whose Lebesgue measure is zero; but not all measurable sets can be built up from points. Thus λ is *more general* than a pointwise function. (Such a function λ is called a *set function.*)

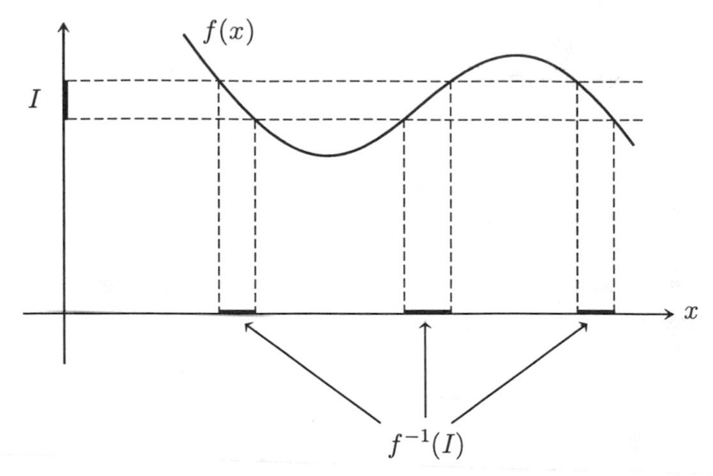

Figure 1.2. Schematic representation of the graph of $f(x)$ and of $f^{-1}(I)$, for an interval I in the y-axis.

In order to define inner products, we must be able to integrate complex-valued functions. For this purpose, the idea of measurability is extended from *sets* $A \subset \mathbf{R}^n$ to *functions* $f : \mathbf{R}^n \to \mathbf{C}$ as follows: A complex-valued function is said to be measurable if its real and imaginary parts are both measurable, so it suffices to define measurability for real-valued functions $f : \mathbf{R}^n \to \mathbf{R}$. Given an arbitrary interval $I \subset \mathbf{R}$, its *inverse image* under f is defined as the set of points in $\mathbf{R}^n$ that get mapped into I, i.e.,

$$f^{-1}(I) \equiv \{u \in \mathbf{R}^n : f(u) \in I\} \subset \mathbf{R}^n. \tag{1.97}$$

Figure 1.2 shows $f^{-1}(I)$ for a function $f : \mathbf{R} \to \mathbf{R}$ and an interval I in the y-axis. Note that f does not need to be one-to-one in order for this to make sense. For example, if $f(t) = 0$ for all $t \in \mathbf{R}^n$, then $f^{-1}(I) = \mathbf{R}^n$ if I contains 0, and $f^{-1}(I)$ is the empty set if I does not contain 0. The function f is said to be *measurable* if for every interval $I \subset \mathbf{R}$, the inverse image $f^{-1}(I)$ is a measurable subset of $\mathbf{R}^n$. Only measurable functions can be integrated in the sense of Lebesgue, subject to extra conditions such as the convergence of improper integrals, etc. The reason for this will become clear below.

Roughly speaking, Lebesgue's definition of the integral starts by partitioning the space of the *dependent variable*, whereas Riemann's definition partitions the space of the *independent variable*. For example, let $f : [a, b] \to \mathbf{R}$ be a measurable function, which we assume to be *nonnegative* for simplicity. Choose an integer $N > 0$ and partition the nonnegative y-axis (i.e., the set of possible *values* of f) into the intervals

$$I_k = [y_k\,, y_{k+1})\,, \quad k = 0, 1, 2, \ldots, \quad \text{where} \quad y_k = \frac{k}{N}\,. \tag{1.98}$$

Then the N-th lower and upper Lebesgue sums approximating $\int_a^b dx\, f(x)$ can be defined as

$$L_N^{\text{lower}} = \sum_{k=0}^{\infty} \lambda(f^{-1}(I_k)) \cdot \frac{k}{N}$$
$$L_N^{\text{upper}} = \sum_{k=0}^{\infty} \lambda(f^{-1}(I_k)) \cdot \frac{k+1}{N}. \tag{1.99}$$

The *Lebesgue integral* $\int_a^b dt\, f(t)$ is defined as the common limit of the lower and upper sums, provided these converge as $N \to \infty$ and their limits are equal. The above equations also make it clear why f needs to be measurable in order for the integral to make sense (otherwise $\lambda(f^{-1}(I_k))$ may not exist). For example, let t be time and let $f(t)$ represent the voltage in a line as a function of time. Then $f^{-1}(I_k)$ is the *set* of all times for which the voltage is in the interval I_k. This set may or may not consist of a single piece. It could be very scattered (if the voltage has much oscillation around the interval I_k) or even empty (if the voltage never reaches the interval I_k). The measure $\lambda(f^{-1}(I_k))$ is simply the *total amount of time spent by the voltage in the interval* I_k*, regardless of the connectivity of* $f^{-1}(I_k)$, and the sums in (1.99) do indeed approximate $\int dt\, f(t)$.

By contrast, Riemann's definition of the same integral goes as follows: Partition the *domain* of f, i.e., the interval $[a, b]$, into small, disjoint pieces (for example, subintervals) A_k, so that $\cup_k A_k = [a, b]$. Then the lower and upper Riemann sums are

$$R^{\text{lower}} = \sum_{k=0}^{\infty} \lambda(A_k) \inf_{x \in A_k} f(x)$$
$$R^{\text{upper}} = \sum_{k=0}^{\infty} \lambda(A_k) \sup_{x \in A_k} f(x), \tag{1.100}$$

where inf ("infimum") and sup ("supremum") mean the least upper bound and greatest lower bound, respectively. The Riemann integral is defined as the common limit of R^{lower} and R^{upper} as the partition of $[a, b]$ gets finer and finer, provided these limits exist and are equal.

The Lebesgue integral can be compared with the Riemann integral as follows: Suppose we want to find the total income of a set of individuals, where person x has income $f(x)$. Riemann's method would be analogous to going from person to person and sequentially adding their incomes. Lebesgue's method, on the other hand, consists of sorting the individuals into many different groups corresponding to their income levels I_k. The set of all people with income level in I_k is $f^{-1}(I_k)$. Then we count the total number of individuals in each group (which corresponds to finding $\lambda(f^{-1}(I_k))$), multiply by the corresonding income levels, and add up. Clearly the second method is more efficient! It also leads to a more general notion of integral. To illustrate that Lebesgue's definition of integrals is indeed broader than Riemann's, consider the function $q:[a,b]\to\mathbf{R}$ defined by

$$q(t)=\begin{cases}1 & \text{if } t \text{ is rational}\\ 0 & \text{if } t \text{ is irrational,}\end{cases} \tag{1.101}$$

for which the Riemann integral does not exist, as we have seen in Section 1.3, because the lower and upper Riemann sums converge to different limits. We first verify that q is measurable. Given any interval $I\subset\mathbf{R}$, there are only four possiblilities for $q^{-1}(I)$: If I contains both 0 and 1, $q^{-1}(I)=[a,b]$; if I contains 1 but not 0, $q^{-1}(I)\equiv\mathbf{Q}[a,b]$ is the set of rationals in $[a,b]$; if I contains 0 but not 1, $q^{-1}(I)\equiv\mathbf{Q}'[a,b]$ is the set of irrationals in $[a,b]$; if I contains neither 0 nor 1, $q^{-1}(I)=\phi$, the empty set. In all four cases, $q^{-1}(I)$ is measurable, hence the function q is measurable. Furthermore, it can be shown that $\mathbf{Q}[a,b]$ is *countable*, i.e., discrete (it can be assembled sequentially from single points); hence it has Lebesgue measure zero. It follows that $\lambda(\mathbf{Q}'[a,b])=b-a$. Given $N\geq 1$, all the points $t\in\mathbf{Q}[a,b]$ get mapped into $1\in I_N$ (since $I_N=[1,\frac{N}{N+1})$), and all the points $t\in\mathbf{Q}'[a,b]$ get mapped into $0\in I_0=[0,\frac{1}{N})$. Since $\lambda(q^{-1}(I_0))=\lambda(\mathbf{Q}'[a,b])=b-a$ and $\lambda(q^{-1}(I_N))=\lambda(\mathbf{Q}[a,b])=0$, we have

$$\begin{aligned} L_N^{\text{lower}} &= (b-a)\cdot\frac{0}{N}+0\cdot\frac{N}{N}=0,\\ L_N^{\text{upper}} &= (b-a)\cdot\frac{1}{N}+0\cdot\frac{N+1}{N}=\frac{b-a}{N}. \end{aligned} \tag{1.102}$$

Hence both Lebesgue sums converge to zero as $N\to\infty$, and

$$\int_a^b dt\, q(t)=0. \tag{1.103}$$

The concept of a measure can be generalized considerably from the above. Let M be an *arbitrary* set. A (positive) *measure* on M is any mapping μ that assigns nonnegative (possibly infinite!) values $\mu(A)$ to certain subsets of $A\subset M$, called *measurable* subsets. The choice of subsets considered to be measurable is also arbitrary, subject only to the following rules already encountered in the discussion of Lebesgue measure:

(a) M itself must be measurable (although possibly $\mu(M)=\infty$);

(b) If A is measurable, then so is its complement $M-A$;

(c) A countable union of measurable sets is measurable.

Such a collection is called a σ*-algebra* of subsets of M. Furthermore, the measure μ must be *countably additive*, meaning that if $\{A_1,A_2,\ldots\}$ is a finite or countably infinite collection of *nonintersecting* measurable sets, then

$$\mu\Big(\bigcup_n A_n\Big)=\sum_n \mu(A_n). \tag{1.104}$$

As in Lebesgue's theory, the idea of measurability is now extended from subsets of M to functions $f : M \to \mathbf{C}$. Again, a complex-valued function is defined to be measurable when its real and imaginary parts are both measurable. A function $f : M \to \mathbf{R}$ is said to be measurable if for every interval $I \subset \mathbf{R}$, the inverse image

$$f^{-1}(I) \equiv \{m \in M : f(m) \in I\} \tag{1.105}$$

is a measurable subset of M. If f is nonnegative, then the lower and upper sums approximating the integral of f over M are defined as in (1.99), but with $\lambda(f^{-1}(I_k))$ replaced by $\mu(f^{-1}(I_k))$. If both sums converge to a common limit $L < \infty$, we say that f is integrable with respect to the measure μ, and we write

$$\int_M d\mu(m)\, f(m) = L. \tag{1.106}$$

A function that can assume both positive and negative values can be written in the form $f(m) = f_+(m) - f_-(m)$, where f_+ and f_- are nonnegative. Then f is said to be integrable if both f_+ and f_- are integrable. Its integral is defined to be $L_+ - L_-$, where $L_\pm$ are the integrals of $f_\pm$.

To illustrate the idea of general measures and integrals, we give two examples.

(a) Let a fluid be distributed in a subset R of $\mathbf{R}^n$ ($n \geq 1$) with density $\rho(x)$. Thus, $\rho : R \to \mathbf{R}$ is a nonnegative function, which we assume to be measurable in the sense of Lebesgue. (That is a safe assumption, since a nonmeasurable function has to be very bizarre). For any measurable subset $A \subset R$, define

$$\mu(A) \equiv \int_A d^n x\, \rho(x), \tag{1.107}$$

which is just the total *mass* of the fluid contained in A. (Note that we write a single integral sign, even though the integral is actually n-dimensional.) Then μ is a measure on $\mathbf{R}^n$, and the corresponding integral is

$$\int_R d\mu(x)\, f(x) = \int_R \rho(x)\, d^n x\, f(x). \tag{1.108}$$

We write $d\mu(x) = \rho(x)\, d^n x$. If $\rho(x) = 1$ for all $x \in R$, then μ reduces to the Lebesgue measure, and (1.108) to the Lebesgue integral. Given $\rho(x)$, it is common to refer to the measure μ in its *differential* form $d\mu(x)$. For example, the continuous wavelet transform will involve a measure on the set of all times $t \in \mathbf{R}$ and all scales $s \neq 0$, defined by the density function $\rho(s,t) = 1/s^2$. We will refer to this measure simply as $ds\, dt/s^2$, rather than in the more elaborate form $\mu(A) = \iint_A ds\, dt/s^2$.

(b) Take $M = \mathbf{Z}$, and let $\{\rho_n : n \in \mathbf{Z}\}$ be a sequence of nonnegative numbers. Given any set of integers, $A \subset \mathbf{Z}$, let $\mu(A) \equiv \sum_{n \in A} \rho_n$. (Note that this sum need not be finite.) Then μ is a measure on $\mathbf{Z}$, and every subset of $\mathbf{Z}$ is measurable. The corresponding integral of a nonnegative function on $\mathbf{Z}$ takes the form

$$\int_{\mathbf{Z}} d\mu(n)\, f(n) = \sum_{n\in\mathbf{Z}} \rho_n\, f(n). \tag{1.109}$$

(The integral on the left is *notation*, and the sum on the right is its definition.) If $\rho_n = 1$ for all n, then μ is called the *counting measure* on $\mathbf{Z}$, since $\mu(A)$ counts the number of elements in A. Note that we may also regard μ as a measure on $\mathbf{R}$ if we define $\mu(A)$ (for $A \subset \mathbf{R}$) as follows: Let $A' = A \cap \mathbf{Z}$ and $\mu(A) \equiv \sum_{n\in A'} \rho_n$. The corresponding integral is then

$$\int_{\mathbf{R}} d\mu(x)\, f(x) = \sum_{n\in\mathbf{Z}} \rho_n\, f(n). \tag{1.110}$$

In fact, $d\mu(x)$ can be expressed as a train of δ-functions:

$$d\mu(x) = \sum_n \rho_n \delta(x-n)\, dx. \tag{1.111}$$

The general idea of measures therefore includes both continuous and discrete analyses as special cases. Given a set M, the choice of a measure on M (along with a σ-algebra of measurable subsets) amounts to a decision about the importance, or *weight*, given to different subsets. Thus, Lebesgue measure is a completely "democratic" way of assigning weight to subsets of $\mathbf{R}$, whereas the measure in Example (b) only assigns weight to sets containing integers. This notion will be useful in the description of *sampling*. If a signal $f : \mathbf{R} \to \mathbf{C}$ is sampled only at a discrete set of points $D \subset \mathbf{R}$ (which is all that is possible in practice), that amounts to a choice of discrete measure on $\mathbf{R}$, i.e., one that assigns a positive measure only to sets $A \subset \mathbf{R}$ that contain points of D.

A set M, together with a class of measurable subsets and a measure defined on these subsets, is called a *measure space*. For a general measure μ on a set M, the measure is related to the integral as follows: Given any measurable set $A \subset M$, define its *characteristic function* X_A by $X_A(m) = 1$ if $m \in A$, $X_A(m) = 0$ if $m \notin A$. Then X_A is measurable and

$$\mu(A) = \int_M d\mu(m)\, X_A(m). \tag{1.112}$$

More generally, the integral of a measurable function f over A (rather than over all of M), where A is a measurable subset of M, is

$$\int_A d\mu(m)\, f(m) \equiv \int_M d\mu(m)\, X_A(m)\, f(m). \tag{1.113}$$

In Chapter 4 we will use the general concept of measures to define *generalized frames*. This concept unifies certain continuous and discrete basis-like systems that are used in signal analysis and synthesis and allows one to view sampling as a change from a continuous measure to a discrete one.

The concept of measure lies at the foundation of probability theory. There, M represents the set of all possible outcomes of an experiment. Measurable sets $A \subset M$ are called *events*, and $\mu(A)$ is the probability of the set of outcomes A. Since *some* outcome is certain to occur, we must have $\mu(M) = 1$, generalizing the finite case discussed earlier. Thus a measure with total "mass" 1 is called a *probability measure*, and measurable functions $f : M \to \mathbf{C}$ are called *random variables* (Papoulis [1984]). For example, any nonnegative Lebesgue-measurable function on $\mathbf{R}^n$ with integral $\int_{\mathbf{R}^n} d^n x\, \rho(x) = 1$ defines a probability measure on $\mathbf{R}^n$ by $d\mu(x) = \rho(x) d^n x$. On the other hand, we could also assign probability 1 to $x = 0$ and probability 0 to all other points. This corresponds to $\rho(x) = \delta(x)$, called the "Dirac measure."

Exercises

1.1. (Linearity) Let $c \in \mathbf{C}$, let $F, G \in \mathbf{C}_M^N$, and let $H \in \mathbf{C}_N^K$. Prove that the functions $cF : \mathbf{C}^M \to \mathbf{C}^N$, $F + G : \mathbf{C}^M \to \mathbf{C}^N$, and $HF : \mathbf{C}^M \to \mathbf{C}^K$ defined below are linear:
(a) $(cF)(u) \equiv c(Fu)$.
(b) $(F + G)(u) \equiv Fu + Gu$.
(c) $(HF)(u) \equiv H(Fu)$.

1.2. (Operators and matrices.) Prove that the matrices representing the operators cF, $F+G$, and HF in Exercise 1.1 with respect to any choice of bases in $\mathbf{C}^M, \mathbf{C}^N$, and $\mathbf{C}^K$ are given by

$$(cF)_m^n = c\,F_m^n, \quad (F+G)_m^n = F_m^n + G_m^n, \quad (HF)_m^k = \sum_{n=1}^{N} H_n^k F_m^n.$$

1.3. Suppose we try to define an "inner product" on $\mathbf{C}^2$ by

$$\langle\, \mathbf{u}, \mathbf{v}\,\rangle = \overline{u^1} v^1.$$

Prove that not every linear functional on $\mathbf{C}^2$ can be represented by a vector, as guaranteed in Theorem 1.6(c). Show this by giving an example of a linear functional that cannot be so represented. What went wrong?

1.4. Let $\mathcal{H}$ be a Hilbert space and let $f \in \mathcal{H}$. Let $f : \mathbf{C} \to \mathcal{H}$ be the corresponding linear map $c \mapsto cf$ (regarding $\mathbf{C}$ as a one-dimensional Hilbert space). Prove that $\|f\|$ is a bound for this mapping and that it is the *lowest* bound.

1.5. (a) Prove that every inner product in $\mathbf{C}$ has the form $\langle c, c' \rangle = p\,\bar{c}\,c'$, for some $p > 0$.
(b) Choose any inner product in $\mathbf{C}$ (i.e., any $p > 0$). Let $H : \mathcal{H} \to \mathbf{C}$ be a bounded linear functional on the Hilbert space $\mathcal{H}$, and show that the vector $h \in \mathcal{H}$ representing H is the *adjoint* of H, now regarded as an operator from $\mathcal{H}$ to the one-dimensional Hilbert space $\mathbf{C}$. Use the characterization (1.56) of adjoint operators.

1.6. (Riesz representation theorem.) Let $F = [\,F_1 \quad \cdots F_N\,] \in \mathbf{C}_N$. Given the inner product (1.25) on $\mathbf{C}^N$, find the components (with respect to the standard basis) of the vector $\mathbf{u} \in \mathbf{C}^N$ representing F, whose existence is guaranteed by Theorem 1.4. That is, find $\mathbf{u} = F^*$.

1.7. (Reciprocal basis.)
(a) Prove that the vectors

$$\mathbf{b}_1 = \begin{bmatrix} 1 \\ 2 \end{bmatrix}, \mathbf{b}_2 = \begin{bmatrix} -1 \\ 1 \end{bmatrix}$$

form a nonorthogonal basis for $\mathbf{R}^2$ with respect to the standard inner product.
(b) Find the corresponding dual basis $\{B^1, B^2\}$ for $\mathbf{R}_2$.
(c) Find the metric operator G (1.43), and use (1.46) to find the reciprocal basis $\{\mathbf{b}^1, \mathbf{b}^2\}$ for $\mathbf{R}^2$.
(d) Sketch the bases $\{\mathbf{b}_1, \mathbf{b}_2\}$ and $\{\mathbf{b}^1, \mathbf{b}^2\}$ in the xy plane, and verify the biorthogonality relations.
(e) Verify the resolutions of unity (1.39) and (1.40).
(f) Express an arbitrary vector $u \in \mathbf{R}^2$ as a superposition of $\{\mathbf{b}_1, \mathbf{b}_2\}$ and also of $\{\mathbf{b}^1, \mathbf{b}^2\}$.

1.8. (Orthonormal bases.) Show that the vectors

$$b_1 = \frac{1}{\sqrt{2}} \begin{bmatrix} 1 \\ i \end{bmatrix}, \quad b_2 = \frac{1}{\sqrt{2}} \begin{bmatrix} 1 \\ -i \end{bmatrix}$$

form an orthonormal basis for $\mathbf{C}^2$ with respect to the standard inner product, and verify that the resolution of unity in the form of (1.43) holds.

1.9. (Star notation in Hilbert space.) Let

$$h(t) = \begin{cases} e^{-2t} & t \geq 1 \\ 0 & t < 1. \end{cases}$$

(a) Show that $h \in L^2(\mathbf{R})$, and compute $\|h\|$.
(b) Find the linear functional $h^* : L^2(\mathbf{R}) \to \mathbf{C}$ corresponding to h. That is, give the expression for $h^* f \in \mathbf{C}$ for any $f \in L^2(\mathbf{R})$.

(c) Let $g(t) = e^{-3t^2}$. Find the operator $F \equiv gh^* : L^2(\mathbf{R}) \to L^2(\mathbf{R})$. That is, give the expression for $Ff \in L^2(\mathbf{R})$ for any f in $L^2(\mathbf{R})$.
(d) Find the operator $G \equiv hg^* : L^2(\mathbf{R}) \to L^2(\mathbf{R})$. How is G related to F?
(e) Compute h^*f and Ff *explicitly* when $f(t) = te^{-t^2}$.

Chapter 2

Windowed Fourier Transforms

Summary: Fourier series are ideal for analyzing periodic signals, since the harmonic modes used in the expansions are themselves periodic. By contrast, the Fourier integral transform is a far less natural tool because it uses periodic functions to expand nonperiodic signals. Two possible substitutes are the windowed Fourier transform (WFT) and the wavelet transform. In this chapter we motivate and define the WFT and show how it can be used to give information about signals simultaneously in the time domain and the frequency domain. We then derive the counterpart of the inverse Fourier transform, which allows us to reconstruct a signal from its WFT. Finally, we find a necessary and sufficient condition that an otherwise arbitrary function of time and frequency must satisfy in order to be the WFT of a time signal with respect to a given window and introduce a method of processing signals simultaneously in time and frequency.

Prerequisites: Chapter 1.

2.1 Motivation and Definition of the WFT

Suppose we want to analyze a piece of music for its frequency content. The piece, as perceived by an eardrum, may be accurately modeled by a function $f(t)$ representing the air pressure on the eardrum as a function of time. If the "music" consists of a single, steady note with fundamental frequency ω_1 (in cycles per unit time), then $f(t)$ is periodic with period $P = 1/\omega_1$ and the natural description of its frequency contents is the Fourier series, since the Fourier coefficients c_n give the amplitudes of the various harmonic frequencies $\omega_n = n\omega_1$ occurring in f (Section 1.4). If the music is a series of such notes or a melody, then it is not periodic in general and we cannot use Fourier series directly. One approach in this case is to compute the Fourier integral transform $\hat{f}(\omega)$ of $f(t)$. However, this method is flawed from a practical point of view: To compute $\hat{f}(\omega)$ we must integrate $f(t)$ over *all* time, hence $\hat{f}(\omega)$ contains the *total* amplitude for the frequency ω in the entire piece rather than the distribution of harmonics in each individual note! Thus, if the piece went on for some length of time, we would need to wait until it was over before computing $\hat{f}$, and then the result would be completely uninformative from a musical point of view. (The same is true, of course, if $f(t)$ represents a speech signal or, in the multidimensional case, an image or a video signal.)

Another approach is to chop f up into approximately single notes and analyze each note separately. This analysis has the obvious drawback of being somewhat arbitrary, since it is impossible to state exactly when a given note ends and the next one begins. Different ways of chopping up the signal may result in widely different analyses. Furthermore, this type of analysis must be tailored to the particular signal at hand (to decide how to partition the signal into notes, for example), so it is not "automatic." To devise a more natural approach, we borrow some inspiration from our experience of hearing. Our ears can hear continuous changes in tone as well as abrupt ones, and they do so without an arbitrary partition of the signal into "notes." We will construct a very simple model for hearing that, while physiologically quite inaccurate (see Roederer [1975], Backus [1977]), will serve mainly as a device for motivating the definition of the windowed Fourier transform.

Since the ear analyzes the frequency distribution of a given signal f in *real time*, it must give information about f simultaneously in the frequency domain and the time domain. Thus we model the output of the ear by a function $\tilde{f}(\omega, t)$ depending on both the frequency ω and the time t. For any fixed value of t, $\tilde{f}(\omega, t)$ represents the frequency distribution "heard" at time t, and this distribution varies with t. Since the ear cannot analyze what has not yet occurred, only the values $f(u)$ for $u \le t$ can be used in computing $\tilde{f}(\omega, t)$. It is also reasonable to assume that the ear has a finite "memory." This means that there is a time interval $T > 0$ such that only the values $f(u)$ for $u \ge t - T$ can influence the output at time t. Thus $\tilde{f}(\omega, t)$ can only depend on $f(u)$ for $t - T \le u \le t$. Finally, we expect that values $f(u)$ near the endpoints $u \approx t - T$ and $u \approx t$ have less influence on $\tilde{f}(\omega, t)$ than values in the middle of the interval. These statements can be formulated mathematically as follows: Let $g(u)$ be a function that vanishes outside the interval $-T \le u \le 0$, i.e., such that $\operatorname{supp} g \subset [-T, 0]$. $g(u)$ will be a weight function, or *window*, which will be used to "localize" signals in time. We allow g to be complex-valued, although in many applications it may be real. For every $t \in \mathbf{R}$, define

$$f_t(u) \equiv \bar{g}(u - t)\, f(u), \tag{2.1}$$

where $\bar{g}(u - t) \equiv \overline{g(u - t)}$. Then $\operatorname{supp} f_t \subset [t - T, t]$, and we think of f_t as a *localized version* of f that depends only on the values $f(u)$ for $t - T \le u \le t$. If g is continuous, then the values $f_t(u)$ with $u \approx t - T$ and $u \approx t$ are small. This means that the above localization is smooth rather than abrupt, a quality that will be seen to be important. We now define the *windowed Fourier transform* (WFT) of f as the Fourier transform of f_t:

$$\begin{aligned} \tilde{f}(\omega, t) \equiv \hat{f}_t(\omega) &= \int_{-\infty}^{\infty} du\; e^{-2\pi i \omega u}\, f_t(u) \\ &= \int_{-\infty}^{\infty} du\; e^{-2\pi i \omega u}\, \bar{g}(u - t)\, f(u). \end{aligned} \tag{2.2}$$

As promised, $\tilde{f}(\omega, t)$ depends on $f(u)$ only for $t - T \le u \le t$ and (if g is continuous) gives little weight to the values of f near the endpoints.

Note: (a) The condition $\operatorname{supp} g \subset [-T, 0]$ was imposed mainly to give a physical motivation for the WFT. In order for the WFT to make sense, as well as for the reconstruction formula (Section 2.3) to be valid, it will only be necessary to assume that $g(u)$ is square-integrable, i.e. $g \in L^2(\mathbf{R})$. (b) In the extreme case when $g(u) \equiv 1$ (so $g \notin L^2(\mathbf{R})$), the WFT reduces to the ordinary Fourier transform. In the following we merely assume that $g \in L^2(\mathbf{R})$.

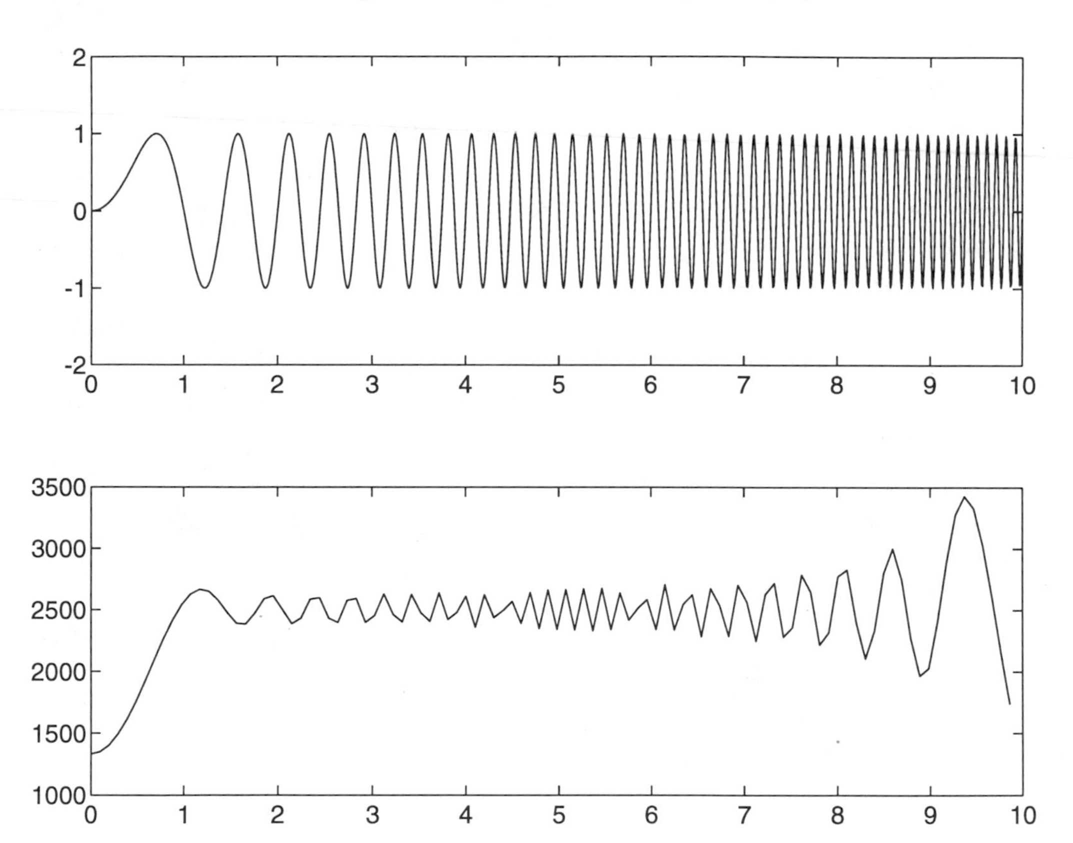

Figure 2.1. *Top*: The chirp signal $f(u) = \sin(\pi u^2)$. *Bottom*: The spectral energy density $|\hat{f}(\omega)|^2$ of f.

If we define

$$g_{\omega,t}(u) \equiv e^{2\pi\, i\omega u}\, g(u - t), \tag{2.3}$$

then $\|g_{\omega,t}\| = \|g\|$; hence $g_{\omega,t}$ also belongs to $L^2(\mathbf{R})$, and the WFT can be expressed as the inner product of f with $g_{\omega,t}$,

$$\tilde{f}(\omega, t) = \langle\, g_{\omega,t}\,, f\,\rangle \equiv g^*_{\omega,t} f, \tag{2.4}$$

which makes sense if both functions are in $L^2(\mathbf{R})$. (See Sections 1.2 and 1.3 for the definition and explanation of the "star notation" $g^* f$.) It is useful to think of $g_{\omega,t}$ as a "musical note" that oscillates at the frequency ω inside the *envelope* defined by $|g(u-t)|$ as a function of u.

Example 2.1: WFT of a Chirp Signal. A *chirp* (in radar terminology) is a signal with a reasonably well defined but steadily rising frequency, such as

$$f(u) = \sin(\pi u^2). \tag{2.5}$$

In fact, the *instantaneous* frequency $\omega_{\text{inst}}(u)$ of f may be defined as the derivative of its phase:

$$2\pi\omega_{\text{inst}}(u) \equiv \partial_u(\pi u^2) = 2\pi u\,. \tag{2.6}$$

Ordinary Fourier analysis hides the fact that a chirp has a well-defined instantaneous frequency by integrating over all of time (or, practically, over a long time period), thus arriving at a very broad frequency spectrum. Figure 2.1 shows $f(u)$ for $0 \le u \le 10$ and its Fourier (spectral) energy density $|\hat{f}(\omega)|^2$ in that range. The spectrum is indeed seen to be very spread out.

We now analyze f using the window function

$$g(u) \equiv \begin{cases} 1+\cos(\pi u) & -1 \le u \le 1 \\ 0 & \text{otherwise}, \end{cases} \tag{2.7}$$

which is pictured in Figure 2.2. (We have centered $g(u)$ around $u = 0$, so it is not causal; but $g(u+1)$ is causal with $\tau = 2$.)

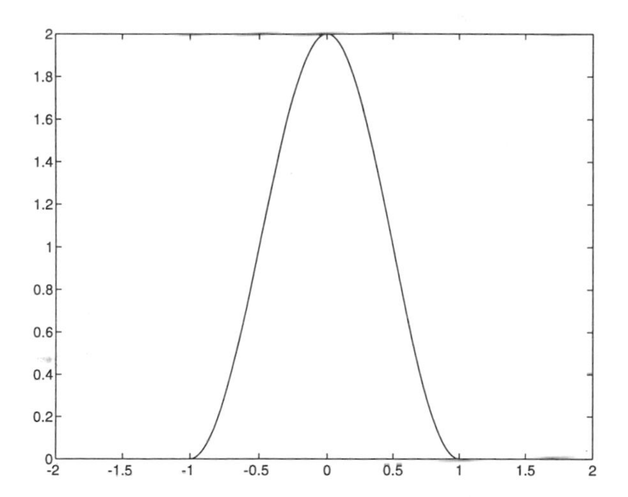

Figure 2.2. The window function $g(u)$.

Figure 2.3 shows the localized version $f_3(u)$ of $f(u)$ and its energy density $|\hat{f}_3(\omega)|^2 = |\tilde{f}(\omega,3)|^2$. As expected, the localized signal has a well-defined (though not *exact!*) frequency $\omega_{\text{inst}}(3) = 3$. It is, therefore, reasonably well localized both

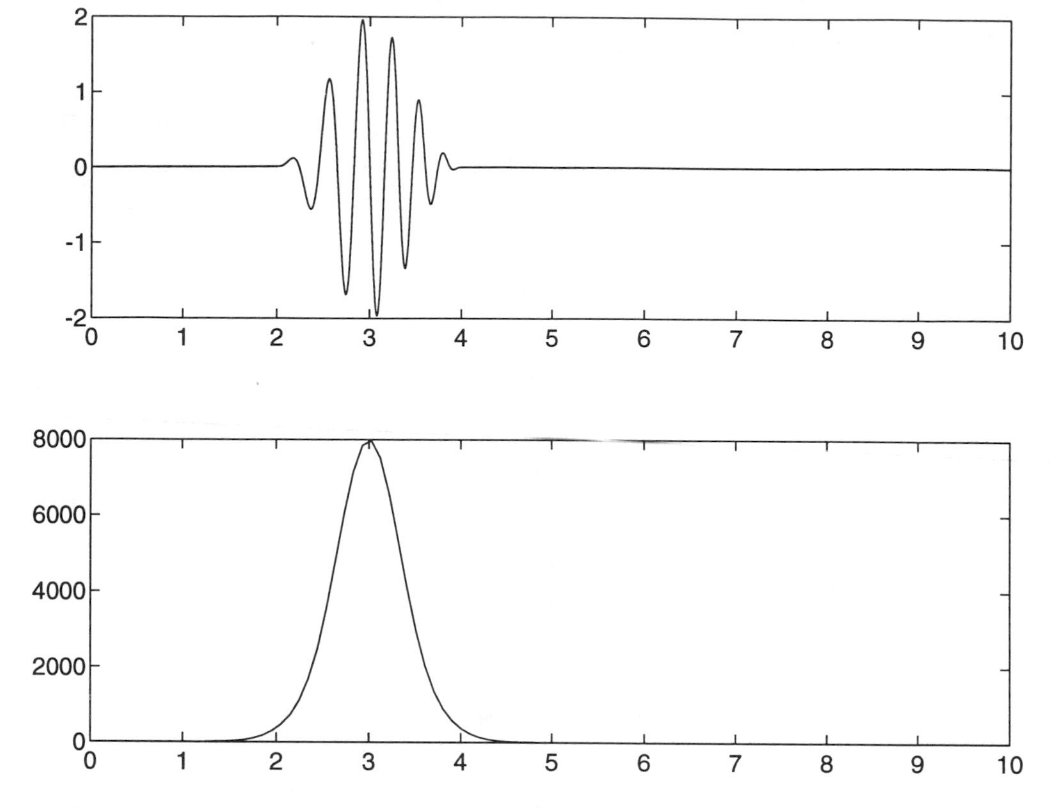

Figure 2.3. *Top*: The localized version $f_3(u)$ of the chirp signal in Figure 2.1 using the window $g(u)$ in Figure 2.2. *Bottom*: The spectral energy density of f_3 showing good localization around the instantaneous frequency $\omega_{\text{inst}}(3) = 3$.

in time and in frequency. Figure 2.4 repeats this analysis at $t = 7$. Now the energy is localized near $\omega_{\text{inst}}(7) = 7$.

Figures 2.5 and 2.6 illustrate frequency resolution. In the top of Figure 2.5 we plot the function

$$\begin{aligned} h(u) &= \text{Re}\left[g_{2,4}(u) + g_{4,6}(u)\right] \\ &= g(u-4)\cos(4\pi u) + g(u-6)\cos(8\pi u), \end{aligned} \tag{2.8}$$

which represents the real part of the sum of two "notes": One centered at $t = 4$ with frequency $\omega = 2$ and the other centered at $t = 6$ with frequency $\omega = 4$. The bottom part of the figure shows the spectral energy density of h. The two peaks are essentially copies of $|\hat{g}(\omega)|^2$ (which is centered at $\omega = 0$) translated to $\omega = 2$ and $\omega = 4$, respectively. This is repeated in Figure 2.6 for $h = \text{Re}\left[g_{2,4} + g_{3,6}\right]$. These two figures show that the window can resolve frequencies down to $\Delta\omega = 2$ but not down to $\Delta\omega = 1$.

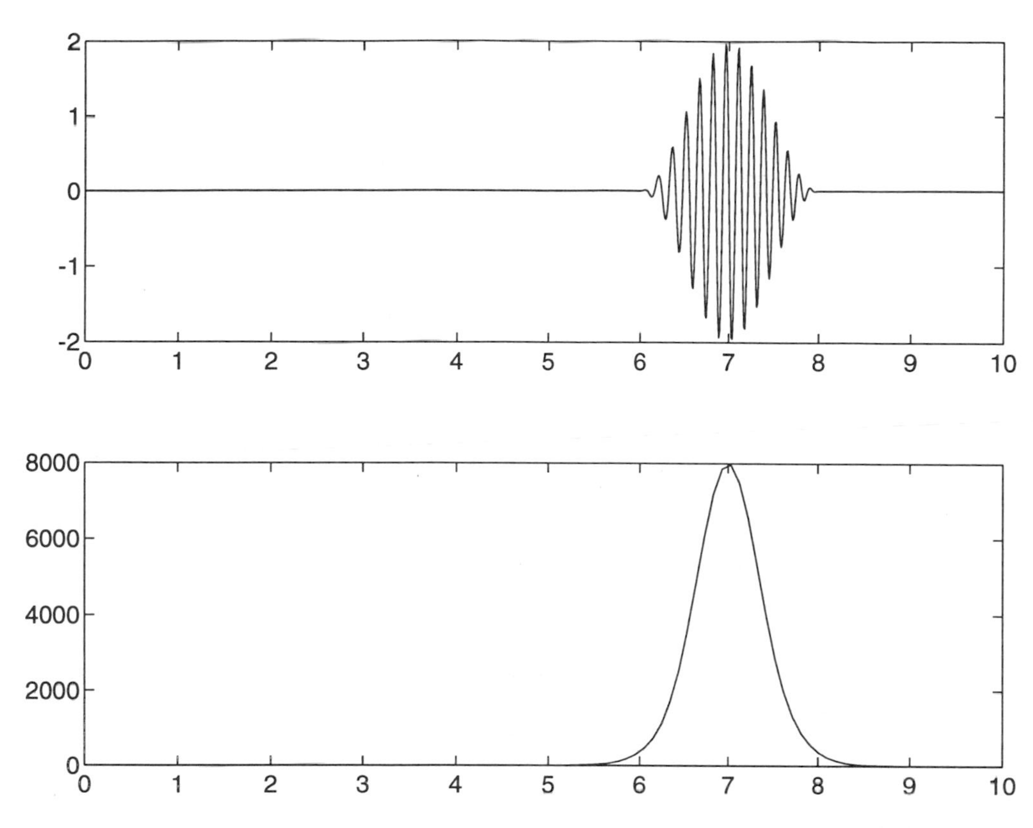

Figure 2.4. *Top*: The localized version $f_7(u)$ of the chirp signal using the window $g(u)$. *Bottom*: The energy density of f_7 showing good localization around the instantaneous frequency $\omega_{\text{inst}}(7) = 7$.

We will see that the vectors $g_{\omega,t}$, parameterized by all frequencies ω and times t, form something analogous to a basis for $L^2(\mathbf{R})$. Note that the inner product $\langle\, g_{\omega,t}\,, f\,\rangle$ is well defined for every choice of ω and t; hence the *values* $\tilde{f}(\omega, t)$ of $\tilde{f}$ are well defined. By contrast, recall from Section 1.3 that the value $f(u)$ of a "function" $f \in L^2(\mathbf{R})$ at any single point u is in general *not* well defined, since f can be modified on a set of measure zero (such as the one-point set $\{u\}$) without changing as an element of $L^2(\mathbf{R})$. The same can be said about the Fourier transform $\hat{f}$ as a function of ω. This remark already indicates that windowed Fourier transforms such as $\tilde{f}(\omega, t)$ are better-behaved than either the corresponding signals $f(t)$ in the time domain or their Fourier transforms $\hat{f}(\omega)$ in the frequency domain. Another example of this can be obtained by applying the Schwarz inequality to (2.4), which gives

$$|\tilde{f}(\omega, t)| = |\langle\, g_{\omega,t}, f\,\rangle| \leq \|g_{\omega,t}\|\, \|f\| = \|g\|\, \|f\|, \tag{2.9}$$

showing that $\tilde{f}(\omega, t)$ *is a bounded function*, since the right-hand side is finite and independent of ω and t. Thus a *necessary* condition for a given function

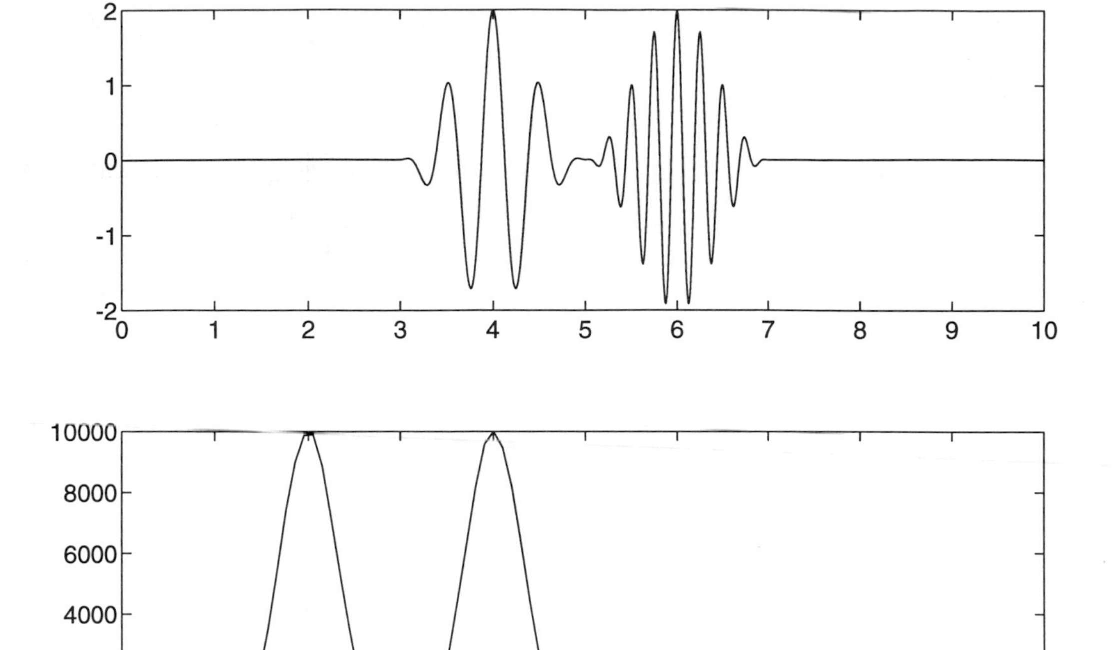

Figure 2.5. *Top*: Plot of $h = \text{Re}\,[g_{2,4} + g_{4,6}]$ for the window in (2.7). *Bottom*: The spectral energy density of h showing good frequency resolution at $\Delta\omega = 2$.

$h(\omega, t)$ to be the WFT of some signal $f \in L^2(\mathbf{R})$ is that h be bounded. However, this condition turns out to be far from *sufficient.* That is, not every bounded function $h(\omega, t)$ is the WFT $\tilde{f}(\omega, s)$ for some time signal $f \in L^2(\mathbf{R})$. In Section 2.3, we derive a condition that is sufficient as well as necessary.

2.2 Time-Frequency Localization

A remarkable aspect of the ordinary Fourier transform is the symmetry it displays between the time domain and the frequency domain, i.e., the fact that the formulas for $f \mapsto \hat{f}$ and $h \mapsto \check{h}$ are identical except for the sign of the exponent. It may appear that this symmetry is lost when dealing with the WFT, since we have treated time and frequency very differently in defining $\tilde{f}$. Actually, the WFT is also completely symmetric with respect to the two domains, as we now show. By Parseval's identity,

$$\tilde{f}(\omega, t) = \langle\, g_{\omega,t}\,, f\,\rangle = \langle\, \hat{g}_{\omega,t}\,, \hat{f}\,\rangle, \tag{2.10}$$

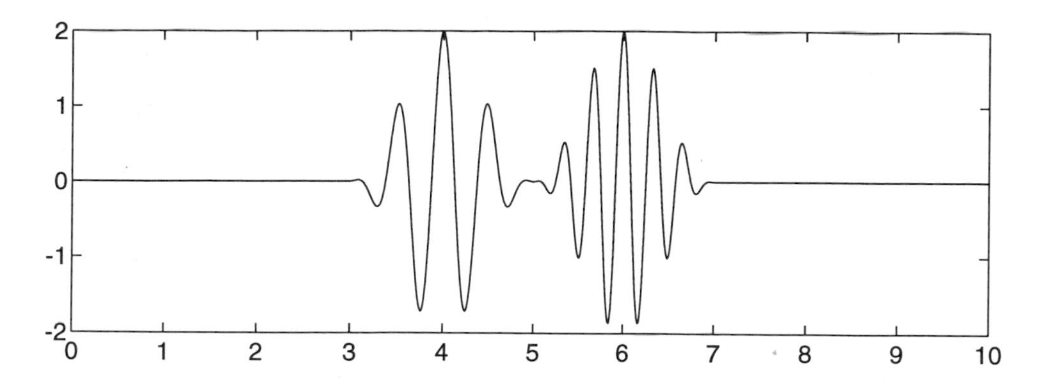

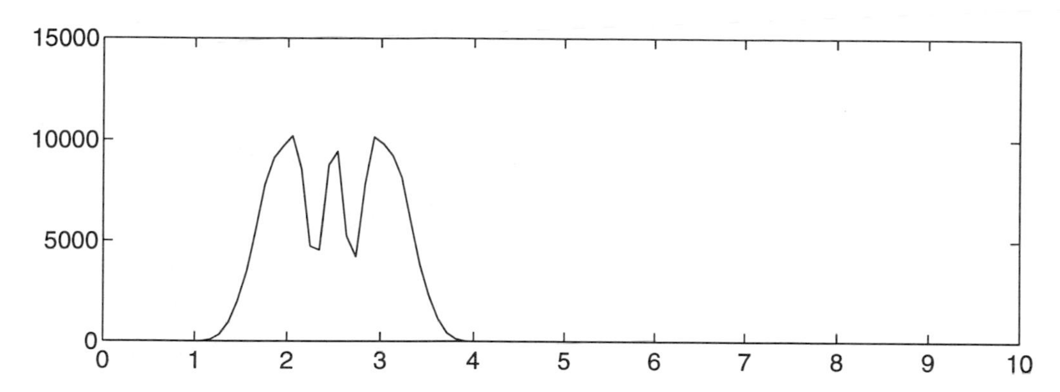

Figure 2.6. *Top*: Plot of $h = \text{Re}\,[g_{2,4} + g_{3,6}]$ for the window in (2.7). *Bottom*: The spectral energy density of h showing poor frequency resolution at $\Delta\omega = 1$.

and the right-hand side is computed to be

$$\begin{aligned} \tilde{f}(\omega, t) &= e^{-2\pi\, i\omega t} \int_{-\infty}^{\infty} d\nu\; e^{2\pi\, i\nu t}\; \overline{\hat{g}}(\nu - \omega)\, \hat{f}(\nu) \\ &= e^{-2\pi\, i\omega t} \left(\overline{\hat{g}}(\nu - \omega)\, \hat{f}(\nu) \right)^{\vee} (t), \end{aligned} \tag{2.11}$$

where $\overline{\hat{g}}(\nu - \omega) \equiv \overline{\hat{g}(\nu - \omega)}$. Equation (2.11) has almost exactly the same form as (2.2) but with the time variable u replaced by the frequency variable ν and the time window $g(u - t)$ replaced by the frequency window $\hat{g}(\nu - \omega)$. (The extra factor $e^{-2\pi\, i\omega t}$ in (2.11) is related to the "Weyl commutation relations" of the *Weyl–Heisenberg group*, which governs translations in time and frequency.) Thus, from the viewpoint of the frequency domain, we begin with the signal $\hat{f}(\nu)$ and "localize" it near the frequency ω by using the window function $\hat{g}$: $\hat{f}_\omega(\nu) \equiv \overline{\hat{g}}(\nu - \omega)\hat{f}(\nu)$; we then take the inverse Fourier transform of $\hat{f}_\omega$ and multiply it by the extra factor $e^{-2\pi\, i\omega t}$, which is a modulation due to the translation of $\hat{g}$ by ω in the frequency domain. If our window g is reasonably well localized in frequency as well as in time, i.e., if $\hat{g}(\nu)$ is small outside a small frequency band in addition to $g(t)$ being small outside a small time interval, then (2.11) shows that the WFT gives a local *time-frequency analysis* of the signal f in the sense that it provides accurate information about f simultaneously in the time domain and in the frequency domain. However, all functions, including windows, obey the *uncertainty principle*, which states that *sharp localizations in time and in frequency are mutually exclusive.* Roughly speaking, if a nonzero function $g(t)$ is small outside a time-interval of length T and its Fourier transform is small outside a frequency band of width Ω, then an inequality of the type $\Omega T \geq c$ must hold for some positive constant $c \sim 1$. The precise value of c depends on how the widths T and Ω of the signal in time and frequency are measured. For example, suppose we normalize g so that $\|g\| = 1$. Let us interpret $|g(t)|^2$ as a "weight distribution" of the window in time (so the total weight in time is $\|g\|^2 = 1$) and $|\hat{g}(\omega)|^2$ as a "weight distribution" of the window in frequency (so the total weight in frequency is $\|\hat{g}\|^2 = \|g\|^2 = 1$, by Plancherel's theorem). The "centers of gravity" of the window in time and frequency are then

$$t_0 \equiv \int_{-\infty}^{\infty} dt \cdot t\,|g(t)|^2, \qquad \omega_0 \equiv \int_{-\infty}^{\infty} d\omega \cdot \omega\,|\hat{g}(\omega)|^2, \tag{2.12}$$

respectively. A common way of defining T and Ω is as the *standard deviations* from t_0 and ω_0:

$$T^2 \equiv \int_{-\infty}^{\infty} dt\ (t - t_0)^2 |g(t)|^2, \qquad \Omega^2 \equiv \int_{-\infty}^{\infty} d\omega\, (\omega - \omega_0)^2 |\hat{g}(\omega)|^2. \tag{2.13}$$

With these definitions, it can be shown that $4\pi\,\Omega \cdot T \geq 1$, so in this case $c = 1/4\pi$.†

Let us illustrate the foregoing with an example. Choose an arbitrary positive constant a and let g be the *Gaussian* window

$$g(t) = (2a)^{1/4}\, e^{-\pi a t^2}. \tag{2.14}$$

Then $\|g\| = 1$,

$$\hat{g}(\omega) = (2/a)^{1/4}\, e^{-\pi\omega^2/a}, \tag{2.15}$$

and

$$t_0 = \omega_0 = 0, \qquad T = \sqrt{\frac{1}{4\pi a}}, \qquad \Omega = \sqrt{\frac{a}{4\pi}}. \tag{2.16}$$

---

† This is the *Heisenberg* form of the uncertainty relation; see Messiah (1961). Contrary to some popular opinion, it is a general property of functions, not at all restricted to quantum mechanics. The connection with the latter is due simply to the fact that in quantum mechanics, if t denotes the position coordinate of a particle, then $2\pi\hbar\omega$ is interpreted as its momentum (where $\hbar$ is Planck's constant), $|f(t)|^2$ and $|\hat{f}(\omega)|^2$ as its probability distributions in space and momentum, and T and $2\pi\hbar\Omega$ as the uncertainties in its position and momentum, respectively. Then the inequality $2\pi\hbar\Omega \cdot T \geq \hbar/2$ is the usual Heisenberg uncertainty relation.

Hence for Gaussian windows, the Heisenberg inequality becomes an *equality*, i.e., $4\pi\,\Omega T = 1$. In fact, equality is attained *only* for such windows and their translates in time or frequency (see Messiah [1961]). Note that Gaussian windows are not causal, i.e., they do not vanish for $t > 0$.

Because of its quantum mechanical origin and connotation, the uncertainty principle has acquired an aura of mystique. It can be stated in a great variety of forms that differ from one another in the way the "widths" of the signal in the time and frequency domains are defined. (There is even a version in which T and Ω are replaced by the *entropies* of the signal in the time and frequency domains; see Zakai [1960] and Białynicki–Birula and Mycielski [1975]). However, all these forms are based on one simple fundamental fact: *The precise measurements of time and frequency are fundamentally incompatible, since frequency cannot be measured instantaneously.* That is, if we want to claim that a signal "has frequency ω_0," then the signal must be observed for at least one period, i.e., for a time interval $\Delta t \geq 1/\omega_0$. (The larger the number of periods for which the signal is observed, the more meaningful it becomes to say that it has frequency ω_0.) Hence we cannot say with certainty exactly *when* the signal has this frequency! It is this basic incompatibility that makes the WFT so subtle and, at the same time, so interesting. The solution offered by the WFT is to observe the signal $f(t)$ over the length of the window $\bar{g}(u-t)$ such that the time parameter t occurring in $\tilde{f}(\omega, t)$ is no longer *sharp* (as was u in $f(u)$) but actually represents a *time interval* (e.g., $[t-T, t]$, if $\operatorname{supp} g \subset [-T, 0]$). As seen from (2.11), the frequency ω occurring in $\tilde{f}(\omega, t)$ is not sharp either but it represents a *frequency band* determined by the spread of the frequency window $\overline{\hat{g}}(\nu - \omega)$. The WFT therefore represents a mutual compromise where both time and frequency acquire an approximate, *macroscopic* significance, rather than an exact, *microscopic* significance. Roughly speaking, *the choice of a window determines the dividing line between time and frequency*: Variations in $f(t)$ over time intervals much longer than T show up in the time behavior of $\tilde{f}(\omega, t)$, while those over time intervals much shorter than T become Fourier-transformed and show up in the frequency behavior of $\tilde{f}(\omega, t)$. For example, an elephant's ear can be expected to have a much longer T-value than that of a mouse. Consequently, what sounds like a rhythm or a flutter (time variation) to a mouse may be perceived as a low-frequency tone by an elephant. (However, we remind the reader that the WFT model of hearing is mainly academic, as it is incorrect from a physiological point of view!)

2.3 The Reconstruction Formula

The WFT is a real-time replacement for the Fourier transform, giving the *dynamical* (time-varying) frequency distribution of $f(t)$. The next step is to find a replacement for the inverse Fourier transform, i.e., to *reconstruct* f from $\tilde{f}$. For this purpose, note that since $\tilde{f}(\omega, t) = \hat{f}_t(\omega)$, we can apply the inverse Fourier transform with respect to the variable ω to obtain

$$\bar{g}(u-t)\, f(u) \equiv f_t(u) = \int_{-\infty}^{\infty} d\omega\; e^{2\pi\, i\omega u}\, \tilde{f}(\omega, t). \tag{2.17}$$

We cannot recover $f(u)$ by dividing by $\bar{g}(u-t)$, since this function may vanish. Instead, we multiply (2.17) by $g(u-t)$ and integrate over t:

$$\int_{-\infty}^{\infty} dt\; |g(u-t)|^2\, f(u) = \int_{-\infty}^{\infty} dt \int_{-\infty}^{\infty} d\omega\; e^{2\pi\, i\omega u}\, g(u-t)\tilde{f}(\omega, t). \tag{2.18}$$

But the left-hand side is just $\|g\|^2\, f(u)$, hence

$$f(u) = C^{-1} \int_{-\infty}^{\infty} dt \int_{-\infty}^{\infty} d\omega\; e^{2\pi\, i\omega u}\, g(u-t)\tilde{f}(\omega, t), \tag{2.19}$$

where we have set $C \equiv \|g\|^2$. This makes sense if g is any nonzero vector in $L^2(\mathbf{R})$, since then $0 < \|g\| < \infty$. By the definition of $g_{\omega,t}$, (2.19) can be written

$$f(u) = C^{-1} \iint d\omega\, dt\; g_{\omega,t}(u)\tilde{f}(\omega, t), \tag{2.20}$$

which is the desired reconstruction formula.

We are now in a position to combine the WFT and its inverse to obtain the analogue of a resolution of unity. To do so, we first summarize our results using the language of vectors and operators (Sections 1.1-1.3). Given a window function $g \in L^2(\mathbf{R})$, we have the following *time-frequency analysis* of a signal $f \in L^2(\mathbf{R})$:

$$\tilde{f}(\omega, t) = \langle\, g_{\omega,t}, f\,\rangle \equiv g_{\omega,t}^{*} f\,. \tag{2.21}$$

The corresponding *synthesis* or reconstruction formula is

$$f = C^{-1} \iint d\omega\, dt\; g_{\omega,t}\, \tilde{f}(\omega, t)\,, \tag{2.22}$$

where we have left the u-dependence on both sides implicit, in the spirit of vector analysis (f and $g_{\omega,t}$ are both vectors in $L^2(\mathbf{R})$). Note that the complex exponentials $e_\omega(u) \equiv e^{2\pi\, i\omega u}$ occurring in Fourier analysis, which oscillate forever, have now been replaced by the "notes" $g_{\omega,t}(u)$, which are *local* in time if g has compact support or decays rapidly. By substituting (2.21) into (2.22), we obtain

$$f = C^{-1} \iint d\omega\, dt\; g_{\omega,t}\, g_{\omega,t}^{*}\, f \tag{2.23}$$

for all $f \in L^2(\mathbf{R})$, hence

$$C^{-1} \iint d\omega\, dt\; g_{\omega,t}\, g_{\omega,t}^{*} = I, \tag{2.24}$$

where I is the identity operator in $L^2(\mathbf{R})$. This is analogous to the resolution of unity in terms of an orthonormal basis $\{\mathbf{b}_n\}$ as given by (1.42) but with the sum $\sum_n$ replaced by the integral $C^{-1}\iint d\omega\, dt$. Equation (2.24) is called a *continuous resolution of unity* in $L^2(\mathbf{R})$, with the "notes" $g_{\omega,t}$ playing a role similar to that of a basis.† This idea will be further investigated and generalized in the following chapters. Note that due to the Hermiticity property (H) of the inner product,

$$f^* g_{\omega,t} = \overline{g^*_{\omega,t} f} = \overline{\tilde{f}(\omega,t)}, \tag{2.25}$$

so that (2.24) implies

$$\begin{aligned} \|f\|^2 \equiv f^* f &= C^{-1} \iint d\omega\, dt\; f^* g_{\omega,t}\, g^*_{\omega,t} f \\ &= C^{-1} \iint d\omega\, dt\; |\tilde{f}(\omega,t)|^2. \end{aligned} \tag{2.26}$$

We may interpret

$$\rho(\omega,t) \equiv C^{-1} |\tilde{f}(\omega,t)|^2$$

as the energy density per unit area of the signal in the time-frequency plane. But area in that plane is measured in cycles, hence $\rho(\omega,t)$ *is the energy density per cycle* of f. Equation (2.26) shows that if a given function $h(\omega,t)$ is to be the WFT of *some* time signal $f \in L^2(\mathbf{R})$, i.e., $h = \tilde{f}$, then h must necessarily be square-integrable in the *joint time-frequency domain.* That is, h must belong to the Hilbert space $L^2(\mathbf{R}^2)$ of all functions with finite norms and inner products defined by

$$\begin{aligned} \|h\|^2_{L^2(\mathbf{R}^2)} &= \iint d\omega\, dt\; |h(\omega,t)|^2, \\ \langle\, h_1, h_2 \,\rangle_{L^2(\mathbf{R}^2)} &= \iint d\omega\, dt\; \bar{h}_1(\omega,t)\, h_2(\omega,t). \end{aligned} \tag{2.27}$$

Equation (2.26) plays a role similar to the Plancherel formula, stating that $\|f\|^2_{L^2(\mathbf{R})} = C^{-1}\|\tilde{f}\|^2_{L^2(\mathbf{R}^2)}$. Since the norm determines the inner product by the polarization identity (1.92), (2.26) implies a counterpart of Parseval's identity:

$$\langle\, f_1, f_2 \,\rangle_{L^2(\mathbf{R})} = C^{-1} \langle\, \tilde{f}_1, \tilde{f}_2 \,\rangle_{L^2(\mathbf{R}^2)} \tag{2.28}$$

for all $f_1, f_2 \in L^2(\mathbf{R})$. (This can also be obtained directly by substituting (2.24) into $f_1^*\, f_2 = f_1^*\, I\, f_2$.)

To simplify the notation, we now normalize g, so that $C = 1$.

---

† Recall that no such resolution of unity existed in terms of the functions e_ω associated with the ordinary Fourier transform, since $e_\omega \notin L^2(\mathbf{R})$; see (1.96) and the discussion below it.

2.4 Signal Processing in the Time-Frequency Domain

The WFT converts a function $f(u)$ of one variable into a function $\tilde{f}(\omega, t)$ of two variables without changing its total energy. This may seem a bit puzzling at first, and it is natural to wonder where the "catch" is. Indeed, not *every* function $h(\omega, t)$ in $L^2(\mathbf{R}^2)$ is the WFT of a time signal. That is, the space of all windowed Fourier transforms of square-integrable time signals,

$$\mathcal{F} \equiv \{\tilde{f} : f \in L^2(\mathbf{R})\}, \tag{2.29}$$

is a *proper subspace* of $L^2(\mathbf{R}^2)$. (This means simply that it is a subspace of $L^2(\mathbf{R}^2)$ but not equal to the latter.) To see this, recall that if $h = \tilde{f}$ for some $f \in L^2(\mathbf{R})$, then h is necessarily *bounded* (2.9). Hence any square-integrable function $h(\omega, t)$ that is unbounded cannot belong to $\mathcal{F}$, and such functions are easily constructed. Thus, being square-integrable is a *necessary* but not *sufficient* condition for $h \in \mathcal{F}$. The next theorem gives an extra condition that is both necessary and sufficient.

Theorem 2.2. *A function $h(\omega, t)$ belongs to $\mathcal{F}$ if and only if it is square-integrable and, in addition, satisfies*

$$h(\omega', t') = \iint d\omega\, dt\, K(\omega', t' \,|\, \omega, t) h(\omega, t)\,, \tag{2.30}$$

where

$$\begin{aligned} K(\omega', t' \,|\, \omega, t) &\equiv g^*_{\omega', t'}\, g_{\omega, t} \equiv \langle\, g_{\omega', t'}\,, g_{\omega, t}\,\rangle = \int_{-\infty}^{\infty} du\ \bar{g}_{\omega', t'}(u)\, g_{\omega, t}(u) \\ &= \int_{-\infty}^{\infty} du\ e^{-2\pi\, i(\omega' - \omega)u}\, \bar{g}(u - t')\, g(u - t). \end{aligned} \tag{2.31}$$

Proof: Our proof will be simple and informal. First, suppose $h = \tilde{f} \in \mathcal{F}$. Then h is square-integrable since $\mathcal{F} \subset L^2(\mathbf{R}^2)$, and we must show that (2.30) holds. Now $h(\omega, t) = \tilde{f}(\omega, t) = g^*_{\omega, t} f$, hence (2.24) implies (recalling that we have set $C = 1$)

$$\begin{aligned} h(\omega', t') = g^*_{\omega', t'}\, I f &= \iint d\omega\, dt\ g^*_{\omega', t'}\, g_{\omega, t}\, g^*_{\omega, t}\, f \\ &= \iint d\omega\, dt\ K(\omega', t' \,|\, \omega, t) h(\omega, t)\,, \end{aligned} \tag{2.32}$$

so (2.30) holds. This proves that the two conditions in the theorem are necessary for $h \in \mathcal{F}$. To prove that they are also sufficient, let $h(\omega, t)$ be any square-integrable function that satisfies (2.30). We will *construct* a signal $f \in L^2(\mathbf{R})$ such that $h = \tilde{f}$. Namely, let

$$f \equiv \iint d\omega\, dt\ g_{\omega, t}\, h(\omega, t)\ . \tag{2.33}$$

(Again, this is a *vector* equation!) Then (2.30) implies

$$\begin{aligned}\|f\|^2 &= \iint d\omega'\, dt' \iint d\omega\, dt\; \bar{h}(\omega', t') \langle\, g_{\omega',t'}\,, g_{\omega,t}\,\rangle\, h(\omega, t)\\ &= \iint d\omega'\, dt' \iint d\omega\, dt\; \bar{h}(\omega', t')\, K(\omega', t' \,|\, \omega, t)\, h(\omega, t) \qquad (2.34)\\ &= \iint d\omega\, dt\; |h(\omega, t)|^2 = \|h\|^2_{L^2(\mathbf{R}^2)} < \infty\,,\end{aligned}$$

which shows that $f \in L^2(\mathbf{R})$. Furthermore, by (2.30),

$$\begin{aligned}\tilde{f}(\omega', t') \equiv g^*_{\omega',t'}\, f &= \iint d\omega\, dt\; g^*_{\omega',t'}\, g_{\omega,t}\, h(\omega, t)\\ &= \iint d\omega\, dt\;\; K(\omega', t' \,|\, \omega, t)\, h(\omega, t) \qquad (2.35)\\ &= h(\omega', t')\,,\end{aligned}$$

hence $h = \tilde{f}$, proving that $h \in \mathcal{F}$. ∎

The function $K(\omega', t' \,|\, \omega, t)$ is called the *reproducing kernel* determined by the window g, and we call (2.30) the associated *consistency condition.*

It is easy to see why not just *any* square-integrable function $h(\omega, t)$ can be the WFT of a time signal: If that were the case, then we could design time signals with arbitrary time-frequency properties and thus violate the uncertainty principle! As will be seen in Chapter 4, reproducing kernels and consistency conditions are naturally associated with a structure we call *generalized frames*, which includes continuous and discrete time-frequency analyses and continuous and discrete wavelet analyses as special cases.

Suppose now that we are given an arbitrary square-integrable function $h(\omega, t)$ that does not necessarily satisfy the consistency condition. Define $f_h(u)$ by blindly substituting h into the reconstruction formula (2.33), i.e.,

$$f_h \equiv \iint d\omega\, dt\;\, g_{\omega,t}\, h(\omega, t)\,. \qquad (2.36)$$

What can be said about f_h?

Theorem 2.3. *f_h belongs to $L^2(\mathbf{R})$, and it is the unique signal with the following property: For any other signal $f \in L^2(\mathbf{R})$,*

$$\|h - \tilde{f}\|_{L^2(\mathbf{R}^2)} > \|h - \tilde{f}_h\|_{L^2(\mathbf{R}^2)}. \qquad (2.37)$$

The meaning of this theorem is as follows: Suppose we want a signal with certain specified properties in both time and frequency. In other words, we look for $f \in L^2(\mathbf{R})$ such that $\tilde{f}(\omega, t) = h(\omega, t)$, where $h \in L^2(\mathbf{R}^2)$ is given. Theorem 2.2 tells us that no such signal can exist unless h satisfies the consistency condition.

The signal f_h defined above comes *closest*, in the sense that the "distance" of its WFT $\tilde{f}_h$ to h is a minimum. We call f_h the *least-squares approximation* to the desired signal. When $h \in \mathcal{F}$, (2.36) reduces to the reconstruction formula. The proof of Theorem 2.3 will be given in a more general context in Chapter 4.

The least-squares approximation can be used to process signals simultaneously in time and in frequency. Given a signal f, we may first compute $\tilde{f}(\omega, t)$ and then modify it in any way desirable, such as by suppressing some frequencies and amplifying others while simultaneously localizing in time. Of course, the modified function $h(\omega, t)$ is generally no longer the WFT of *any* time signal, but its least-squares approximation f_h comes closest to being such a signal, in the above sense. Another aspect of the least-squares approximation is that even when we do not purposefully tamper with $\tilde{f}(\omega, t)$, "noise" is introduced into it through round-off error, transmission error, human error, etc. Hence, by the time we are ready to reconstruct f, the resulting function h will no longer belong to $\mathcal{F}$. ($\mathcal{F}$, being a subspace, is a very *thin* set in $L^2(\mathbf{R}^2)$, like a plane in three dimensions. Hence any random change is almost certain to take h out of $\mathcal{F}$.) The "reconstruction formula" in the form (2.36) then automatically yields the least-squares approximation to the original signal, given the incomplete or erroneous information at hand. This is a kind of built-in stability of the WFT reconstruction, related to *oversampling*. It is typical of reconstructions associated with generalized frames, as explained in detail in Chapter 4.

Exercises

2.1. Prove (2.11) by computing $\hat{g}_{\omega,t}(\omega')$.

2.2. Prove (2.15) and (2.16). *Hint: If* $b > 0$ *and* $u \in \mathbf{R}$, *then*

$$\int_{-\infty}^{\infty} dt\; e^{-\pi b(t+iu)^2} = b^{-1/2}. \tag{2.38}$$

2.3. (a) Show that translation in the time domain corresponds to *modulation* in the frequency domain, and that translation in the frequency domain corresponds to modulation in the time domain, according to

$$\begin{aligned} (f(t-t_0))^\wedge(\omega) &= e^{-2\pi i\omega t_0}\,\hat{f}(\omega), \\ (h(\omega-\omega_0))^\vee(t) &= e^{2\pi i\omega_0 t}\,\check{h}(t). \end{aligned} \tag{2.39}$$

(b) Given a function f_1 such that $f_1(t) \approx 0$ for $|t - t_0| \geq T/2$ and $|\hat{f}_1(\omega)| \approx 0$ for $|\omega - \omega_0| \geq \Omega/2$, use (2.39) to construct a function $f(t)$ such that $f(t) \approx 0$ for $|t| \geq T/2$ and $\hat{f}(\omega) \approx 0$ for $|\omega| \geq \Omega/2$. How can this be used to extend the qualitative explanation of the uncertainty principle given at the end of Section 2.2 to functions centered about arbitrary times t_0 and frequencies ω_0?

2.4. Let $f(t) = e^{2\pi i\alpha t}$, where $\alpha \in \mathbf{R}$. Note that $f \notin L^2(\mathbf{R})$. Nevertheless, the WFT of f with respect to the Gaussian window $g(t) = e^{-\pi t^2}$ is well defined. Use (2.38) to compute $\tilde{f}(\omega, t)$, and show that $|\tilde{f}(\omega, t)|^2$ is maximized when $\omega = \alpha$. Interpret this in view of the fact that $\tilde{f}(\omega, t)$ represents the frequency distribution of f "at" the (macroscopic) time t.

Chapter 3

Continuous Wavelet Transforms

Summary: The WFT localizes a signal simultaneously in time and frequency by "looking" at it through a window that is translated in time, then translated in frequency (i.e., *modulated* in time). These two operations give rise to the "notes" $g_{\omega,t}(u)$. The signal is then reconstructed as a superposition of such notes, with the WFT $\tilde{f}(\omega, t)$ as the coefficient function. Consequently, any features of the signal involving time intervals much shorter than the width T of the window are *underlocalized* in time and must be obtained as a result of constructive and destructive interference between the notes, which means that "many notes" must be used and $\tilde{f}(\omega, t)$ must be spread out in frequency. Similarly, any features of the signal involving time intervals much longer than T are *overlocalized* in time, and their construction must again use "many notes," with $\tilde{f}(\omega, t)$ spread out in time. This can make the WFT an inefficient tool for analyzing regular time behavior that is either very rapid or very slow relative to T. The wavelet transform solves both of these problems by replacing modulation with *scaling* to achieve frequency localization.

Prerequisites: Chapters 1 and 2.

3.1 Motivation and Definition of the Wavelet Transform

Recall how the *ordinary* Fourier transform and its inverse are used to reproduce a function $f(t)$ as a superposition of the complex exponentials $e_\omega(t) = e^{2\pi\, i\omega t}$. The Fourier transform $\hat{f}(\omega) \equiv e_\omega^* f$ gives the coefficient function that must be used in order that the (continuous) superposition $\int d\omega\, e_\omega(t)\, \hat{f}(\omega)$ be equal to $f(t)$. Now $e_\omega(t)$ is spread out over all of time, since $|e_\omega(t)| \equiv 1$. How can a "sharp" feature of the signal, such as a spike, be reproduced by superposing such e_ω's? The extreme case is that of the "delta function" or *impulse* $\delta(t)$, which can be viewed as a limit of spikes with ever-increasing height and sharpness, such as

$$\delta_\varepsilon(t) = \frac{1}{\pi} \frac{\varepsilon}{t^2 + \epsilon^2}, \qquad \varepsilon > 0. \tag{3.1}$$

Note that $\int_{-\infty}^{\infty} dt\, \delta_\varepsilon(t) = 1$ for all $\varepsilon > 0$, and as $\varepsilon \to 0^+$, then $\delta_\varepsilon(t) \to 0$ when $t \neq 0$ and $\delta_\varepsilon(t) \to \infty$ when $t = 0$. Therefore, if $f(t)$ is a "nice" function (e.g., continuous and bounded), then $\lim_{\varepsilon \to 0^+} \int_{-\infty}^{\infty} dt\, \delta_\varepsilon(t)\, f(t) = f(0)$. For such functions, one therefore writes[†]

$$\int_{-\infty}^{\infty} dt\ \delta(t)\, f(t) = f(0) \tag{3.2}$$

and, in particular, $\int_{-\infty}^{\infty} dt\ \delta(t) = 1$. Hence the Fourier transform of δ is

$$\hat{\delta}(\omega) = \int_{-\infty}^{\infty} dt\ e^{-2\pi\, i\omega t}\, \delta(t) \equiv 1, \tag{3.3}$$

and

$$\delta(t) = \int_{-\infty}^{\infty} d\omega\ e^{2\pi\, i\omega t}\ \hat{\delta}(\omega) = \int_{-\infty}^{\infty} d\omega\, e^{2\pi\, i\omega t}\,. \tag{3.4}$$

That is, complex exponentials with *all* frequencies ω must be combined with the same amplitude $\hat{\delta}(\omega) \equiv 1$ in order to produce $\delta(t)$. This is an extreme example of the uncertainty principle (Section 2.2): $\delta(t)$ has width $T = 0$, therefore its Fourier transform must have width $\Omega = \infty$. What is the *intuitive* meaning of (3.4)? Clearly, when $t = 0$, the right-hand side gives $\delta(0) = \infty$ as desired. When $t \neq 0$, the right-hand side must vanish, since $\delta(t) = 0$. This can only be due to *destructive interference* between the different complex exponentials with different frequencies. Thus we can summarize (3.4) by saying that it is due entirely to *interference* (constructive when $t = 0$, destructive otherwise)! To obtain an ordinary spike with positive width $T \sim \varepsilon$ like $\delta_\varepsilon(t)$, we must combine exponentials in a frequency band of finite width $\Omega \sim 1/\varepsilon$. *The narrower the spike, the more high-frequency exponentials must enter into its construction in order to achieve the desired sharpness of the spike. But then, since these exponentials oscillate forever, we need even more exponentials in order cancel the previous ones before and after the spike occurs!* This shows the tremendous inefficiency of ordinary Fourier analysis in dealing with *local* behavior, and it is clear that this inefficiency is due entirely to its nonlocal nature: We are trying to compose local variations in time by using the nonlocal exponentials e_ω. The windowed Fourier transform solves this problem only partially. $\tilde{f}(\omega, t)$ is used as a coefficient function to superimpose the "notes" $g_{\omega,t}$ in order to reconstruct f. When the window $g(u)$ has width T, then the notes $g_{\omega,t}(u)$ have the same width. A sharp spike in $f(u)$ of width $\Delta u \ll T$, located near $u = u_0$, can only be produced by superimposing notes $g_{\omega,t}(u)$ with $t \approx u_0$ and many different frequencies ω. We must again rely on constructive and destructive interference

---

[†] Equation (3.2) must be viewed as "symbolic" in the sense that $\delta(t)$ cannot actually be defined as a *function*. If it were an ordinary function, it would have to vanish for all $t \neq 0$; hence the integral in (3.2) would vanish. Objects such as δ, which were introduced by P.A.M. Dirac in the context of quantum mechanics (see Messiah [1961]), now belong to a well-established branch of mathematics called the *theory of distributions* or *generalized functions*. See Gel'fand and Shilov (1964).

among these notes to produce the spike, since its width is much narrower than that of the notes. The situation is not quite as bad as in ordinary Fourier analysis, since the destructive interference among the notes is only needed within a time interval of order T around $u = u_0$ rather than for all time. Still, this shows that the WFT can be a rather inefficient tool when very short time intervals are of interest.

A similar situation occurs when very *long and smooth* features of the signal are to be reproduced by the WFT, i.e., ones characterized by time intervals $\Delta u \gg T$. In this case, many low-frequency notes must be combined over the time interval Δu. (High-frequency notes produce rough behavior rather than smooth behavior.) Hence $\tilde{f}(\omega, t)$ must be spread out in time. The point uniting both cases ($\Delta u \ll T$ and $\Delta u \gg T$) is that *the WFT introduces a scale into the analysis and reconstruction of signals, namely the width of the window. Features much shorter and much longer than this scale can only be produced by combining "many notes," even though these features may be very simple in that they contain little information. Roughly speaking, features with time scales much shorter than T are synthesized in the frequency domain (i.e., by combining many notes with similar values of t but different frequencies), while features with time scales much longer than T are synthesized in the time domain (by combining many notes with different values of t).*

Recall that the WFT was inspired as a model (however inaccurate) for the process of hearing. Since an ear *does* have a characteristic "response time" (which is related, among other things, to its physical size), it is not necessarily bad to introduce such a time interval T into the analysis. However, it is legitimate to wonder if some way can be found to process signals locally without prejudice to scale. Wavelet analysis is precisely such a *scale-independent* method. Again, one begins with a (complex-valued) window function $\psi(t)$, this time called a *mother wavelet* or *basic wavelet.* As before, ψ introduces a scale (its width) into the analysis. Since we want to avoid commitment to any particular scale, we use not only ψ but *all possible scalings of* ψ. This can be done as follows: Fix an arbitrary $p \geq 0$, and for any real number $s \neq 0$, define

$$\psi_s(u) \equiv |s|^{-p}\, \psi\left(\frac{u}{s}\right). \tag{3.5}$$

To visualize ψ_s, imagine that we already have a graph of $\psi(u)$, with u measured in minutes, and then decide to replot ψ with u measured in *seconds.* Clearly, the new graph will look much like the old one but be *stretched* by a factor of 60. In fact, it is nothing but the graph of $\psi(u/60)$. For general $s > 1$, $\psi_s(u)$ is a version of ψ *stretched by the factor* s in the horizontal direction. Similarly, if $0 < s < 1$, then ψ_s is a version of ψ *compressed* in the horizontal direction. If $s = -1$, then ψ_s is a *reflected* version of ψ. If $-1 < s < 0$, then ψ_s is a reflected *and* compressed version of ψ. Finally, if $s < -1$, then ψ_s is a reflected and stretched version of ψ. We refer to s as a *scale factor.*

The factor $|s|^{-p}$ in (3.5) has a similar effect on the vertical direction, i.e., on the dependent variable. If $p > 0$, then ψ is *compressed* along the vertical whenever it is stretched along the horizontal and it is stretched along the vertical whenever compressed along the horizontal. (If $p < 0$, then ψ is simultaneously stretched or compressed in both directions, but only conventions with $p \geq 0$ are used.) For example, if $p = 1$, then the integral $\int du\,|\psi(u)|$ is unaffected by the scaling operation. If $p = 1/2$, then the integral of $|\psi(u)|$ is affected but the L^2 norm is not, i.e., $\|\psi_s\| = \|\psi\|$. We shall see that *the actual value of p is completely irrelevant to the basic theory.* Different choices are made in the literature: Chui (1992a), Daubechies (1992), and Meyer (1993a, b) use $p = 1/2$; Mallat and Zhong (1992) use $p = 1$. When dealing with orthonormal bases of wavelets (Chapter 8), the choice $p = 0$ is sometimes convenient. This results in a confusing proliferation of notations and square roots of 2 (much like the different conventions for the Fourier transform). In order to show the irrelevance of the choice of p and unify the different conventions, we leave p arbitrary for the time being. This will also have the benefit of tracing the role of p in the reconstruction formula. In Chapters 6–8, we bow to the prevailing convention and set $p = 1/2$.

Time localization of signals is achieved by "looking" at them through translated versions of ψ_s. If $\psi(u)$ is supported on an interval of length T near $u = 0$, then $\psi_s(u)$ is supported on an interval of length $|s|T$ near $u = 0$ and the function

$$\psi_{s,t}(u) \equiv \psi_s(u - t) = |s|^{-p}\,\psi\left(\frac{u - t}{s}\right) \tag{3.6}$$

is supported on an interval of length $|s|T$ near $u = t$. In Figure 3.1, we illustrate scaled and shifted versions of the wavelet $\psi(u) = u\,e^{-u^2}$.

We assume that ψ belongs to $L^2(\mathbf{R})$. Then so does $\psi_{s,t}$, since

$$\|\psi_{s,t}\|^2 = |s|^{-2p}\int_{-\infty}^{\infty} du\,\left|\psi\left(\frac{u - t}{s}\right)\right|^2 = |s|^{1-2p}\,\|\psi\|^2. \tag{3.7}$$

The functions $\psi_{s,t}(u)$ are the *wavelets* generated by ψ and will play a role similar to that played earlier by the "notes" $g_{\omega,t}(u)$. The *continuous wavelet transform* (CWT) of a signal $f(t)$ is defined as

$$\tilde{f}(s,t) \equiv \int_{-\infty}^{\infty} du\,\bar{\psi}_{s,t}(u)\,f(u) = \langle\,\psi_{s,t}\,, f\,\rangle = \psi_{s,t}^*\,f\,. \tag{3.8}$$

The inner product on the right exists since both f and $\psi_{s,t}$ belong to $L^2(\mathbf{R})$.

The wavelet transform $\tilde{f}(s,t)$ can be given a more or less direct interpretation, depending on the wavelet used: As a function of t at a fixed value of s, it represents the *detail* contained in the signal $f(t)$ at the scale s. This idea will be explained through examples in Section 3.4.

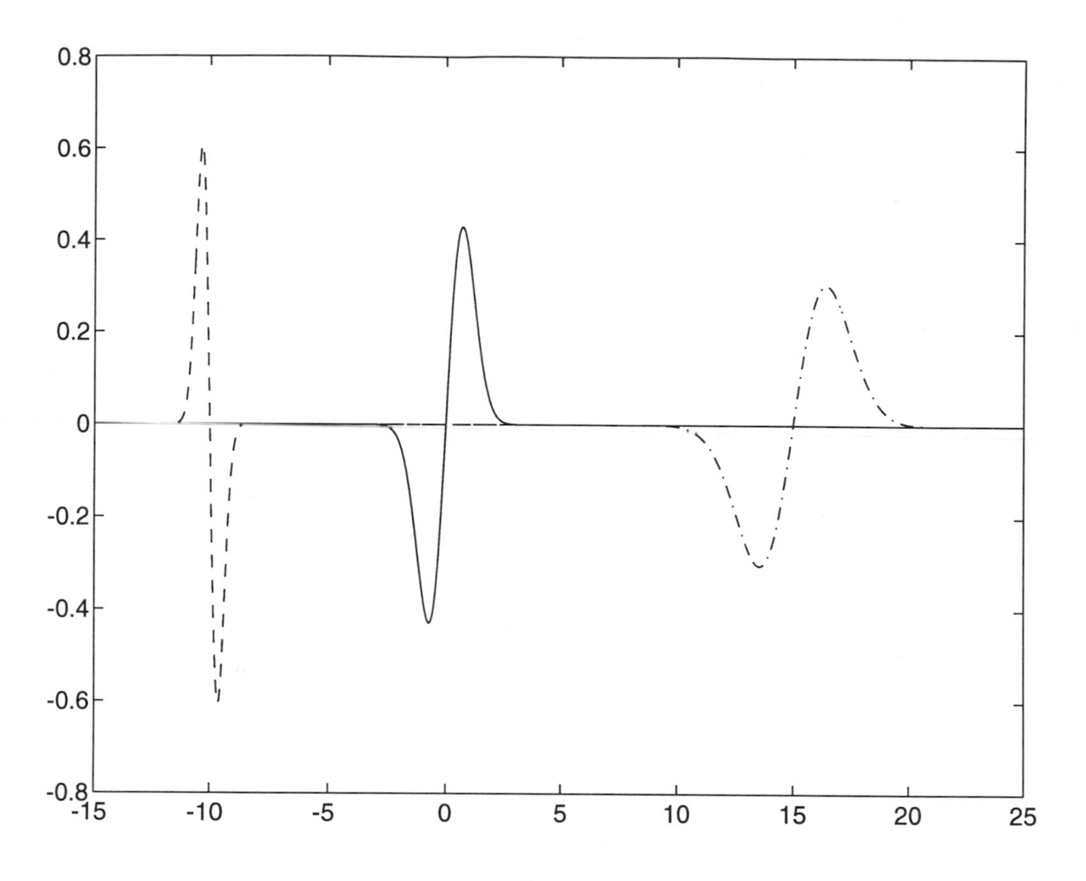

Figure 3.1. Graphs of $\psi(u) = u\,e^{-u^2}$ (*center*) and its scaled and shifted versions $\psi_{2,15}(u)$ (*right*) and $\psi_{-.5,-10}(u)$ (*left*). Note that $\psi_{-.5,-10}$ is reflected because $s < 0$.

The use of scaled and translated versions of a single function was proposed by Morlet (1983) for the analysis of seismic data. (See also Grossmann and Morlet (1984).) As explained in Chapters 9–11, a similar method is useful (and for similar reasons) in the analysis of radar and sonar data. For a good general review of both continuous and discrete wavelet transforms, see Heil and Walnut (1989).

3.2 The Reconstruction Formula

Our next step is to reconstruct $f(u)$ from its CWT $\tilde{f}(s,t)$. The key is to apply Parseval's identity to (3.8),

$$\tilde{f}(s,t) = \langle\, \psi_{s,t}\,, f\,\rangle = \langle\, \hat{\psi}_{s,t}\,, \hat{f}\,\rangle, \tag{3.9}$$

and to compute

$$\hat{\psi}_{s,t}(\omega) = |s|^{1-p}\, e^{-2\pi\, i\omega t}\, \hat{\psi}(s\omega). \tag{3.10}$$

Thus (3.9) becomes

$$\begin{aligned} \tilde{f}(s,t) &= |s|^{1-p} \int_{-\infty}^{\infty} d\omega\, e^{2\pi\, i\omega t}\, \overline{\hat{\psi}}(s\omega)\, \hat{f}(\omega) \\ &= |s|^{1-p} \left(\overline{\hat{\psi}}(s\omega)\, \hat{f}(\omega) \right)^{\vee} (t). \end{aligned} \tag{3.11}$$

That is, as a function of t, $\tilde{f}(s,t)$ is the inverse Fourier transform of $|s|^{1-p}\, \overline{\hat{\psi}}(a\omega)\, \hat{f}(\omega)$ with respect to the variable ω, with s playing the role of a parameter throughout. Thus, applying the Fourier transform with respect to t to both sides gives

$$\int_{-\infty}^{\infty} dt\; e^{-2\pi\, i\omega t}\; \tilde{f}(s,t) = |s|^{1-p}\, \overline{\hat{\psi}}(s\omega)\, \hat{f}(\omega). \tag{3.12}$$

If we can recover $\hat{f}(\omega)$, then $f(t)$ can be found. We cannot simply divide (3.12) by $\overline{\hat{\psi}}(s\omega)$, since this function may vanish. Instead, we multiply both sides of the equation by $\hat{\psi}(s\omega)$ and integrate with respect to the scale parameter s. However, it will be necessary to use a *weight function* $w(s)$, which is presently unknown but will be determined by the reconstruction. In later chapters, when studying discrete wavelet analysis, we will need to deal separately with positive scales ($s > 0$) and negative scales ($s < 0$). For this reason, we begin by integrating only over $s > 0$. This has the added advantage that the roles of positive and negative scales in the reconstruction will be clear from the beginning. Thus, (3.12) implies

$$\begin{aligned} \int_{0}^{\infty} ds\, w(s) \int_{-\infty}^{\infty} dt\; e^{-2\pi\, i\omega t}\; \hat{\psi}(s\omega)\, \tilde{f}(s,t) \\ = \int_{0}^{\infty} ds\, w(s)\, s^{1-p}\, |\hat{\psi}(s\omega)|^{2}\, \hat{f}(\omega)\,. \end{aligned} \tag{3.13}$$

Define

$$Y(\omega) \equiv \int_{0}^{\infty} ds\, w(s)\, s^{1-p}\, |\hat{\psi}(s\omega)|^{2} \tag{3.14}$$

and assume, for the moment, that both $Y(\omega)$ and its reciprocal are bounded (except possibly on a set of measure zero), i.e., that

$$A \leq Y(\omega) \leq B \qquad \text{a.e.} \tag{3.15}$$

for some constants A, B with $0 < A \le B < \infty$. (The implications of this assumption will be studied below.) Then by (3.13),

$$\begin{aligned} \hat{f}(\omega) &= Y(\omega)^{-1} \int_0^\infty ds\, w(s) \int_{-\infty}^\infty dt\, e^{-2\pi i\omega t}\, \hat{\psi}(s\omega)\, \tilde{f}(s,t) \\ &= Y(\omega)^{-1} \int_0^\infty ds\, w(s)\, s^{p-1} \int_{-\infty}^\infty dt\, \hat{\psi}_{s,t}(\omega)\, \tilde{f}(s,t) \\ &= \int_0^\infty ds\, w(s)\, s^{p-1} \int_{-\infty}^\infty dt\, \hat{\psi}^{s,t}(\omega)\, \tilde{f}(s,t), \end{aligned} \tag{3.16}$$

where we have defined a new set of "wavelets" $\psi^{s,t}(u)$ through their Fourier transforms $\hat{\psi}^{s,t}(\omega)$ by

$$\hat{\psi}^{s,t}(\omega) \equiv Y(\omega)^{-1} \hat{\psi}_{s,t}(\omega) = s^{1-p}\, Y(\omega)^{-1}\, e^{-2\pi i\omega s}\, \hat{\psi}(s\omega). \tag{3.17}$$

Note that

$$|\hat{\psi}^{s,t}(\omega)| \le A^{-1}\, |\hat{\psi}_{s,t}(\omega)| \tag{3.18}$$

by (3.15), hence by (3.7), $\hat{\psi}^{s,t} \in L^2(\mathbf{R})$, with

$$\|\psi^{s,t}\|^2 \le A^{-2}\, \|\psi_{s,t}\|^2 = |s|^{1-2p} A^{-2}\, \|\psi\|^2. \tag{3.19}$$

In the time domain the new wavelets have the form

$$\begin{aligned} \psi^{s,t}(u) &= s^{1-p} \int_{-\infty}^\infty d\omega\, e^{2\pi i\omega(u-t)}\, Y(\omega)^{-1}\, \hat{\psi}(s\omega) \\ &\equiv \psi^s(u-t), \end{aligned} \tag{3.20}$$

where

$$\begin{aligned} \psi^s(u) &\equiv s^{1-p} \int_{-\infty}^\infty d\omega\, e^{2\pi i\omega u}\, Y(\omega)^{-1}\, \hat{\psi}(s\omega) \\ &= s^{-p} \int_{-\infty}^\infty d\omega\, e^{2\pi i\omega u/s}\, Y(\omega/s)^{-1}\, \hat{\psi}(\omega). \end{aligned} \tag{3.21}$$

The original signal can now be recovered by taking the inverse Fourier transform of (3.16):

$$f(u) = \int_0^\infty ds\, w(s)\, s^{p-1} \int_{-\infty}^\infty dt\, \psi^{s,t}(u)\, \tilde{f}(s,t). \tag{3.22}$$

In vector notation (i.e., dropping u and viewing f and $\psi^{s,t}$ as vectors in $L^2(\mathbf{R})$),

$$f = \int_0^\infty ds\, w(s)\, s^{p-1} \int_{-\infty}^\infty dt\, \psi^{s,t}\, \tilde{f}(s,t). \tag{3.23}$$

This gives f as a superposition of the new vectors $\psi^{s,t}$, with $\tilde{f}(s,t)$ as the coefficient function.

The reconstruction formula (3.23) is not satisfactory as it stands since it requires us to compute a whole new family of vectors $\psi^{s,t}$. This would not be altogether bad if the entire family could be obtained from a single "mother wavelet" by translations and dilations, as was the case for the original wavelets $\psi_{s,t}$. For then we would only need to compute the new mother wavelet and use it to generate the new family. Equation (3.20) shows that translations are no problem, so we concentrate on dilations. By (3.21), a *sufficient* condition for the new family to have a mother is that

$$Y(\omega/s) = Y(\omega) \quad \text{for (almost) all} \quad \omega \in \mathbf{R}, s > 0, \tag{3.24}$$

since then $\psi^s(u) = \psi^1(u/s)$ and, by (3.21),

$$\psi^{s,t}(u) = s^{-p}\,\psi^1\left(\frac{u-t}{s}\right), \tag{3.25}$$

where ψ^1 is ψ^s with $s = 1$. Hence ψ^1 generates the new family just as ψ generated the old one. Equation (3.24) means that $Y(\omega)$ *must be a constant for all* $\omega > 0$, *and a (possibly different) constant for all* $\omega < 0$. (Fix $\omega \neq 0$, and let s vary from 0 to ∞.) We can now use the freedom of choosing the weight function $w(s)$ to arrange for (3.24) to hold. Since ω enters the definition of $Y(\omega)$ only via the product $s\omega$, we require that $w(s)$ be such that

$$ds\, w(s)\, s^{1-p} = \frac{ds}{s}\,. \tag{3.26}$$

This works because, for any $\omega > 0$, the change of variables $s\omega = \xi$ gives

$$Y(\omega) = \int_0^\infty \frac{ds}{s}\,|\hat{\psi}(s\omega)|^2 = \int_0^\infty \frac{d\xi}{\xi}\,|\hat{\psi}(\xi)|^2 \equiv C_+\,, \tag{3.27}$$

and for any $\omega < 0$, the change of variables $s\omega = -\xi$ gives

$$Y(\omega) = \int_0^\infty \frac{ds}{s}\,|\hat{\psi}(s\omega)|^2 = \int_0^\infty \frac{d\xi}{\xi}\,|\hat{\psi}(-\xi)|^2 \equiv C_-\,. \tag{3.28}$$

Equation (3.15) therefore holds for all $\omega \neq 0$ (hence almost everywhere in ω) if and only if

$$0 < C_\pm = \int_0^\infty \frac{d\xi}{\xi}\,|\hat{\psi}(\pm\xi)|^2 < \infty. \tag{3.29}$$

This is called an *admissibility condition* for the mother wavelet ψ. As already noted, the choice of $w(s)$ dictated by (3.26), i.e., $w(s) = s^{p-2}$, not only guarantees the existence of a mother wavelet for the new family but also makes $Y(\omega)$ piecewise constant. That, in turn, simplifies the computation of the new mother wavelet, since now (3.21), with $s = 1$, becomes

$$\begin{aligned}\psi^1(u) &= C_-^{-1}\int_{-\infty}^0 d\omega\; e^{2\pi i\omega u}\,\hat{\psi}(\omega) + C_+^{-1}\int_0^\infty d\omega\; e^{2\pi i\omega u}\,\hat{\psi}(\omega)\\ &\equiv C_-^{-1}\psi_-(u) + C_+^{-1}\psi_+(u),\end{aligned} \tag{3.30}$$

where $\psi_\pm$ are the *positive- and negative-frequency components* of ψ, obtained by taking the inverse Fourier transform of $\hat{\psi}(\omega)$ over just the positive and negative frequencies, respectively. (Such functions, called *analytic signals*, are studied in Section 3.3.) ψ^1 will be called the *reciprocal* mother wavelet of ψ, and $\{\psi^{s,t}\}$, the wavelet family *reciprocal* to $\{\psi_{s,t}\}$. It is easy to show that ψ^1 is also admissible in the sense of (3.29), and that the reciprocal mother wavelet of ψ^1 is ψ. We summarize our results so far.

Theorem 3.1. *Let ψ be admissible in the sense of (3.29), and let $\{\psi^{s,t}\}$ be the wavelet family reciprocal to $\{\psi_{s,t}\}$. Then any signal $f \in L^2(\mathbf{R})$ can be reconstructed from its continuous wavelet transform $\tilde{f}(s,t) = \psi_{s,t}^* f$ by*

$$f(u) = \int_0^\infty ds \cdot s^{2p-3} \int_{-\infty}^\infty dt \; \psi^{s,t}(u)\, \tilde{f}(s,t). \tag{3.31}$$

This gives a resolution of unity in $L^2(\mathbf{R})$:

$$\int_0^\infty ds \cdot s^{2p-3} \int_{-\infty}^\infty dt \; \psi^{s,t}\, \psi_{s,t}^* = I. \tag{3.32}$$

If ψ is such that $C_- = C_+ \equiv C/2$, then (3.30) gives $\psi^1 = (2/C)\psi$, so no computation is involved even in finding ψ^1, except for calculating C. This is the case, for example, if $\psi(t)$ is *real-valued*, since then $\hat{\psi}(-\omega) = \overline{\hat{\psi}(\omega)}$. However, sometimes it is convenient to use wavelets whose Fourier transforms vanish for $\omega < 0$ or $\omega > 0$; then (3.29) cannot be satisfied since either $C_- = 0$ or $C_+ = 0$. A glance at (3.11) shows that if $\hat{\psi}(\omega) = 0$ for $\omega < 0$, then for $s > 0$, the transform $\tilde{f}(s,t)$ depends only on $\hat{f}(\omega)$ with $\omega > 0$. Hence it is *a priori* impossible to recover the negative-frequency part of f by integrating only over positive scales. To recover the whole signal, it is in general necessary to integrate over negative scales as well. (This point will be further elaborated in Section 3.3.) Repeating the derivation of Theorem 3.1 but integrating over all $s \neq 0$ with the weight function $w(s) = |s|^{p-2}$, we get the following result:

Theorem 3.2. *Let ψ be admissible in the sense that $0 < C < \infty$, where*

$$C \equiv \int_{-\infty}^\infty \frac{d\xi}{|\xi|} |\hat{\psi}(\xi)|^2 = C_- + C_+ \,. \tag{3.33}$$

Then $f \in L^2(\mathbf{R})$ can be recovered from $\tilde{f}$ by

$$f(u) = C^{-1} \iint_{\mathbf{R}^2} |s|^{2p-3}\, ds\, dt\, \psi_{s,t}(u)\, \tilde{f}(s,t)\,, \tag{3.34}$$

and we have the corresponding resolution of unity in $L^2(\mathbf{R})$:

$$C^{-1} \iint_{\mathbf{R}^2} |s|^{2p-3}\, ds\, dt\, \psi_{s,t}\, \psi_{s,t}^* = I. \tag{3.35}$$

We emphasize that (3.34) is more general than (3.31). However, if $0 < C_\pm < \infty$, it is preferable to use the reconstruction in Theorem 3.1 since it involves only half the integration and is therefore easier to implement.

An immediate consequence of (3.35) is

$$\begin{aligned} \|f\|^2 &= C^{-1} \iint_{\mathbf{R}^2} |s|^{2p-3} ds\, dt\, f^* \psi_{s,t}\, \psi_{s,t}^* f \\ &= C^{-1} \iint_{\mathbf{R}^2} |s|^{2p-3} ds\, dt\, |\tilde{f}(s,t)|^2. \end{aligned} \tag{3.36}$$

The function

$$\rho(s,t) \equiv C^{-1} |s|^{2p-3}\, |\tilde{f}(s,t)|^2$$

can therefore be interpreted as the *energy density of the signal in the time-scale plane.*

Let us denote the space of all wavelet transforms (with respect to a fixed mother ψ) by $\mathcal{F}$:

$$\mathcal{F} \equiv \{\tilde{f} : f \in L^2(\mathbf{R})\}. \tag{3.37}$$

Then (3.36) shows that $\mathcal{F}$ is a subspace of the space of all measurable functions $h(s,t)$ that are square-integrable with respect to the weight function $|s|^{2p-3}$, i.e., for which the norm

$$\|h\|^2 \equiv C^{-1} \iint_{\mathbf{R}^2} |s|^{2p-3} ds\, dt\, |h(s,t)|^2 < \infty. \tag{3.38}$$

The set of all such functions forms a Hilbert space (with the obvious inner product, obtained by polarizing (3.38)). Let us denote this Hilbert space by $\mathcal{L}$. Then (3.36) states that $\|f\|^2_{L^2(\mathbf{R})} = \|\tilde{f}\|^2_{\mathcal{L}}$. This shows that the map $f \mapsto \tilde{f}$ defines an *operator* $T : L^2(\mathbf{R}) \to \mathcal{L}$ that satisfies $\|f\|^2_{L^2(\mathbf{R})} = \|Tf\|^2_{\mathcal{L}}$, analogous to Plancherel's formula in Fourier analysis. Equation (3.37) states that the *range* of T is $\mathcal{F}$. Applying the polarization identity to (3.36) gives the analogue of Parseval's identity:

$$\langle f, g \rangle_{L^2(\mathbf{R})} = \langle Tf, Tg \rangle_{\mathcal{L}} \quad \text{for all } f, g \in L^2(\mathbf{R}).$$

That is, the operator T preserves inner products, hence "lengths" and "angles." Such an operator is called an *isometry.* But T is only a *partial* isometry because its range $\mathcal{F}$ is not all of $\mathcal{L}$. This is shown next.

Not every function $h \in \mathcal{L}$ can be the CWT of some signal $f \in L^2(\mathbf{R})$. For if $h = \tilde{f}$, then by (3.35)

$$\begin{aligned} h(s',t') &= \psi_{s',t'}^* f = \psi_{s',t'}^*\, I f \\ &= C^{-1} \iint_{\mathbf{R}^2} |s|^{2p-3} ds\, dt\, \psi_{s',t'}^*\, \psi_{s,t}\, \psi_{s,t}^* f \\ &= \iint_{\mathbf{R}^2} |s|^{2p-3} ds\, dt\, K(s',t' \,|\, s,t)\, h(s,t), \end{aligned} \tag{3.39}$$

where

$$\begin{aligned} K(s',t'\,|\,s,t\,) &\equiv C^{-1}\,\langle\,\psi_{s',t'}\,,\psi_{s,t}\,\rangle = C^{-1}\,\langle\,\hat{\psi}_{s',t'}\,,\hat{\psi}_{s,t}\,\rangle \\ &= C^{-1}\,|s's|^{1-p}\int_{-\infty}^{\infty} d\omega\; e^{2\pi\, i\omega(t'-t)}\;\overline{\hat{\psi}}(s'\omega)\,\hat{\psi}(s\omega) \end{aligned} \tag{3.40}$$

by (3.10). $K(s',t'\,|\,s,t\,)$ is called the *reproducing kernel* associated with ψ. As in the case of the WFT, it can be shown that the condition (3.39) is not only necessary but also sufficient for a function $h \in \mathcal{L}$ to belong to $\mathcal{F}$. Again, this condition is related to the linear dependence among the wavelets $\psi_{s,t}$. There is also a "least-squares approximation" f_h determined by functions $h(s,t\,)$ that do not necessarily belong to $\mathcal{F}$. Rather than restate all these facts explicitly for the CWT, we postpone the discussion to the next chapter, where they will be seen to emerge from the structure of *generalized frames*, of which the WFT and CWT, as well as all their discrete versions, are special cases.

3.3 Frequency Localization and Analytic Signals†

The CWT $\tilde{f}$ represents a *time-scale analysis* of the signal. This is different from, but closely related to, the time-frequency analysis discussed in Chapter 2. In this section we study the relation between these two analyses. In addition, we introduce an effective way to treat *real-valued* signals by considering only their positive-frequency components, the so-called *analytic signals* introduced in D. Gabor (1946).

The admissibility conditions of Theorems 3.1 and 3.2 require that $\hat{\psi}(\omega) \to 0$ as $\omega \to 0$. If $\hat{\psi}(\omega)$ is continuous (which is the case, for example, if $\int du\,|\psi(u)| < \infty$), then it follows that $\hat{\psi}(0) = 0$, i.e.,

$$\int_{-\infty}^{\infty} du\;\psi(u) = 0. \tag{3.41}$$

That is, *a wavelet must actually be a "small wave"!* It cannot just be a bump function like a Gaussian, but it must "wiggle" around the time axis. Since $\hat{\psi}(\omega)$ is square-integrable, it must also decay as $|\omega| \to \infty$. If $\hat{\psi}(\omega)$ decays rapidly as $|\omega| \to \infty$ *and* as $\omega \to 0$, then $\hat{\psi}(\omega)$ will be small outside of a "frequency band" $\alpha \le |\omega| \le \beta$ for some fixed $0 < \alpha < \beta$. Then (3.10) shows that $\hat{\psi}_{s,t}(\omega) \approx 0$ outside the frequency band $\alpha/|s| \le |\omega| \le \beta/|s|$, and (3.11) shows that $\tilde{f}(s,t\,)$ contains information about $\hat{f}(\omega)$ mainly in this band, with $\hat{\psi}(s\omega)$ acting as the localizing window in frequency. Just as the WFT uses modulation in the time domain to *translate* the window in frequency, so does the CWT use scaling in the time domain to *scale* the window in frequency. From a physical standpoint, the

† This section is optional.

scaling of frequencies seems to be more natural: In music, going up an octave always involves *doubling* the frequency, rather than shifting it by a constant additive term. Of course, since all scales $s \neq 0$ are used, the reconstruction is highly redundant. Ideally, the entire frequency spectrum should be covered by *discrete* scalings of ψ. This is, in fact, exactly what happens in discrete wavelet analysis.

If the mother wavelet ψ is real-valued, then its Fourier transform is reflection-symmetric, i.e., $\overline{\hat{\psi}(\omega)} = \hat{\psi}(-\omega)$. Hence the effective "frequency band" of ψ mentioned above is symmetric about the origin. Sometimes it is more convenient to work with mother wavelets that have only positive frequencies, i.e., for which $\hat{\psi}(\omega) = 0$ for $\omega < 0$. Such wavelets are necessarily complex-valued (i.e., not real-valued) since $\hat{\psi}(\omega)$ cannot be reflection-symmetric. To see the advantage offered by these one-sided wavelets, consider the analogous situation in ordinary Fourier analysis, where the theory can be formulated either in terms of complex exponentials *or* in terms of sines and cosines. Here, the function $e_1(u) \equiv e^{2\pi\, iu}$ plays the role of a "mother wavelet" in the following sense: It generates all the other complex exponentials by $e_\omega(u) = e_1(\omega u)$, i.e., by *dilations* with scaling factor $s = \omega^{-1} \neq 0$ (take $p = 0$ in (3.5)). Since $\hat{e}_1(\omega) = \delta(\omega - 1)$, the frequency spectrum of e_1 is concentrated at $\omega = 1$, i.e., on an infinitely narrow "band" of positive frequencies. This is, of course, why Fourier analysis gives *precise* frequency information. However, this precision has a price: e_1 is not square-integrable, and translations do not yield anything interesting since $e_\omega(u-t) = e^{-2\pi\, i\omega t}\, e_\omega(u)$. (This would not be the case if the spectrum of e_1 had some thickness, like that of ψ.) Hence the complex exponentials e_ω are analogous to the wavelets $\psi_{s,t}$, and $e_\omega^* f = \hat{f}(\omega)$ is analogous to the CWT $\psi_{s,t}^* f = \tilde{f}(s,t)$. By contrast, the spectra of the functions $\sin(2\pi t)$ and $\cos(2\pi t)$ are less well localized in frequency since they are concentrated at $\omega = \pm 1$. This mixing of positive and negative frequencies makes the orthogonality relations for the sine and cosine functions more complicated, resulting in greater complication for the formulas expressing an arbitrary function as a superposition of sines and cosines. The innate symmetry of Fourier analysis, which depends on allowing negative as well as positive frequencies, remains hidden in the sine–cosine formulation.

Thus, to obtain a simplified form of the CWT, let us try using "one-sided" wavelets. We have already encountered such wavelets in Theorem 3.1, where ψ_+ and ψ_- contained only positive and negative frequencies, respectively. Functions with a positive frequency spectrum were introduced into signal analysis by D. Gabor in his fundamental paper on communication theory (Gabor [1946]). He called them "analytic signals" because they can be extended analytically to the upper-half complex time plane (see also Chapter 9). Similarly, functions with a negative frequency spectrum can be extended analytically to the lower-half complex time plane. We therefore call a function $f \in L^2(\mathbf{R})$ an *upper analytic signal* if $\hat{f}(\omega) = 0$ for $\omega < 0$ or a *lower analytic signal* if $\hat{f}(\omega) = 0$ for $\omega > 0$. Clearly, every $f \in L^2(\mathbf{R})$ can be written uniquely as a sum $f = f_+ + f_-$ of an upper and a lower analytic signal, namely

$$f_+(u) = \int_0^\infty d\omega\; e^{2\pi\, i\omega u}\, \hat{f}(\omega), \qquad f_-(u) = \int_{-\infty}^0 d\omega\; e^{2\pi\, i\omega u}\, \hat{f}(\omega), \tag{3.42}$$

and $\langle\, f_+, f_-\,\rangle = 0$ by Parseval's identity. The upper and lower analytic signals form subspaces of $L^2(\mathbf{R})$, which we denote by $L^2_+(\mathbf{R})$ and $L^2_-(\mathbf{R})$. Then the above decomposition means that $L^2(\mathbf{R})$ is the *orthogonal sum* of $L^2_+(\mathbf{R})$ and $L^2_-(\mathbf{R})$. The operators

$$P_\pm : L^2(\mathbf{R}) \to L^2(\mathbf{R}) \quad \text{defined by} \quad P_\pm f = f_\pm \tag{3.43}$$

are the *orthogonal projections* onto $L^2_\pm(\mathbf{R})$.

When dealing with real-valued signals $f(u)$, no information is lost by considering only their positive-frequency parts $f_+(u)$. Since $\hat{f}(-\omega) = \overline{\hat{f}(\omega)}$, it follows that $f_-(u) = \overline{f_+(u)}$, hence $f(u) = 2\mathrm{Re}\, f_+(u)$. However, sometimes it is convenient or even necessary to work with complex-valued signals,† and then it is natural to consider both the positive- and the negative-frequency parts of f, which are now independent.

A function $\psi(u)$ that is admissible in the sense of Theorem 3.2 (i.e., $0 < C < \infty$) will be called an *upper analytic wavelet* if $\psi \in L^2_+(\mathbf{R})$. (Note that then $C = C_+$, since $C_- = 0$.) A version of the continuous wavelet transform corresponding to the complex exponential form of the Fourier transform can now be stated as follows:

Theorem 3.3. *Let ψ be an upper analytic wavelet, and let f be an arbitrary signal in $L^2(\mathbf{R})$. Then the wavelet transform $\tilde{f}(s,t)$ of $f \in L^2(\mathbf{R})$ with respect to ψ contains only positive-frequency information about f if $s > 0$ and only negative-frequency information if $s < 0$. Specifically,*

$$\tilde{f}(s,t) = \begin{cases} \tilde{f}_+(s,t) & \text{if } s > 0 \\ \tilde{f}_-(s,t) & \text{if } s < 0, \end{cases} \tag{3.44}$$

† This is the case in quantum mechanics, for example, where the state of a system is described by a complex-valued wave function whose overall phase has no physical significance. A similar situation can occur in signal analysis, where a real signal is often written as the real part of a complex-valued function. An overall phase factor amounts to a constant phase shift, which has no effect on the information carried by the signal.

where $\tilde{f}_{\pm}$ *are the wavelet transforms of the upper and lower analytic signals* $f_{\pm}$ *of* f. *Furthermore,*

$$f_{\pm} = C^{-1} \iint_{\mathbf{R}^2_{\pm}} |s|^{2p-3}\, ds\, dt\, \psi_{s,t}\, \tilde{f}(s,t), \tag{3.45}$$

where $\mathbf{R}^2_{\pm} = \{(s,t) \in \mathbf{R}^2 : \pm s > 0\}$, *and the orthogonal projections onto the subspaces* $L^2_{\pm}(\mathbf{R})$ *of upper and lower analytic signals are given by*

$$P_{\pm} \equiv C^{-1} \iint_{\mathbf{R}^2_{\pm}} |s|^{2p-3}\, ds\, dt\, \psi_{s,t}\, \psi^*_{s,t}\,. \tag{3.46}$$

Proof: Let $s > 0$. Then by (3.11),

$$\begin{aligned} \tilde{f}(s,t) &= s^{1-p} \int_{-\infty}^{\infty} d\omega\; e^{2\pi\, i\omega t}\; \overline{\hat{\psi}}(s\omega)\, \hat{f}(\omega) \\ &= s^{1-p} \int_{0}^{\infty} d\omega\; e^{2\pi\, i\omega t}\; \overline{\hat{\psi}}(s\omega)\, \hat{f}(\omega) = \tilde{f}_{+}(s,t), \end{aligned} \tag{3.47}$$

since $\hat{\psi}(s\omega)$ vanishes for $\omega < 0$. The proof that $\tilde{f}(s,t) = \tilde{f}_{-}(s,t)$ for $s < 0$ is similar. By Theorem 3.2, the integral over $\mathbf{R}^2_{+}$ in (3.45) gives f_{+} since $\tilde{f}(s,t) = \tilde{f}_{+}(s,t)$ for $s > 0$. The proof that f_{-} is obtained by integrating over $\mathbf{R}^2_{-}$ is similar. Equation (3.46) follows from (3.45) since $\tilde{f}(s,t) = \psi^*_{s,t}\, f$. ■

Note that since $(P_{+} + P_{-})f = f_{+} + f_{-} = f$, it follows that $P_{+} + P_{-} = I$. When the expressions (3.46) are inserted for $P_{\pm}$, we get the resolution of unity in Theorem 3.2. Hence the choice of an upper analytic wavelet for ψ amounts to a *refinement* of Theorem 3.2 due to the fact that ψ does not mix positive and negative frequencies.

Multidimensional generalizations of analytic signals can also be defined. They are a natural objects in signal theory, quantum mechanics, electromagnetics, and acoustics; see Chapters 9–11.

3.4 Wavelets, Probability Distributions, and Detail

In Section 3.1 we mentioned that as a function of t for fixed $s \neq 0$, the wavelet transform $\tilde{f}(s,t)$ can be interpreted as the "detail" contained in the signal at the scale s. This view will become especially useful in the discrete case (Chapters 6–8) in connection with *multiresolution analysis.* We now give some simple examples of this interpretation. Suppose we begin with a *probability distribution* $\phi(u)$ *with zero mean and unit variance.* That is, $\phi(t)$ is a nonnegative function satisfying

$$\int_{-\infty}^{\infty} du\, \phi(u) = 1, \quad \int_{-\infty}^{\infty} du\, \cdot u\, \phi(u) = 0, \quad \int_{-\infty}^{\infty} du \cdot u^2\, \phi(u) = 1. \tag{3.48}$$

Assume that ϕ is at least n times differentiable (where $n \geq 1$) and its $(n-1)$-st derivative satisfies

$$\lim_{u \to \pm\infty} \phi^{(n-1)}(u) = 0, \tag{3.49}$$

and let

$$\psi^n(u) \equiv (-1)^n \, \phi^{(n)}(u). \tag{3.50}$$

Then

$$\int_{-\infty}^{\infty} du \, \psi^n(u) = (-1)^n \left[\phi^{(n-1)}(\infty) - \phi^{(n-1)}(-\infty) \right] = 0. \tag{3.51}$$

Thus ψ^n satisfies the admissibility condition and can be used to define a CWT. For $s \neq 0$ and $t \in \mathbf{R}$, let

$$\phi_{s,t}(u) \equiv |s|^{-1} \phi\left(\frac{u-t}{s}\right), \qquad \psi^n_{s,t}(u) \equiv |s|^{-1} \psi^n\left(\frac{u-t}{s}\right). \tag{3.52}$$

Then $\phi_{s,t}$ is a probability distribution with mean t and variance s^2 (standard deviation $|s|$) and $\psi^n_{s,t}$ is the wavelet family of ψ^n (take $p = 1$ in (3.5)). Let

$$\overline{f}(s,t) \equiv \phi_{s,t}{}^* f, \qquad \tilde{f}_n(s,t) \equiv \psi^n_{s,t}{}^* f. \tag{3.53}$$

$\overline{f}(s,t)$ is a *local average* of f at t, taken at the scale s. $\tilde{f}_n(s,t)$ is the CWT of f with respect to ψ^n. But (3.50) implies that

$$\psi^n_{s,t}(u) = (-1)^n \, s^n \, \partial_u^n \, \phi_{s,t}(u) = s^n \, \partial_t^n \, \phi_{s,t}(u), \tag{3.54}$$

where ∂_u and ∂_t denote the partial derivatives with respect to u and t, respectively. Thus

$$\begin{aligned} \tilde{f}_n(s,t) &\equiv \int_{-\infty}^{\infty} du \, \psi^n_{s,t}(u) f(u) \\ &= s^n \, \partial_t^n \int_{-\infty}^{\infty} du \, \phi_{s,t}(u) f(u) = s^n \, \partial_t^n \, \overline{f}(s,t). \end{aligned} \tag{3.55}$$

Hence the CWT of f with respect to ψ^n is proportional (by the factor s^n) to the n-th derivative of the average of f at scale s. Now the n-th derivative $f^{(n)}(t)$ may be said to give the *exact* n-th order detail of f at t, i.e., the *n-th order detail at scale zero.* For example, $f'(t)$ measures the rapidity with which f is changing (first-order detail) and $f''(t)$ measures the concavity (second-order detail). Then (3.55) shows that *$\tilde{f}_n(s,t)$ is the n-th order detail of f at scale s*, in the sense that it is proportional to the n-th derivative of $\overline{f}(s,t)$.

As an explicit example, consider the well-known *normal distribution* of zero mean and unit variance,

$$\phi(u) = \frac{e^{-u^2/2}}{\sqrt{2\pi}}, \tag{3.56}$$

which indeed satisfies (3.48). ϕ is infinitely differentiable, so we can take n to be *any* positive integer. The wavelets with $n = 1$ and $n = 2$ are

$$\begin{aligned} \psi^1(u) &= -\phi'(u) = \frac{u \cdot e^{-u^2/2}}{\sqrt{2\pi}} \\ \psi^2(u) &= \phi''(u) = \frac{(u^2 - 1)e^{-u^2/2}}{\sqrt{2\pi}}. \end{aligned} \tag{3.57}$$

$-\psi^2$ is known as the *Mexican hat function* (see Daubechies [1992]) because of the shape of its graph. Figure 3.2 shows a Mexican hat wavelet, a test signal consisting of a trapezoid with two different slopes and a third-order chirp

$$f(u) = \cos(u^3),$$

and sections of $\tilde{f}(s,t)$ for decreasing values of $s > 0$. The property (3.55) is clearly visible: $\tilde{f}(s,t)$ is the second derivative of a "moving average" of f performed with a translated and dilated Gaussian. Note also that as $s \to 0^+$, there is a shift from lower to higher frequencies in the chirp.

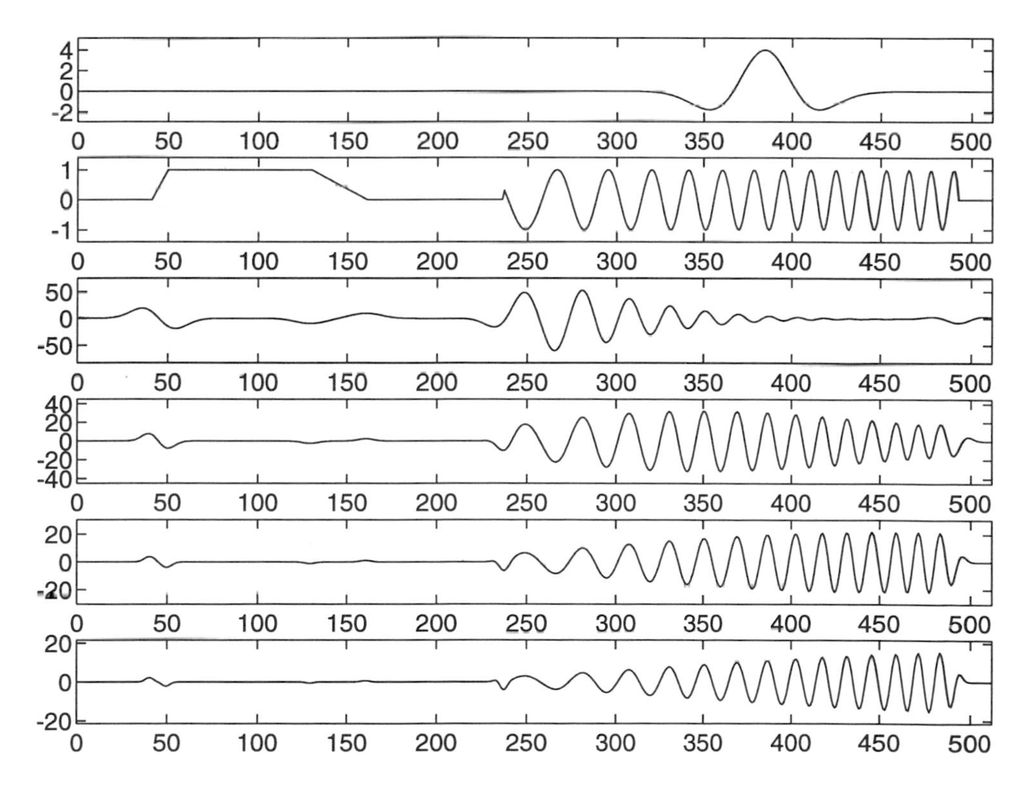

Figure 3.2. From top to bottom: Mexican hat wavelet; a signal consisting of a trapezoid and a third order chirp; $\tilde{f}(s,t)$ with decreasing values of s, going from coarse to fine scales. This figure was produced by Hans Feichtinger with MATLAB using his irregular sampling toolbox (IRSATOL).

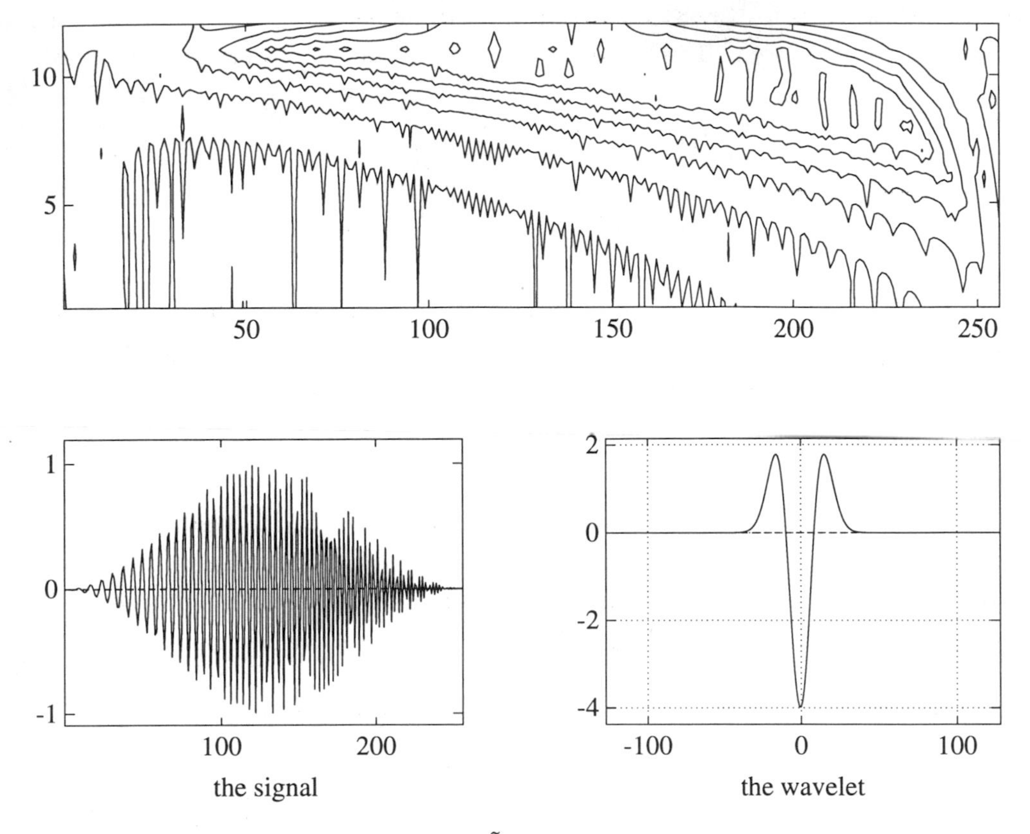

Figure 3.3. Contour plot of $\log|\tilde{f}(s,t)|$ for a windowed third-order chirp $f(u) = \cos(u^3)$ (shown below), using a Mexical hat wavelet. This figure was produced by Hans Feichtinger with MATLAB using his irregular sampling toolbox (IRSATOL).

We can compute the corresponding transforms $\tilde{f}_1(s,t)$ and $\tilde{f}_2(s,t)$ of a Gaussian "spike" of width $\sigma > 0$ centered at time zero,

$$f(u) = \frac{e^{-u^2/2\sigma^2}}{\sigma\sqrt{2\pi}} \equiv N_\sigma(u), \tag{3.58}$$

which is itself the normal distribution with zero mean and standard deviation σ. Then it can be shown that

$$\bar{f}(s,t) = \frac{e^{-t^2/2\sigma(s)^2}}{\sigma(s)\sqrt{2\pi}} \equiv N_{\sigma(s)}(t), \tag{3.59}$$

where $\sigma(s) \equiv \sqrt{\sigma^2 + s^2}$. Hence by (3.55),

$$\tilde{f}_1(s,t) = sN'_{\sigma(s)}(s) = -\frac{st}{\sigma(s)^3\sqrt{2\pi}} \exp\left[-\frac{t^2}{2\sigma(s)^2}\right] \tag{3.60}$$

and

$$\tilde{f}_2(s,t) = s^2 N''_{\sigma(s)}(t) = \frac{s^2[t^2 - \sigma(s)^2]}{\sigma(s)^5\sqrt{2\pi}} \exp\left[-\frac{t^2}{2\sigma(s)^2}\right]. \tag{3.61}$$

The simplicity of this result is related to the magic of Gaussian integrals and cannot be expected to hold for other kinds of signals or when analyzing with other kinds of wavelets. Nevertheless, the picture of the wavelet transform $\tilde{f}(s,t)$ as a derivative of a local average of f, suggested by (3.55), is worth keeping in mind as a rough explanation for the way the wavelet transform works. For a wavelet that cannot be written as a derivative of a nonnegative function, the admissibility condition nevertheless makes $\tilde{f}$ look like a "generalized" derivative of a local average.

Exercises

3.1. Prove (3.10).

3.2. Show that the dual mother wavelet ψ^1 defined in (3.30) satisfies

$$C^1_\pm \equiv \int_0^\infty \frac{d\xi}{\xi}\,|\hat{\psi}^1(\pm\xi)|^2 = 1/C_\pm\,, \tag{3.62}$$

where $C_\pm$ are defined by (3.27) and (3.28). Hence ψ^1 is admissible in the sense of (3.29).

3.3. Prove Theorem 3.2.

3.4. Prove (3.45) for f_-.

3.5. Prove (3.59), (3.60) and (3.61).

3.6. On a computer, find $\tilde{f}(s,t)$ as a function of t at various scales, for the chirp $f(u) = \cos(u^2)$, using the wavelet ψ^1 of (3.57). Experiment with a range of scales and times. Interpret your results in terms of frequency localization, and also in terms of (3.35).

3.7. Repeat Problem 3.6 using the wavelet ψ^2 of (3.57).

3.8. Repeat Problem 3.6 with the third-order chirp $f(u) = \cos(u^3)$.

3.9. Repeat Problem 3.6 with $f(u) = \cos(u^3)$ and using the wavelet ψ^2 of (3.57).

Chapter 4

Generalized Frames: Key to Analysis and Synthesis

Summary: In this chapter we develop a general method of analyzing and reconstructing signals, called the theory of *generalized frames*. The windowed Fourier transform and the continuous wavelet transform are both special cases. So are their manifold discrete versions, such as those described in the next four chapters. In the discrete case the theory reduces to a well-known construction called (ordinary) frames. The general theory shows that the results obtained in Chapters 2 and 3 are not isolated but are part of a broad structure. One immediate consequence is that certain types of theorems (such as reconstruction formulas, consistency conditions, and least-square approximations) do not have to be proved again and again in different settings; instead, they can be proved once and for all in the setting of generalized frames. Since the field of wavelet analysis is so new, it is important to keep a broad spectrum of options open concerning its possible course of development. The theory of generalized frames provides a tool by which many different wavelet-like analyses can be developed, studied, and compared.

Prerequisites: Chapters 1, 2, and 3.

4.1 From Resolutions of Unity to Frames

Let us recall the essential facts related to the windowed Fourier transform and the continuous wavelet transform. The WFT analizes a signal by taking its inner product with the *notes* $g_{\omega,t}(u) \equiv e^{2\pi\, i\omega u}\, g(u-t)$, i.e., $\tilde{f}(\omega,t) = g_{\omega,t}^{*} f$. The reconstruction fomula (2.19) may be written in the form

$$f = \iint_{\mathbf{R}^2} d\omega\, dt\; g^{\omega,t}\, \tilde{f}(\omega,t), \tag{4.1}$$

where $g^{\omega,t} \equiv C^{-1} g_{\omega,t}$ is the *reciprocal family* of notes. (In this case, it is actually unnecessary to introduce the reciprocal family, since it is related so simply to the original family; but in the general case studied here, this relation may not be as simple.) The combined analysis and synthesis gives a resolution of unity in $L^2(\mathbf{R})$ in terms of the pair of reciprocal families $\{g_{\omega,t}\,, g^{\omega,t}\}$:

$$\iint_{\mathbf{R}^2} d\omega\, dt\; g^{\omega,t}\, g_{\omega,t}^{*} = I. \tag{4.2}$$

Similarly, the continuous wavelet transform of Theorem 3.1 analyzes signals by using the *wavelets* $\psi_{s,t}(u) \equiv s^{-p}\, \psi((u-t)/s)$ with $s > 0$, i.e., $\tilde{f}(s,t) = \psi_{s,t}^{*} f$.

Its reconstruction formula, (3.31), is

$$f = \iint_{\mathbf{R}^2_+} s^{2p-3} ds\, dt\, \psi^{s,t}\, \tilde{f}(s,t), \tag{4.3}$$

where $\{\psi^{s,t}\}$ is the reciprocal family of wavelets defined by (3.25) and (3.30). Combined with the analysis, this gives a resolution of unity in $L^2(\mathbf{R})$ in terms of $\{\psi_{s,t}\,, \psi^{s,t}\}$:

$$\iint_{\mathbf{R}^2_+} s^{2p-3} ds\, dt\, \psi^{s,t}\, \psi^*_{s,t} = I. \tag{4.4}$$

Equations (4.2) and (4.4) are reminiscent of the discrete resolution of unity in $\mathbf{C}^N$ given by (1.40) in terms of a pair of reciprocal bases $\{\mathbf{b}_n, b^n\}$,

$$\sum_{n=1}^{N} \mathbf{b}^n\, \mathbf{b}^*_n = I, \tag{4.5}$$

but with the following important differences:

(a) The family $\{\mathbf{b}_n\}$ is finite, whereas each of the above families contains an infinite number of vectors. This is to be expected, since $L^2(\mathbf{R})$ is infinite-dimensional.

(b) $\{\mathbf{b}_n\}$ are parameterized by the *discrete* variable n, whereas the above families are parameterized by *continuous* variables: $(\omega, t) \in \mathbf{R}^2$ for $\{g_{\omega,t}\}$ and $(s,t) \in \mathbf{R}^2_+$ for $\{\psi_{s,t}\}$.

(c) Unlike the basis $\{\mathbf{b}_n\}$, each of the families $\{g_{\omega,t}\}$ and $\{\psi_{s,t}\}$ are *linearly dependent* in a certain sense, as explained below. Hence *they are not bases.* Yet, the resolutions of unity make it possible to compute with them in a way similar to that of computing with bases.

The arguments are almost identical for $\{g_{\omega,t}\}$ and $\{\psi_{s,t}\}$, so we concentrate on the latter. To see in what sense the vectors in $\{\psi_{s,t}\}$ are linearly dependent, we proceed as follows: Taking the adjoint of (4.4), we obtain

$$\iint_{\mathbf{R}^2_+} s^{2p-3} ds\, dt\, \psi_{s,t}\, (\psi^{s,t})^* = I, \tag{4.6}$$

which states that the original family $\{\psi_{s,t}\}$ is reciprocal to *its* reciprocal $\{\psi^{s,t}\}$. Applying this to $\psi_{s',t'}$ gives

$$\begin{aligned} \psi_{s',t'} &= \iint_{\mathbf{R}^2_+} s^{2p-3} ds\, dt\, \psi_{s,t}\, (\psi^{s,t})^*\, \psi_{s',t'} \\ &= \iint_{\mathbf{R}^2_+} s^{2p-3} ds\, dt\, \psi_{s,t}\, \overline{K(s',t' \,|\, s,t)}, \end{aligned} \tag{4.7}$$

where

$$K(s',t' \,|\, s,t) \equiv \langle\, \psi_{s',t'}\,, \psi^{s,t}\, \rangle = \overline{\langle\, \psi^{s,t}\,, \psi_{s',t'}\, \rangle}. \tag{4.8}$$

But it is easy to see that $K(s', t' \mid s, t)$ does not, in general, vanish when $s' \neq s$ or $t' \neq t$; hence (4.7) shows that the family $\{\psi_{s,t}\}$ is linearly dependent, in the sense that each of its vectors can be expressed as a *continuous* linear superposition of the others. Usually, linear dependence is defined in terms of *discrete* linear combinations. But then, the usual definition of *completeness* also involves discrete linear combinations. To discuss basis-like systems having a continuous set of labels we need to generalize the idea of linear dependence as well as the idea of completeness. Both these generalizations occur simultaneously when discrete linear combinations are replaced with continuous ones. But to take a continuous linear combination we must *integrate*, and that necessarily involves introducing a measure. The reader unfamiliar with measure theory may want to read Section 1.5 at this point, although that will not be necessary in order to understand this chapter.

The linear dependence of the $\psi_{s,t}$'s, in the above sense, is precisely the cause of the consistency condition that any CWT $\tilde{f}$ must satisfy: Taking the inner product of both sides of (4.7) with $f \in L^2(\mathbf{R})$, we get

$$\begin{aligned} \tilde{f}(s', t') \equiv \psi_{s',t'}^* f &= \iint_{\mathbf{R}_+^2} s^{2p-3} ds\, dt\; K(s', t' \mid s, t)\, \psi_{s,t}^*\, f \\ &= \iint_{\mathbf{R}_+^2} s^{2p-3} ds\, dt\; K(s', t' \mid s, t)\, \tilde{f}(s, t). \end{aligned} \tag{4.9}$$

That is, since the *values* of $\tilde{f}$ are given by taking inner products with the family $\{\psi_{s,t}\}$ and the latter are linearly dependent, so must be the values of $\tilde{f}$. The situation is exactly the same for the WFT.

This shows that resolutions of unity are a more flexible tool than bases: A basis gives rise to a resolution of unity, but not every resolution of unity comes from a basis. On the other hand, all of our results on the WFT and the CWT are a direct consequence of the two resolutions of unity given by (4.2) and (4.4).

To obtain a resolution of unity, we need a pair of reciprocal families. In practice, we begin with a single family of vectors and then attempt to *compute* a reciprocal. For a reciprocal family to exist, the original family must satisfy certain conditions that are, in fact, a generalization of the conditions that it be a basis. Aside from technical complications, the procedure is quite similar to the one in the finite-dimensional case. We therefore recommend that the reader review Section 1.2, especially in connection with the metric operator G (1.43). In the general case, we need the following:

(a) For the analysis ($f \mapsto \tilde{f}$), we need a Hilbert space $\mathcal{H}$ of possible signals, a set M of *labels*, and a family of vectors $h_m \in \mathcal{H}$ labeled by $m \in M$. For the WFT, $\mathcal{H} = L^2(\mathbf{R})$, $M = \mathbf{R}^2$ is the set of all frequencies and times $m = (\omega, t)$, and $h_m = g_{\omega,t}$. For the CWT, $\mathcal{H} = L^2(\mathbf{R})$, $M = \mathbf{R}^2_+$ is the set of all scales and times $m = (s, t)$, and $h_m = \psi_{s,t}$.

(b) For the synthesis ($\tilde{f} \mapsto f$), we need to construct a family of vectors $\{h^m\}$ in $\mathcal{H}$ that is "reciprocal" to $\{h_m\}$ in an appropriate sense. Furthermore, we need a way to *integrate* over the labels in M in order to reconstruct the signal as a superposition of the vectors h^m. This will be given by a *measure* μ on M, as described in Section 1.5. That means that given any "reasonable" (i.e., measurable) subset $A \subset M$, we can assign to it a "measure" $\mu(A)$, $0 \le \mu(A) \le \infty$. As explained in Section 1.5, this leads to the possibility of integrating measurable functions $F : M \to \mathbf{C}$, the integral being written as $\int_M d\mu(m)\, F(m)$.

The measure spaces we need for the WFT and the CWT are the following.

1. For the WFT, $M = \mathbf{R}^2$ and $\mu(A) =$ area of A, where A is any measurable subset of $\mathbf{R}^2$. This is just the Lebesgue measure on $\mathbf{R}^2$, and the associated integral is the Lebesgue integral

$$\int_M d\mu(m)\, F(m) = \iint_{\mathbf{R}^2} d\omega\, dt\; F(\omega, t). \tag{4.10}$$

2. For the CWT, $M = \mathbf{R}^2_+$ and

$$\mu(A) = \iint_A s^{2p-3} ds\, dt \tag{4.11}$$

for any measurable subset $A \subset \mathbf{R}^2_+$. The associated integral is

$$\int_M d\mu(m)\, F(m) = \iint_{\mathbf{R}^2_+} s^{2p-3} ds\, dt\; F(s, t). \tag{4.12}$$

Given an arbitrary measure space M with measure μ and a measurable function $g : M \to \mathbf{C}$, write

$$\|g\|^2_{L^2} \equiv \int_M d\mu(m)\, |g(m)|^2. \tag{4.13}$$

Since $|g(m)|^2$ is measurable and nonnegative, either the integral exists and is finite, or $\|g\|^2_{L^2} = +\infty$. Let $L^2(\mu)$ be the set of all g's for which $\|g\|^2_{L^2} < \infty$. Then $L^2(\mu)$ *is a Hilbert space* with inner product given by

$$\langle g_1, g_2 \rangle_{L^2} = \int_M d\mu(m)\, \overline{g_1(m)}\, g_2(m), \tag{4.14}$$

which is the result of polarizing the norm (4.13) (Section 1.3). For example, when $M = \mathbf{R}^n$ and μ is the Lebesgue measure on $\mathbf{R}^n$, then $L^2(\mu)$ is just $L^2(\mathbf{R}^n)$; when $M = \mathbf{Z}^n$ and μ is the counting measure ($\mu(A) =$ number of elements in A), then $L^2(\mu) = \ell^2(\mathbf{Z}^n)$.

We are now ready to give the general definition.

Definition 4.1. *Let $\mathcal{H}$ be a Hilbert space and let M be a measure space with measure μ. A* generalized frame *in $\mathcal{H}$ indexed by M is a family of vectors $\mathcal{H}_M \equiv \{h_m \in \mathcal{H} : m \in M\}$ such that*

(a) *For every $f \in \mathcal{H}$, the function $\tilde{f} : M \to \mathbf{C}$ defined by*

$$\tilde{f}(m) \equiv \langle h_m , f \rangle_{\mathcal{H}} \tag{4.15}$$

is measurable.

(b) *There is a pair of constants $0 < A \leq B < \infty$ such that for every $f \in \mathcal{H}$,*

$$A\|f\|^2_{\mathcal{H}} \leq \|\tilde{f}\|^2_{L^2} \leq B\|f\|^2_{\mathcal{H}}. \tag{4.16}$$

The vectors $h_m \in \mathcal{H}_M$ are called *frame vectors*, (4.16) is called the *frame condition*, and A and B are called *frame bounds*. Note that $\tilde{f} \in L^2(\mu)$ whenever $f \in \mathcal{H}$, by (4.16). The function $\tilde{f}(m)$ will be called the *transform* of f with respect to the frame.

In the sequel we will often refer to generalized frames simply as *frames*, although in the literature the latter usually presumes that the label set M is *discrete* (see below).

In the special case that $A = B$, the frame is called *tight*. Then the frame condition reduces to

$$A\|f\|^2_{\mathcal{H}} = \|f\|^2_{L^2}, \tag{4.17}$$

which plays a similar role in the general analysis as that played in Fourier analysis by the Plancherel formula, stating that no information is lost in the transformation $f \mapsto \tilde{f}$, which will allow f to be reconstructed from $\tilde{f}$. However, we will see that reconstruction is possible even when $A < B$, although it is a bit more difficult. In certain cases it is difficult or impossible to obtain tight frames with desired properties, but frames with $A < B$ may still exist.

Given a frame, we may assume without any loss of generality that all the frame vectors are *normalized*, i.e., $\|h_m\| = 1$ for all $m \in M$. For if that is not the case to begin with, define another frame, essentially equivalent to $\mathcal{H}_M$, as follows: First, delete any points from M for which $h_m = 0$, and denote the new set by M'. For $m \in M'$, define the new frame vectors by $h'_m = h_m/\|h_m\|$, so that $\|h'_m\| = 1$. The transform of $f \in \mathcal{H}$ with respect to the new frame vectors is

$$\tilde{f}'(m) \equiv \langle h'_m, f \rangle_{L^2} = \|h_m\|^{-1} \tilde{f}(m), \qquad m \in M', \tag{4.18}$$

hence

$$\|\tilde{f}\|^2_{L^2} \equiv \int_M d\mu(m) \, |\tilde{f}(m)|^2 = \int_{M'} d\mu(m) \, \|h_m\|^2 \, |\tilde{f}'(m)|^2, \tag{4.19}$$

where we have used the fact that the integrand vanishes when $m \notin M'$ (since then $\tilde{f}(m) = \langle 0, f \rangle = 0$). We now define a measure μ' on M' by $d\mu'(m) =$

$\|h_m\|^2 \, d\mu(m)$, i.e.,

$$\mu'(A) \equiv \int_A d\mu'(m) \, \|h_m\|^2 \tag{4.20}$$

for any measurable $A \subset M'$. Then (4.19) states that

$$\|\tilde{f}\|^2_{L^2(\mu)} = \int_{M'} d\mu'(m) \, |\tilde{f}'(m)|^2 \equiv \|\tilde{f}'\|^2_{L^2(\mu')}, \tag{4.21}$$

where the right-hand side denotes the norm in $L^2(\mu')$. The frame condition thus becomes

$$A\|f\|^2 \le \|\tilde{f}'\|^2_{L^2(\mu')} \le B\|f\|^2, \tag{4.22}$$

showing that $\mathcal{H}'_{M'} \equiv \{h'_m : m \in M'\}$ is also a frame, with identical frame bounds as $\mathcal{H}_M$. If $\|h_m\| = 1$ for all $m \in M$, we say that the frame is *normalized.*

However, sometimes frame vectors have a "natural" normalization, not necessarily unity. This occurs, for example, in the wavelet formulations of electromagnetics and acoustics (Chapters 9 and 11), where f represents an electromagnetic wave, M is complex space-time, and $\tilde{f}(m)$ is *analytic* in m. Then the normalized version $\|h_m\|^{-1}\tilde{f}(m)$ of $\tilde{f}$ is no longer analytic. Thus we do *not* assume that $\|h_m\| = 1$ in general, but we leave the normalization of h_m as a built-in freedom in the general theory of frames. This will also mean that when M is *discrete*, our generalized frames reduce to the usual (discrete) frames when μ is the "counting measure" on M (see (4.63)).

Since $\tilde{f} \in L^2(\mu)$ for every $f \in \mathcal{H}$, the mapping $f \mapsto \tilde{f}$ defines a function $T : \mathcal{H} \to L^2(\mu)$ (i.e., $T(f) \equiv \tilde{f}$), and T is clearly linear, hence it is an operator. T is called the *frame operator* of $\mathcal{H}_M$ (Daubechies [1992]). To get some intuitive understanding of T, let us normalize f (i.e., $\|f\| = 1$). Then the Schwarz inequality implies that

$$|\tilde{f}(m)|^2 = |\langle h_m, f \rangle|^2 \le \|h_m\|^2 \equiv H(m),$$

with equality if and only if f is a multiple of h_m. This states that $|\tilde{f}(m)|^2$ is dominated by the function $H(m)$, which is uniquely determined by the frame. Furthermore, for any *fixed* m_0, $|\tilde{f}(m)|^2$ can attain its maximum value $H(m_0)$ at m_0 if and only if f is a multiple of h_{m_0}. (But then $|\tilde{f}(m)|^2 < H(m)$ for every $m \ne m_0$, if we exclude the trivial case when h_m is a multiple of h_{m_0}.) The number $|\tilde{f}(m)|^2$ therefore measures how much of the "building block" h_m is contained in the signal f. Thus T *analyzes* the signal f in terms of the building blocks h_m. For this reason, we also refer to T as the *analyzing operator* associated with the frame $\mathcal{H}_M$. Synthesis or reconstruction amounts to finding an inverse map $\tilde{f} \mapsto f$. We now show that *the frame condition can be interpreted precisely as stating that this inverse exists as a (bounded) operator.* Note that by the definition of the adjoint operator $T^* : L^2(\mu) \to \mathcal{H}$,

$$\|\tilde{f}\|^2_{L^2} = \|Tf\|^2_{L^2} = \langle Tf, Tf \rangle_{L^2} = \langle f, T^*Tf \rangle_{\mathcal{H}}. \tag{4.23}$$

The frame condition can be written in terms of the operator $G \equiv T^*T : \mathcal{H} \to \mathcal{H}$, *without any reference to the "auxiliary" Hilbert space* $L^2(\mu)$, as follows: We have $\langle f, Gf \rangle = f^*Gf$ and $\|f\|^2 = f^*f$, with all norms and inner products understood to be in $\mathcal{H}$. Then the frame condition states that

$$Af^*f \leq f^*Gf \leq Bf^*f, \qquad f \in \mathcal{H}. \tag{4.24}$$

Note that the operator G is *self-adjoint*, i.e., $G^* = G$, hence all of its eigenvalues are real (Section 1.3). If H is any Hermitian operator on $\mathcal{H}$ such that $f^*Hf \geq 0$ for all $f \in \mathcal{H}$, then all the eigenvalues of H are necessarily nonnegative. We then say that the *operator* H itself is nonnegative and write this as an *operator inequality* $H \geq 0$. (This only makes sense when H is self-adjoint, since otherwise its eigenvalues are complex, in general.) Since (4.24) can be written as $f^*(G - AI)f \geq 0$ and $f^*(BI - G)f \geq 0$ (where I is the identity operator), it is equivalent to the operator inequalities $G - AI \geq 0$, $BI - G \geq 0$, or

$$AI \leq G \leq BI. \tag{4.25}$$

This operator inequality states that all the eigenvalues λ of G satisfy $A \leq \lambda \leq B$. It is the frame condition in *operator form*, and it is equivalent to (4.16). We call G the *metric operator* of the frame (see also Section 1.2).

The operator inequality (4.25) has a simple interpretation: $G \leq BI$ *means that the operator G is bounded (since $B < \infty$), and $G \geq AI$ means that G also has a bounded inverse G^{-1} (since $A > 0$).* In fact, G^{-1} satisfies the operator inequalities

$$B^{-1}I \leq G^{-1} \leq A^{-1}I. \tag{4.26}$$

Now for any $f, g \in \mathcal{H}$,

$$\begin{aligned} f^*Gg = \langle f, T^*Tg \rangle_{\mathcal{H}} = \langle Tf, Tg \rangle_{L^2} &= \int_M d\mu(m)\, \overline{\tilde{f}(m)}\, \tilde{g}(m) \\ = \int_M d\mu(m)\, \overline{h_m^* f}\, h_m^* g &= \int_M d\mu(m)\, f^* h_m\, h_m^* g. \end{aligned} \tag{4.27}$$

Hence we may express G as an integral over M of the operators $h_m h_m^* : \mathcal{H} \to \mathcal{H}$:

$$G = \int_M d\mu(m)\, h_m h_m^*, \tag{4.28}$$

which generalizes (1.43). It is now clear that the frame condition generalizes the idea of resolutions of unity, since it reduces to a resolution of unity (i.e., $G = I$) when $A = B = 1$. When $G \neq I$, we will need to construct a reciprocal family $\{h^m\}$ in order to obtain a resolution of unity and the associated reconstruction.

Frames were first introduced in Duffin and Schaeffer (1952). Generalized frames as described here appeared in Kaiser (1990a).

4.2 Reconstruction Formula and Consistency Condition

Suppose now that we have a frame. We know that for every signal $f \in \mathcal{H}$, the transform $\tilde{f}(m)$ belongs to $L^2(\mu)$. Two questions must now be answered:

(a) Given an *arbitrary* function $g(m)$ in $L^2(\mu)$, how can we tell whether $g = \tilde{f}$ for some signal $f \in \mathcal{H}$?

(b) If $g(m) = \tilde{f}(m)$ for some $f \in \mathcal{H}$, how do we reconstruct f?

The answer to (a) amounts to finding the *range* of the operator $T : \mathcal{H} \to L^2(\mu)$, i.e., the subspace

$$\mathcal{F} \equiv \{\tilde{f} : f \in \mathcal{H}\} \subset L^2(\mu). \tag{4.29}$$

The answer to (b) amounts to finding a *left inverse* of T, i.e., an operator $S : L^2(\mu) \to \mathcal{H}$ such that $ST = I$, the identity on $\mathcal{H}$. For if such an operator S can be found, then $\tilde{f} = Tf$ implies $S\tilde{f} = f$, which is the desired reconstruction. We now show that the key to both questions is finding the (two-sided) inverse of the metric operator G, whose existence is guaranteed by the frame condition (see (4.25) and (4.26)). Suppose, therefore, that G^{-1} is known explicitly. Then the operator $S \equiv G^{-1}T^*$ maps vectors in $L^2(\mu)$ to $\mathcal{H}$, i.e.,

$$L^2(\mu) \xrightarrow{\quad T^* \quad} \mathcal{H} \xrightarrow{\quad G^{-1} \quad} \mathcal{H}\,. \tag{4.30}$$

Furthermore, $ST = G^{-1}T^*T = G^{-1}G = I$, which proves that S is a left inverse of T. Hence the reconstruction is given by $f = S\tilde{f}$. We call S the *synthesizing operator* associated with the analyzing operator T. To find the range of T, consider the operator $P = TS : L^2(\mu) \to L^2(\mu)$. One might think that P is the identity operator on $L^2(\mu)$, i.e., that a left inverse is necessarily also a right inverse. This is not true in general. In fact, we have the following important result:

Theorem 4.2. *$P \equiv TS = TG^{-1}T^*$ is the orthogonal projection operator to the range $\mathcal{F}$ of T in $L^2(\mu)$.*

Proof: Any vector $g \in L^2(\mu)$ can be written uniquely as a sum $g = \tilde{f} + g_\perp$, where $\tilde{f} \in \mathcal{F}$ and $g_\perp$ belongs to the *orthogonal complement* $\mathcal{F}^\perp$ of $\mathcal{F}$ in $L^2(\mu)$. The statement that P is the orthogonal projection onto $\mathcal{F}$ in $L^2(\mu)$ means that $Pg = \tilde{f}$. To show this, note that since $\tilde{f} \in \mathcal{F}$, it can be written as $\tilde{f} = Tf$ for some $f \in \mathcal{H}$, hence $P\tilde{f} = PTf = TSTf = Tf = \tilde{f}$. Furthermore, $g_\perp \in \mathcal{F}^\perp$ means that $\langle g_\perp, \tilde{f}' \rangle_{L^2} = 0$ for every $\tilde{f}' \in \mathcal{F}$. Since $\tilde{f}' = Tf'$ for some $f' \in \mathcal{H}$, this means

$$0 = \langle g_\perp, Tf' \rangle_{L^2} = \langle T^* g_\perp, f' \rangle \tag{4.31}$$

for every $f' \in \mathcal{H}$. Hence $T^* g_\perp$ must be the zero vector in $\mathcal{H}$ (choosing $f' = T^* g_\perp$ gives $\|T^* g_\perp\|^2 = 0$). Thus $P g_\perp = T G^{-1} T^* g_\perp = 0$, and

$$Pg = P(\tilde{f} + g_\perp) = \tilde{f} + 0 = \tilde{f}, \tag{4.32}$$

proving the theorem. ■

We are now ready to define the reciprocal family. Let

$$h^m \equiv G^{-1} h_m\,, \quad m \in M. \tag{4.33}$$

The *reciprocal frame* of $\mathcal{H}_M$ is the collection

$$\mathcal{H}^M \equiv \{h^m : m \in M\}.$$

Then (4.28) implies

$$\int_M d\mu(m)\, h^m\, h_m^* = G^{-1} \int_M d\mu(m)\, h_m\, h_m^* = G^{-1} G = I, \tag{4.34}$$

which is the desired resolution of unity in terms of the pair $\mathcal{H}_M$ and $\mathcal{H}^M$. Taking the adjoint of (4.34), we also have

$$\int_M d\mu(m)\, h_m\, (h^m)^* = I, \tag{4.35}$$

which states that $\mathcal{H}_M$ is also reciprocal to $\mathcal{H}^M$. Since

$$\int_M d\mu(m)\, h^m\, h^{m\,*} = G^{-1} \int_M d\mu(m)\, h_m\, h_m^*\, G^{-1} = G^{-1} G G^{-1} = G^{-1} \tag{4.36}$$

and $B^{-1} \le G^{-1} \le A^{-1}$, it follows that $\mathcal{H}^M \equiv \{h^m\}$ also forms a (generalized) frame, with frame bounds B^{-1}, A^{-1} and metric operator G^{-1}.

We can now give explicit formulas for the actions of the operators T^*, S, and P. To simplify the notation, the inner product in $\mathcal{H}$ will be denoted without a subscript, i.e., $\langle f, f' \rangle \equiv \langle f, f' \rangle_{\mathcal{H}}$. Some of the proofs below rely implicitly on the boundedness of various operators and linear functionals.

Theorem 4.3. *The adjoint $T^* : L^2(\mu) \to \mathcal{H}$ of the analyzing operator T is given by*

$$(T^* g)(m) = \int_M d\mu(m)\, h_m\, g(m), \tag{4.37}$$

where the equality holds in the "weak" sense in $\mathcal{H}$, i.e., the inner products of both sides with an arbitrary vector $f \in \mathcal{H}$ are equal.

Proof: For any $f \in \mathcal{H}$ and $g \in L^2(\mu)$,

$$\begin{aligned} \langle f, T^* g \rangle = \langle Tf, g \rangle_{L^2} &= \int_M d\mu(m)\, \overline{h_m^* f}\, g(m) \\ &= \int_M d\mu(m)\, f^*\, h_m\, g(m) = \left\langle f, \int_M d\mu(m)\, h_m\, g(m) \right\rangle. \end{aligned} \tag{4.38}$$
■

Theorem 4.4. *The synthesizing operator $S = G^{-1}T^* : L^2(\mu) \to \mathcal{H}$ is given by*

$$Sg = \int_M d\mu(m)\, h^m\, g(m). \tag{4.39}$$

In particular, $f \in \mathcal{H}$ can be reconstructed from $\tilde{f} \in \mathcal{F}$ by

$$f = S\tilde{f} = \int_M d\mu(m)\, h^m\, \tilde{f}(m). \tag{4.40}$$

Proof: For $g \in L^2(\mu)$, (4.37) implies

$$\begin{aligned} Sg = G^{-1} \int_M d\mu(m)\, h_m\, g(m) &= \int_M d\mu(m)\, G^{-1} h_m\, g(m) \\ &= \int_M d\mu(m)\, h^m\, g(m). \quad \blacksquare \end{aligned} \tag{4.41}$$

Theorem 4.5. *The orthogonal projection $P : L^2(\mu) \to L^2(\mu)$ to the range $\mathcal{F}$ of T is given by*

$$(Pg)(m') = \int_M d\mu(m)\, K(m' \,|\, m)\, g(m), \tag{4.42}$$

where

$$K(m' \,|\, m) = \langle\, h_{m'}\,, h^m \,\rangle = \langle\, h_{m'}\,, G^{-1} h_m \,\rangle. \tag{4.43}$$

In particular, a function $g \in L^2(\mu)$ belongs to $\mathcal{F}$ if and only if it satisfies the consistency condition

$$g(m') = \int_M d\mu(m)\, K(m' \,|\, m)\, g(m). \tag{4.44}$$

Proof: By Theorem 4.4,

$$(Pg)(m') = (TSg)(m') = \langle\, h_{m'}\,, Sg \,\rangle = \int_M d\mu(m)\, \langle\, h_{m'}\,, h^m \,\rangle\, g(m). \tag{4.45}$$

Since $g \in \mathcal{F}$ if and only if $g = Pg$, (4.44) follows. $\blacksquare$

The function $K(m' \,|\, m)$ is called the *reproducing kernel* of $\mathcal{F}$ associated with the frame $\mathcal{H}_M$. For the WFT and the CWT, it reduces to the reproducing kernels introduced in Chapters 2 and 3. Equation (4.44) also reduces to the consistency conditions derived in Chapters 2 and 3. For background on reproducing kernels, see Hille (1972).

A *basis* in $\mathcal{H}$ is a frame $\mathcal{H}_M$ whose vectors happen to be *linearly independent* in the sense that if $g \in L^2(\mu)$ is such that

$$\int_M d\mu(m)\, h_m\, g(m) = 0 \quad \text{in} \quad \mathcal{H}, \tag{4.46}$$

then $g = 0$ as an element of $L^2(\mu)$, i.e., the set $\{m \in M : g(m) \neq 0\}$ has zero measure (Section 1.5). By (4.37), this means that $T^*g = 0$ implies $g = 0$. That is equivalent to saying that T^* is one-to-one, since $T^*(g_1 - g_2) = T^*g_2 - T^*g_2 = 0$ implies $g_1 - g_2 = 0$. We state this result as a theorem for emphasis.

Theorem 4.6. *A generalized frame is a basis if and only if the operator* $T^* : L^2(\mu) \to \mathcal{H}$ *is one-to-one.*

In this book we consider only *separable* Hilbert spaces, meaning Hilbert spaces whose bases are countable. Hence a generalized frame $\mathcal{H}_M$ can be a basis only if M is countable.

4.3 Discussion: Frames versus Bases

When the frame vectors h_m are linearly dependent, the transform $\tilde{f}(m)$ carries redundant information. This corresponds to *oversampling* the signal. Although it might seem inefficient, such redundancy has certain advantages. For example, it can detect and correct errors, which is impossible when only the minimal information is given. Oversampling has proven its worth in modern digital recording technology. In the context of frames, the tendency for correcting errors is expressed in the least-squares approximation, examples of which were already discussed in Chapters 2 and 3. We now show that this is a general property of frames.

Theorem 4.7: Least-Squares Approximation. *Let $g(m)$ be an arbitrary function in $L^2(\mu)$ (not necessarily in $\mathcal{F}$.) Then the unique signal $f \in \mathcal{H}$ which minimizes the "error"*

$$\|g - \tilde{f}\|_{L^2}^2 \equiv \int_M d\mu(m)\, |g(m) - \tilde{f}(m)|^2 \tag{4.47}$$

is given by $f = f_g \equiv Sg = \int_M d\mu(m)\, h^m\, g(m)$.

Proof: The unique $f \in \mathcal{H}$ which minimizes $\|g - \tilde{f}\|_{L^2}$ is the orthogonal projection of g to $\mathcal{F}$ (see Figure 4.1), which is $Pg = TSg = Tf_g = \tilde{f}_g$. ■

Some readers will object to using linearly dependent vectors, since that means that the coefficients used in expanding signals are not unique. In the context of frames, this nonuniqueness can be viewed as follows: Recall that reconstruction is achieved by applying the operator $SG^{-1}T^* : L^2(\mu) \to \mathcal{H}$ to the transform $\tilde{f}$ of f. Since $T^*g_\perp = 0$ whenever $g_\perp$ belongs to the orthogonal complement $\mathcal{F}^\perp$ of $\mathcal{F}$ in $L^2(\mu)$ (4.31), it follows that $Sg_\perp = 0$. Hence by Theorem 4.4,

$$f = S\tilde{f} = S(\tilde{f} + g_\perp) = \int_M d\mu(m)\, h^m \left[\tilde{f}(m) + g_\perp(m)\right] \tag{4.48}$$

for all $g_\perp \in \mathcal{F}^\perp$. If the frame vectors are linearly independent, then $\mathcal{F} = L^2(\mu)$, hence $\mathcal{F}^\perp = \{0\}$ (the trivial space containing only the zero vector) and the representation of f is unique, as usual. If the frame vectors are linearly dependent, then $\mathcal{F}^\perp$ is nontrivial and there is an infinite number of representations for f (one for each $g_\perp \in \mathcal{F}^\perp$). In fact, (4.48) gives *the most general* coefficient

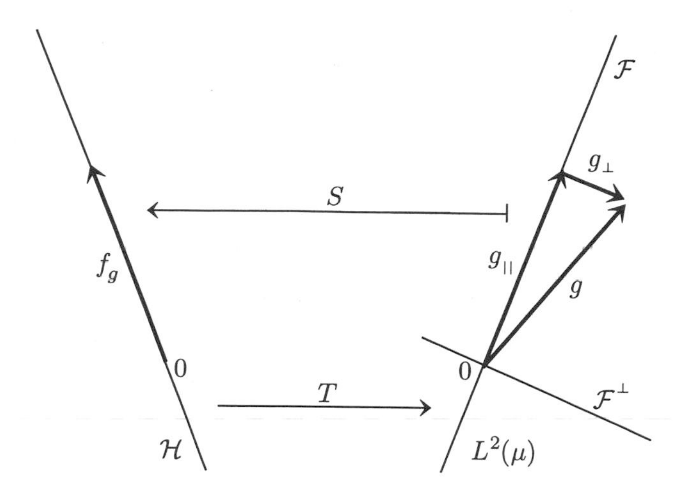

Figure 4.1. Schematic representation of $\mathcal{H}$, the analyzing operator T, its range $\mathcal{F}$, the synthesizing operator S, and the least-squares approximation f_g.

function that can be used to superimpose the h^m's, if the result is to be a given signal f. However, the coefficient function $g = \tilde{f}$ (i.e., $g_\perp = 0$) is very special, as shown next.

Theorem 4.8: Least-Energy Representation. *Of all possible coefficient functions $g \in L^2(\mu)$ for $f \in \mathcal{H}$, the function $g = \tilde{f}$ is unique in that it minimizes the "energy" $\|g\|^2_{L^2}$.*

Proof: Since $\tilde{f}$ and $g_\perp$ are orthogonal in $L^2(\mu)$, we have

$$\|g\|^2_{L^2} = \|\tilde{f} + g_\perp\|^2_{L^2} = \|\tilde{f}\|^2_{L^2} + \|g_\perp\|^2_{L^2} \geq \|\tilde{f}\|^2_{L^2}, \tag{4.49}$$

with equality if and only if $g_\perp = 0$. ■

The coefficient function $g = \tilde{f}$ is also distinguished by the fact that it is the only one satisfying the consistency condition (Theorem 4.5). Furthermore, this choice of coefficient function is bounded, since $|\tilde{f}(m)| \leq \|h_m\| \, \|f\| = \|f\|$, whereas general elements of $L^2(\mu)$ are unbounded.

Theorems 4.7 and 4.8 are similar in spirit, though their interpretations are quite different. In Theorem 4.7, we fixed $g \in L^2(\mu)$ and compared all possible signals $f \in \mathcal{H}$ such that $Tf \approx g$, whereas in Theorem 4.8, we fixed $f \in \mathcal{H}$ and compared all possible $g \in L^2(\mu)$ such that $Sg = f$.

4.4 Recursive Reconstruction

The above constructions made extensive use of the reciprocal frame vectors $h^m = G^{-1}h_m$, hence they depend on our ability to compute the reciprocal metric operator G^{-1}. We now examine methods for determining or approximating G^{-1}. The operator form of the frame condition can be written as

$$-\tfrac{1}{2}(B-A)\,I \le G - \tfrac{1}{2}(B+A)\,I \le \tfrac{1}{2}(B-A)\,I. \tag{4.50}$$

Setting $\delta \equiv (B-A)/(B+A)$ and $\alpha = 2/(B+A)$, we have

$$-\delta I \le I - \alpha G \le \delta I. \tag{4.51}$$

Now $0 < A \le B < \infty$ implies that $0 \le \delta < 1$, and (4.51) implies that the eigenvalues ε of the Hermitian operator $E \equiv I - \alpha G$ all satisfy $|\varepsilon| \le \delta < 1$. If G were represented by an ordinary (i.e., finite-dimensional) matrix, we could use (4.51) to prove that the following infinite series for G^{-1} converges:

$$G^{-1} = \alpha(I-E)^{-1} = \alpha \sum_{n=0}^{\infty} E^n. \tag{4.52}$$

The same argument works even in the infinite-dimensional case. The smaller δ is, the faster the convergence. Now δ is precisely a measure of the *tightness* of the frame, since $\delta = 0$ if and only if the frame is tight and $\delta \to 1^-$ as $A \to 0^+$ and/or $B \to \infty$ (loose frame). The above computation of G^{-1} works best, therefore, when the frame is as tight as possible. Frames with $0 < \delta \ll 1$ are called *snug.* Any frame has a unique set of best frame bounds, defined as $A_{\text{best}} \equiv \sup A$ and $B_{\text{best}} \equiv \inf B$, where the least upper bound "sup" and greatest lower bound "inf" are taken over all lower and upper frame bounds A and B, respectively; see Daubechies (1992).

In practice, one can only compute a finite number of terms in (4.52). A finite approximation to G^{-1} is given by

$$H_N \equiv \alpha \sum_{n=0}^{N} E^n. \tag{4.53}$$

Hence H_N can be computed recursively from H_{N-1} by

$$H_N = \alpha I + E H_{N-1} = \alpha I + (I - \alpha G)\, H_{N-1}. \tag{4.54}$$

The reciprocal frame vectors are then approximated by

$$h_N^m \equiv H_N h_m = \alpha h_m + E h_{N-1}^m, \tag{4.55}$$

which can be used to approximate the reconstruction formula:

$$f_N \equiv H_N G f = H_N \int_M d\mu(m)\, h_m h_m^* f = \int_M d\mu(m)\, h_N^m\, \tilde{f}(m). \tag{4.56}$$

Since $H_N G \to I$ as $N \to \infty$, it follows that $f_N \to f$. More precisely, it can be shown (Daubechies (1992), p. 62) that

$$\|f - f_N\| \le \delta^{N+1}\, \|f\|, \tag{4.57}$$

so the convergence is exponentially fast and is very rapid for small values of δ. A practical way to compute f_N is by using (4.54):

$$f_N = (\alpha I + EH_{N-1})Gf = \alpha Gf + Ef_{N-1} = f_0 + Ef_{N-1}. \tag{4.58}$$

Recent work by Gröchenig (1993) makes it possible to "accelerate" the above algorithm.

4.5 Discrete Frames

We now consider the case when M is *discrete* or *countable.* This means that the elements of M can be "counted," i.e., labeled by the positive integers: $M = \{m_1, m_2, m_3, \ldots\}$. Examples of discrete sets are the set $\mathbf{N}$ of positive integers, the set $\mathbf{Z}$ of all integers (let $m_1 = 0$, $m_2 = 1$, $m_3 = -1$, $m_4 = 2$, $m_5 = -2$, etc.), the set $\mathbf{Q}$ of all rational numbers, and the set $\mathbf{Z}^n \subset \mathbf{R}^n$ of all n-tuples of integers. In the next two chapters we will encounter discrete sets of frequencies and times associated with the windowed Fourier transform,

$$P_{\nu,\tau} = \{(k\nu, n\tau) : k, n \in \mathbf{Z}\} \subset \mathbf{R}^2, \tag{4.59}$$

where ν is a fixed positive frequency interval and τ is a fixed positive time interval, and discrete sets of scales and times associated with the wavelct transform,

$$S_{\sigma,\tau} = \{(\pm\sigma^k, n\sigma^k\tau) : k, n \in \mathbf{Z}\} \subset \mathbf{R}^2, \tag{4.60}$$

where σ is a fixed positive scale factor and τ is a fixed positive time interval.

If M is discrete, we assume that *every* subset $A \subset M$ is measurable. It then follows that every function $g : M \to \mathbf{C}$ is measurable, and the integral over M becomes a *sum*:

$$\int_M d\mu(m)\, g(m) = \sum_{m \in M} \mu_m\, g(m), \tag{4.61}$$

where $\mu_m \equiv \mu(\{m\})$ is the measure of the one-point set $\{m\}$. In general, $0 \le \mu_m \le \infty$. However, we may assume without loss of generality that $0 < \mu_m < \infty$ for every $m \in M$. For if $\mu_{m_0} = 0$ for some m_0, then the value $g(m_0)$ cannot contibute to the sum in (4.61), and we may delete m_0 from M without any loss. On the other hand, if $\mu_{m_0} = \infty$ for some m_0, then every integrable function $g(m)$ must vanish at m_0, and we can again delete m_0. With this provision, the frame condition (4.16) becomes

$$A\|f\|^2 \le \sum_{m \in M} \mu_m\, |\tilde{f}(m)|^2 \le B\|f\|^2. \tag{4.62}$$

When $\mu_m = 1$ for all $m \in M$, μ is called the *counting measure* on M since $\mu(A)$ is simply the number of elements in $A \subset M$. Then the frame condition reduces to

$$A\|f\|^2 \le \sum_{m\in M} |\tilde{f}(m)|^2 \le B\|f\|^2. \tag{4.63}$$

This is the classical definition of frames as originally introduced by Duffin and Schaeffer (1952); see also Young (1980) and Daubechies (1992). It is for this reason that we have refered to the family introduced in Definition 4.1 as a *generalized* frame. When M is discrete and μ is the counting measure on M, then our definition reduces to the classical one.

However, the counting measure is not always the natural choice in the discrete case. In Chapters 5 and 6, we will start with a continuous tight frame $\mathcal{H}_M$ associated with the WFT or CWT and arrive at a discrete "subframe" by a discretization procedure, which amounts to *sampling* $\tilde{f}(m)$ on a discrete subset $\Gamma \subset M$. Each sample point $m \in \Gamma$ then represents a certain region $A_m \subset M$ (with $m \in A_m$), and it is natural to choose $\mu_m = \mu(A_m)$. When this is done, $\mu_m \ne 1$, in general. Although we could *force* μ_m to be unity by simply choosing the counting measure on Γ (which leads to different frame constants, assuming that we still *have* a frame), it will be seen that a smooth transition to the continuous case as $\mu(A_m) \to 0$ can only be obtained by choosing $\mu_m = \mu(A_m)$.

The next two theorems tell us under what conditions a discrete frame can be a basis and an orthonormal basis.

Theorem 4.9. *A discrete frame is a basis if and only if its reproducing kernel is given by*

$$K(k\,|\,m) \equiv h_k^*\, h^m = \mu_m^{-1}\, \delta_k^m\,,$$

for all $k, m \in M$. *Then* $\{\mu_m h^m\}$ *is the basis biorthogonal to* $\{h_m\}$.

Proof: In the discrete case, the resolutions of unity (4.34) and (4.35) become

$$\sum_{m\in M} \mu_m\, h^m\, h_m^* = I, \qquad \sum_{m\in M} \mu_m\, h_m\, (h^m)^* = I. \tag{4.64}$$

Thus, using the second identity, we have

$$f = \sum_{m\in M} \mu_m\, h_m\, (h^m)^* f = \sum_{m\in M} \mu_m\, h_m\, \tilde{f}^{\#}(m), \tag{4.65}$$

where $\tilde{f}^{\#}(m) \equiv \langle\, h^m, f\,\rangle$. That proves that the frame vectors h_m span $\mathcal{H}$ (which is true for *any* frame). Hence they form a basis if and only if they are linearly independent. Now

$$h_k = \sum_{m\in M} \mu_m\, h_m\, (h^m)^* h_k = \sum_{m\in M} \mu_m\, h_m\, \overline{K(k\,|\,m)}. \tag{4.66}$$

If the frame vectors are linearly independent, it follows that $K(k\,|\,m) = \mu_m^{-1}\, \delta_k^m$. On the other hand, if $K(k\,|\,m) = \mu_m^{-1}\, \delta_k^m$, then the consistency condition (4.44), which now reads

$$g(k) = \sum_{m \in M} \mu_m \, K(k \,|\, m) \, g(m), \tag{4.67}$$

reduces to an identity. Hence Theorem 4.7 shows that $\mathcal{F} = L^2(\mu)$, and thus $\mathcal{F}^\perp = \{0\}$. Suppose that a linear combination of the h_m's vanishes:

$$\sum_{m \in M} h_m c^m = 0, \tag{4.68}$$

where $\sum_m |c^m|^2 < \infty$. Then the function $g(m) \equiv c^m$ belongs to $L^2(\mu)$, and for any $f \in \mathcal{H}$,

$$\langle \tilde{f}, g \rangle_{L^2} = \sum_{m \in M} \mu_m \, f^* h_m \, g(m) = 0 \tag{4.69}$$

(since $\overline{\tilde{f}(m)} = f^* h_m$). This proves that g is orthogonal to every $\tilde{f} \in \mathcal{F}$, i.e., that $g \in \mathcal{F}^\perp = \{0\}$. Hence $g = 0$, i.e., $c^m = 0$ for all $m \in M$. Therefore the frame vectors are linearly independent and they form a basis. The condition $h_k^* \, h^m = \mu_m^{-1} \, \delta_k^m$ is equivalent to the biorthogonality condition between $\{\mu_m h^m\}$ and $\{h_m\}$. ■

Note that (4.65) is an expansion of f in terms of the frame vectors h_m, with $\tilde{f}^\#(m)$ as the coefficient function. The usual reconstruction formula,

$$f = \sum_{m \in M} \mu_m \, h^m h_m^* f = \sum_{m \in M} \mu_m \, h^m \, \tilde{f}(m), \tag{4.70}$$

is obtained by using the first of the identities (4.64) and is related to (4.65) by exchanging the frame $\mathcal{H}_M$ with its reciprocal frame $\mathcal{H}^M$. The coefficient functions $\tilde{f}^\#$ and $\tilde{f}$ in (4.65) and (4.70) are equal if and only if $h^m = h_m$ for all m, which is true if and only if $G = I$.

Theorem 4.10. *Let $\mathcal{H}_M$ be a discrete, normalized frame ($\|h_m\| = 1$) with $\mu_m = \gamma > 0$ (constant) for all $m \in M$. Then its upper frame bound satisfies $B \geq \gamma$. Furthermore, the* best *upper frame bound is $B_{\text{best}} = \gamma$ if and only if $\mathcal{H}_M$ is an orthonormal basis. In that case, $G = \gamma I$, and $\mathcal{H}_M$ is tight.*

Proof: Let us not assume, to begin with, that the frame is normalized, and let $N_k \equiv \|h_k\|^2 > 0$. Applying the (discrete) frame condition (4.62) to $f = h_k$, we obtain

$$AN_k \leq \sum_{m \in M} \gamma \, |\langle \, h_m, h_k \, \rangle|^2 = \gamma \, N_k^2 + N_k \mathcal{O}_k \leq BN_k, \tag{4.71}$$

where

$$\mathcal{O}_k \equiv N_k^{-1} \sum_{m \neq k} \gamma \, |\langle \, h_m, h_k \, \rangle|^2 \geq 0. \tag{4.72}$$

Hence

$$A \leq \gamma \, N_k + \mathcal{O}_k \leq B, \tag{4.73}$$

so $B \geq \gamma N_k$ for every k. Since we have assumed the frame to be normalized, $B \geq \gamma$ as claimed. Furthermore, $B_{\text{best}} = \gamma$ if and only if $\mathcal{O}_k = 0$ for every k, which means that $\langle h_k, h_m \rangle = 0$ whenever $k \neq m$. This combines with $\|h_k\| = 1$ to give $\langle h_k, h_m \rangle = \delta_m^k$. Since the h_m's also span $\mathcal{H}$, they form an orthonormal basis. Thus $\sum_m h_m h_m^* = I$, and $G = \sum_m \gamma h_m h_m^* = \gamma I$. ■

The number $\mathcal{O}_m$ defined in (4.72) provides a numerical measure of the extent to which the particular vector h_m is orthogonal to the rest of the frame: It is orthogonal if and only if $\mathcal{O}_m = 0$, and it is "almost orthogonal" if $0 < \mathcal{O}_m \ll N_m$. Similarly, the single number

$$\mathcal{O} \equiv \sup_m \mathcal{O}_m \tag{4.74}$$

measures the extent to which the entire frame is orthogonal (not necessarily orthonormal, unless it is normalized). $\mathcal{O}$ will be called the *orthogonality index* of the frame. If $\mathcal{H}_M$ is normalized and tight with $G = CI$, then (4.73) shows that $\mathcal{O}_m = C - \gamma$ for all $m \in M$. In particular, $\mathcal{O}_m$ is independent of m.

Corollary 4.11. *Let $\mathcal{H}_M$ be a discrete, normalized frame with counting measure. Then $\mathcal{H}_M$ is self-reciprocal (i.e., $G = I$) if and only if it is an orthonormal basis.*

We now describe a method for making a self-reciprocal frame from an arbitrary frame. It involves computing $G^{-1/2}$. Just as a positive number posesses a unique positive square root, so does a positive (finite-dimensional) matrix H. $H^{1/2}$ can be defined as the unique positive matrix whose square is H; its eigenvalues are the square roots of the eigenvalues of H. This carries over to bounded positive operators such as the reciprocal metric operator G^{-1}. In fact, $G^{-1/2}$ can be computed by the same technique used to compute G^{-1} in (4.52):

$$\begin{aligned} G^{-1/2} &= \sqrt{\alpha}\,(I - E)^{-1/2} \\ &= \sqrt{\alpha}\left[I + \frac{1}{2}E + \frac{1\cdot 3}{2\cdot 4}E^2 + \frac{1\cdot 3\cdot 5}{2\cdot 4\cdot 6}E^3 + \cdots\right], \end{aligned} \tag{4.75}$$

and the convergence is uniform, since $-\delta I \leq E \leq \delta I$ and $0 \leq \delta < 1$. Again, the series converges rapidly if the frame is snug ($\delta \ll 1$) and the resulting operator satisfies

$$B^{-1/2} I \leq G^{-1/2} \leq A^{-1/2} I. \tag{4.76}$$

Given $G^{-1/2}$, define

$$\breve{h}_m \equiv G^{-1/2}\, h_m\,. \tag{4.77}$$

Then

$$\begin{aligned}\breve{G} &\equiv \sum_{m\in M} \mu_m\, \breve{h}_m\, \breve{h}_m^* \\ &= \sum_{m\in M} \mu_m\, G^{-1/2}\, h_m\, h_m^*\, G^{-1/2} \\ &= G^{-1/2}\, G\, G^{-1/2} = I,\end{aligned} \tag{4.78}$$

where we have used $(G^{-1/2}\, h_m)^* = h_m^*(G^{-1/2})^* = h_m^* G^{-1/2}$. Hence $\breve{h}^m = \breve{h}_m$, so $\breve{\mathcal{H}}_M$ is self-reciprocal. In view of Theorem 4.10 and Corollary 4.11, one might think at first that $\hat{\mathcal{H}}_M$ is an orthonormal basis, but this is generally not the case. Assume $\mu_m = \gamma$ for all m. Setting $\breve{N}_k = \|\breve{h}_k\|^2$ and $\hat{\mathcal{O}}_k = \breve{N}_k^{-1} \sum_{m\neq k} \gamma\, |\langle\, \breve{h}_m, \breve{h}_k\,\rangle|^2$, the counterpart of (4.73) states that $\gamma\, \breve{N}_k + \breve{\mathcal{O}}_k = 1$ for all $k \in M$. In order for $\breve{\mathcal{H}}_M$ to be an orthonormal basis, we need $\breve{N}_k = \gamma^{-1}$ for all $k \in M$, and this is not true in general. In fact,

$$\breve{N}_k = \langle\, G^{-1/2} h_k\,, G^{-1/2} h_k\,\rangle = h_k^*\, G^{-1} h_k. \tag{4.79}$$

In general, $\breve{N}_k \neq N_k$, so $\breve{\mathcal{H}}_M$ is not normalized even if $\mathcal{H}_M$ is. However, when $\mathcal{H}_M$ is a discrete subframe of certain continuous frames associated with the WFT, then (as it will be shown in Section 5.5) $\breve{N}_k = N_k$, i.e., multiplication by $G^{-1/2}$ does not change the norm of h_k. Thus if $\mathcal{H}_M$ is normalized and $\mu_m = \gamma$ for all m, then $\breve{G} = I$ implies that

$$1 = \|\breve{h}_m\|^2 = \sum_{k\in M} \gamma\, |\langle\, \breve{h}_k, \breve{h}_m\,\rangle|^2 = \gamma + \breve{\mathcal{O}}_m \tag{4.80}$$

for all $m \in M$. Hence $\breve{\mathcal{O}}_m$ is independent of m, and the orthogonality index of $\breve{\mathcal{H}}_M$ is $\mathcal{O} = 1 - \gamma$. Note that this also proves $\gamma \leq 1$.

4.6 Roots of Unity: A Finite-Dimensional Example

Since the idea of frames is not widely known or appreciated, we now give some examples of discrete frames. The basic ideas can be seen most clearly in the simplest *finite-dimensional* case $\mathcal{H} = \mathbf{R}^2$, which also has the advantage that no complicated questions of convergence, etc. need to be dealt with; moreover, the results can be easily visualized. It will be convenient to associate vectors in $\mathbf{R}^2$ with complex numbers by the rule

$$\mathbf{a} \equiv \begin{bmatrix} a_1 \\ a_2 \end{bmatrix} \leftrightarrow \alpha \equiv a_1 + i a_2 \in \mathbf{C}. \tag{4.81}$$

(This is just the usual correspondence between real vectors (x, y) and complex numbers $x + iy$.) The standard inner product on $\mathbf{R}^2$ can then be written as $\langle\, \mathbf{a}, \mathbf{b}\,\rangle = \mathrm{Re}\,(\bar{\alpha}\,\beta)$, where $\mathbf{a} \leftrightarrow \alpha$ and $\mathbf{b} \leftrightarrow \beta$. Let p be any integer ≥ 2, and define $\mathbf{g}_k$, $k = 0, 1, \ldots, p-1$ as the vectors in $\mathbf{R}^2$ corresponding to $\gamma_k = \Omega^k$, where $\Omega \equiv e^{2\pi i/p}$. ($\gamma_0, \ldots, \gamma_{p-1}$ are just the p-th roots of unity.) Then

$$\mathbf{g}_k = \begin{bmatrix} \cos(2\pi k/p) \\ \sin(2\pi k/p) \end{bmatrix} \equiv \begin{bmatrix} c_k \\ s_k \end{bmatrix}, \tag{4.82}$$

and

$$\langle \mathbf{g}_m, \mathbf{g}_k \rangle = \mathrm{Re}\,(\bar{\gamma}_m \gamma_k) = \cos(2\pi(m-k)/p). \tag{4.83}$$

The metric operator is

$$G = \sum_{k=0}^{p-1} \mathbf{g}_k \mathbf{g}_k^* = \sum_{k=0}^{p-1} \begin{bmatrix} c_k^2 & s_k c_k \\ s_k c_k & s_k^2 \end{bmatrix}. \tag{4.84}$$

To compute this, note that $\gamma_k^2 = 2c_k^2 - 1 + 2ic_k s_k$, hence

$$2\sum_{k=0}^{p-1} c_k^2 - p + 2i\sum_{k=1}^{p} c_k s_k = \sum_{k=0}^{p-1} \Omega^{2k} = \frac{\Omega^{2p}-1}{\Omega^2 - 1} = 0. \tag{4.85}$$

Thus

$$\sum_{k=0}^{p-1} c_k^2 = \frac{p}{2} \quad \text{and} \quad \sum_{k=0}^{p-1} c_k s_k = 0, \tag{4.86}$$

which gives

$$G = (p/2)I, \tag{4.87}$$

a tight frame. The factor $p/2$ can be interpreted as the *redundancy* of the frame (see Chapters 5 and 6). This makes sense since we have p normalized frame vectors in two (real) dimensions. The numbers $\mathcal{O}_k$ defined in the last section are therefore given by $1 + \mathcal{O}_k = p/2$. The orthogonality-symmetry of the frame (independence of $\mathcal{O}_k$ from k) is geometrically evident in this case, since the frame vectors are evenly distributed around the unit circle in $\mathbf{R}^2$ (Figure 4.2). The orthogonality index is $\mathcal{O} = \mathcal{O}_k = (p-2)/2$, which increases (meaning less orthogonality) with increasing p. However, it must be stressed that orthogonality-symmetry is not quite the same as complete geometric symmetry (but almost!), since we can change any one of the above vectors $\mathbf{g}_k$ to $-\mathbf{g}_k$ (leaving all the others the same) without affecting $\mathcal{O}_k$. More generally, given any orthogonality-symmetric *complex* frame $\mathcal{H}_M$, we can multiply each h_m by an arbitrary phase factor ($h_m \to e^{i\phi_m} h_m$) without affecting $\mathcal{O}_m$.

Exercises

4.1. This is a simple, though somewhat lengthy, exercise designed to guide you through the various stages of analysis and synthesis using a *finite-dimensional* frame. By the time you are done, you should have a thorough and concrete understanding of the fundamentals. Since the "analysis" and

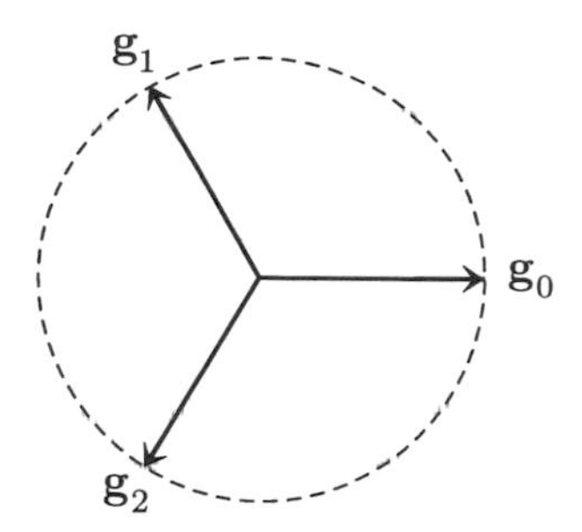

Figure 4.2. The p-th complex roots of unity as an example of a tight frame in $\mathbf{R}^2$ with redundancy $p/2$. Here $p = 3$.

"synthesis" are finite-dimensional, they display all the *algebraic* aspects without getting entangled in the (often difficult) analytical subtleties, such as worrying whether sums or integrals converge (and, if so, in what exact sense), exchanging orders of summation, etc. To simplify matters even further and allow visualization in three dimensions, we assume that all vectors and functions are *real.*

Let $\mathcal{H} = \mathbf{R}^2$, with the standard inner product: $\langle u, v \rangle = u^1 v^1 + u^2 v^2$. For the measure space, choose $M = \{1, 2, 3\}$ with μ the counting measure. That is, $\mu(1) = \mu(2) = \mu(3) = 1$. Let $\mathcal{H}_M = \{h_1, h_2, h_3\}$, where $h_1 = \begin{bmatrix} 1 \\ 0 \end{bmatrix}$, $h_2 = \begin{bmatrix} 0 \\ 1 \end{bmatrix}$, $h_3 = \begin{bmatrix} a \\ b \end{bmatrix}$ with $a, b \in \mathbf{R}$.

(a) *Auxiliary Hilbert space*: Let $L^2(\mu)$ denote the space of square-integrable *real-valued* functions on M. Show that $L^2(\mu) = \mathbf{R}^3$ with the standard inner product. (Use the definition of the integral given in Section 1.5 to find $\langle f, g \rangle \equiv \int_M d\mu(m)\, f(m) g(m)$ for arbitrary functions $f, g : M \to \mathbf{R}$.)

(b) *Analysis*: Given $u \in \mathcal{H}$, find its transform $\tilde{u} \in \mathbf{R}^3$.

(c) *Analyzing operator*: Find the matrix $T : \mathbf{R}^2 \to \mathbf{R}^3$ that represents the mapping $u \mapsto \tilde{u}$.

(d) *Metric operator*: Find $G = T^*T$ and verify that $G = \sum_m h_m\, h_m^*$.

(e) *Frame property*: Prove that $\|\tilde{u}\|^2_{\mathbf{R}^3} = \|u\|^2_{\mathbf{R}^2} + (\langle h_3, u \rangle)^2$, and show that this implies $\|u\|^2_{\mathbf{R}^2} \le \|\tilde{u}\|^2_{\mathbf{R}^3} \le (1 + \|h_3\|^2)\|u\|^2_{\mathbf{R}^2}$. This establishes that $\mathcal{H}_M$ is a frame.

(f) *Best frame bounds*: The frame bounds $A = 1, B = 1 + \|h_3\|^2$ above are the best possible, i.e., it is not possible to find frame bounds A', B' with either $A' > A$ and/or $B' < B$. Show this in two different ways: (i) by producing vectors $u, u' \in \mathbf{R}^2$ such that $\|\tilde{u}\|^2_{\mathbf{R}^3} = \|u\|^2_{\mathbf{R}^2}$ and $\|\tilde{u}'\|^2_{\mathbf{R}^3} = (1 + \|h_3\|^2)\|u'\|^2_{\mathbf{R}^2}$; (ii) by finding the eigenvalues of G. The best frame bounds are the lowest and highest eigenvalues.

(g) *Reciprocal frame*: Find the vectors $h^m = G^{-1}h_m$, and verify that

$$\sum_m h^m\, h_m^* = I, \quad \sum_m h_m\, h^{m\,*} = I, \tag{4.88}$$

and

$$\sum_m h^m\, h^{m\,*} = G^{-1}. \tag{4.89}$$

Use the first identity in (4.88) to expand an arbitrary $u \in \mathbf{R}^2$ as a superposition of h^1 and h^2 and the second identity to expand the same u as a superposition of h_1 and h_2. Sketch the vectors h_m and h^m for $a = 1$ and $b = 2$, and also for $a = b = 1$.

(h) *Synthesizing operator*: Find the synthesizing operator $S = G^{-1}T^*$ and verify that if a vector $g \in \mathbf{R}^3$ is the transform of some $u \in \mathbf{R}^2$, then $u = Sg$.

(i) *Projection operator*: Find the projection operator $P = TS : \mathbf{R}^3 \to \mathbf{R}^3$ onto the range of T.

(j) *Consistency condition and reproducing kernel*: Verify that an arbitrary vector $g \in \mathbf{R}^3$ can be the transform of some $u \in \mathbf{R}^2$ (i.e., $g \in \mathcal{F}$) if and only if $Pg = g$. Find the reproducing kernel, which can be written as a 3×3 matrix.

(k) *Least-squares approximation*: Given an arbitrary vector $g \in \mathbf{R}^3$, compute $u_g \equiv Sg$. Show that $u_g = S\tilde{u}_g$, and draw a sketch to give a geometric explanation of the statement that u_g is the unique solution to the "least-squares problem": Find $u \in \mathbf{R}^2$ such that $\|g - \tilde{u}\|^2$ is a minimum.

(l) *Least-energy property*: Find the conditions a vector $g_\perp \in \mathbf{R}^3$ must satisfy in order that $g_\perp \in \mathcal{F}^\perp$, and verify that a given $u \in \mathbf{R}^2$ can be written as $u = S(\tilde{u} + g_\perp)$ for an arbitrary $g_\perp \in \mathcal{F}^\perp$. Draw a sketch to give a geometric explanation of the statement that of all such "coefficient functions" $g = \tilde{u} + g_\perp$, the choice $g_\perp = 0$ has the least "energy" $\|g\|^2_{\mathbf{R}^3}$.

4.2. Repeat parts (a)–(j) of Problem 4.1 using $\mathcal{H} = \mathbf{C}^2$ with the standard inner product, $M = \{1, 2\}$ with $\mu =$ counting measure, and $\mathcal{H}_M = \{h_1, h_2\}$, where $h_1 = \begin{bmatrix} 1 \\ 0 \end{bmatrix}$ and $h_2 = \begin{bmatrix} a \\ b \end{bmatrix}$ with $a, b \in \mathbf{C}$, normalized to $\|h_2\| = 1$. This time, you must let $L^2(\mu)$ be the space of *complex-valued* functions on M, since otherwise the analyzing operator T could not be linear (try it!). For which values of a and b is $\mathcal{H}_M$ *not* a frame? Interpret this.

Chapter 5

Discrete Time-Frequency Analysis and Sampling

Summary: The reconstruction formula for windowed Fourier transforms is highly redundant since it uses *all* the notes $g_{\omega,t}$ to recover the signal and these notes are linearly dependent. In this chapter we prove a reconstruction formula using only a discrete subset of notes. Although still redundant, this reconstruction is much more efficient and can be approximated numerically by ignoring notes with very large time or frequency parameters. The present reconstruction is a generalization of the well-known Shannon sampling theorem, which underlies digital recording technology. We discuss its advantages over the latter, including the possibility of cutting the frequency spectrum of a signal into a number of "subbands" and processing these subbands in parallel.

Prerequisites: Chapters 1 and 2. An understanding of frames (Chapter 4) is helpful but not necessary.

5.1 The Shannon Sampling Theorem

A signal $f(t)$ is called *band-limited* if its Fourier transform $\hat{f}(\omega)$ vanishes outside of some bounded interval, say $-\Omega \le \omega \le \Omega$. The least such Ω is then called the *bandwidth* of f. Since the frequency content of f is limited, $f(t)$ can be expected to vary slowly, its precise degree of slowness being governed by Ω: The smaller Ω is, the slower the variation. In turn, we expect that a slowly varying signal can be *interpolated* from a knowledge of its values at a discrete set of points, i.e., by *sampling*. The slower the variation is, the less frequently the signal needs to be sampled. This is the intuition behind the *Shannon sampling theorem* (Shannon [1949]), which states that the interpolation can, in fact, be made *exact*. We first state and prove the theorem, and then we discuss its intuitive meaning in greater detail.

Theorem 5.1. *Let $f \in L^2(\mathbf{R})$, and let $\hat{f}(\omega) = 0$ for $|\omega| > \Omega$. Then $f(t)$ can be reconstructed from its samples at the times $t_n = n/2\Omega$, $n \in \mathbf{Z}$, by the following interpolation formula:*

$$f(t) = \sum_{n \in \mathbf{Z}} \frac{\sin[2\pi\Omega(t - t_n)]}{2\pi\Omega(t - t_n)} f(t_n). \tag{5.1}$$

Proof: Note that

$$\begin{aligned} \|\hat{f}\|^2_{L^2([-\Omega,\Omega])} &\equiv \int_{-\Omega}^{\Omega} d\omega\, |\hat{f}(\omega)|^2 = \int_{-\infty}^{\infty} d\omega\ \ |\hat{f}(\omega)|^2 \\ &= \|\hat{f}\|^2_{L^2(\mathbf{R})} = \|f\|^2_{L^2(\mathbf{R})} < \infty \end{aligned} \tag{5.2}$$

by Plancherel's theorem, hence $\hat{f} \in L^2([-\Omega, \Omega])$. Thus $\hat{f}(\omega)$ can be expanded in a Fourier series in the interval $-\Omega \le \omega \le \Omega$:

$$\hat{f}(\omega) = \frac{1}{2\Omega} \sum_{n\in\mathbf{Z}} e^{-2\pi\, i\omega t_n}\ c_n\,, \tag{5.3}$$

where $t_n = n/2\Omega$ and

$$c_n = \int_{-\Omega}^{\Omega} d\omega\ e^{2\pi\, i\omega t_n}\, \hat{f}(\omega) = \int_{-\infty}^{\infty} d\omega\ \ e^{2\pi\, i\omega t_n}\, \hat{f}(\omega) = f(t_n). \tag{5.4}$$

That is, the Fourier coefficients in (5.3) are *samples* of $f(t)$.† Thus

$$\begin{aligned} f(t) &= \int_{-\infty}^{\infty} d\omega\ \ e^{2\pi\, i\omega t}\, \hat{f}(\omega) = \int_{-\Omega}^{\Omega} d\omega\ e^{2\pi\, i\omega t}\, \hat{f}(\omega) \\ &= \frac{1}{2\Omega} \sum_{n\in\mathbf{Z}} \int_{-\Omega}^{\Omega} d\omega\ e^{2\pi\, i\omega(t-t_n)}\ f(t_n) \\ &= \sum_{n\in\mathbf{Z}} \frac{\sin\left[2\pi\Omega(t-t_n)\right]}{2\pi\Omega(t-t_n)}\ f(t_n), \end{aligned} \tag{5.5}$$

where the equality holds in the sense of $L^2(\mathbf{R})$. But since band-limited functions actually extend to the complex time plane as entire functions (see (5.8)), the equality is also pointwise. ■

To get a better intuitive understanding of the sampling theorem, note that $f(t)$ is a superposition of the complex exponentials $e_\omega(t) \equiv e^{2\pi\, i\omega t}$ with $|\omega| \le \Omega$, and each of these satisfies

$$\left| \frac{d}{dt} e_\omega(t) \right| = 2\pi|\omega||e_\omega(t)| \le 2\pi\Omega. \tag{5.6}$$

Consequently, it can be shown that

$$|f'(t)| \le 2\pi\Omega\, C_f\,, \quad \text{where} \quad C_f \equiv \max_{t\in\mathbf{R}} |f(t)| < \infty. \tag{5.7}$$

† Note that the Fourier series (5.3) is "backwards," since it expresses a function in the *frequency* domain as a superposition of periodic functions of frequency. The discrete times t_n play the role of the discrete harmonic frequencies in the usual Fourier series. The fact that sampling a function in the time domain leads to periodic repetitions in the frequency domain is called *aliasing*; see Papoulis (1977), and Chapter 7.

This is a special case of *Bernstein's inequality* (Achieser (1956), p. 138). Let us normalize f so that $C_f = 1$. Then (5.7) states that $f(t)$ cannot grow or decay at a rate faster than $2\pi\Omega$. Similarly, the n-th derivative of f is bounded by $|f^{(n)}(t)| \le (2\pi\Omega)^n$. It therefore makes sense that if f is sampled sufficiently frequently, i.e., if we measure $f(t_n)$ at sufficiently close times t_n , then we should be able to recover $f(t)$ for all t by interpolation, since no "surprises" (such as sharp spikes) can occur between the samples if the latter are sufficiently dense. The time interval between samples in Shannon's theorem is $\Delta t \equiv t_{n+1} - t_n = 1/2\Omega$; hence the *sampling rate* is $R \equiv 1/\Delta t = 2\Omega$, in samples per unit time. (Note how the bandwidth, in *cycles* per unit time, is related to the sampling rate, in *samples* per unit time: R is the width of the support of $\hat{f}$. This relation will be generalized in Section 5.3.) R is the theoretical *minimum* for the sampling rate if we want to recover $f(t)$ completely. It is called the *Nyquist rate.* For a given signal with bandwidth Ω, we can also apply Shannon's theorem with $\Omega' > \Omega$, since then $\hat{f}(\omega) = 0$ for $|\omega| > \Omega'$. This leads to the higher sampling rate $R' = 2\Omega' > R$, which amounts to *oversampling.* In practice, signals need to be sampled at a higher rate than R for at least two reasons: First, a "fudge factor" must be built in, since the actual sampling and interpolation cannot exactly match the theoretical one; second, oversampling has a tendency to reduce errors. For example, the highest audible frequency is approximately 18,000 cycles per second (Hertz), depending on the listener. Theoretically, therefore, an audio signal should be sampled at least 36,000 times per second in order not to lose any information. The actual rate on compact disk players is usually 44,000 samples per second.

Another property of band-limited functions, related to the above sampling theorem, is that they can actually be extended to the entire complex time plane as *analytic functions.* In fact, the substitution $t \to t + iv \in \mathbf{C}$ in the Fourier representation of $f(t)$ gives

$$f(t+iv) = \int_{-\Omega}^{\Omega} d\omega \; e^{2\pi\, i\omega(t+iv)} \; \hat{f}(\omega), \tag{5.8}$$

and this integral can be shown to converge, defining $f(t+iv)$ as an entire function. (When f is not band-limited, i.e., $\Omega \to \infty$, the integral diverges in general because the exponential $e^{-2\pi\,\omega v}$ can grow without bounds.) The bandwidth Ω is related to the growth rate of $f(t+iv)$ in the complex plane. Since analytic functions are determined by their values on certain discrete sets (such as sequences with accumulation points), it may not seem surprising that $f(t)$ is determined by its discrete samples $f(t_n)$. However, the set $\{t_n\}$ has no accumulation point. The sampling theorem is actually a consequence of the above-mentioned growth rate of $f(t+iv)$.

5.2 Sampling in the Time-Frequency Domain

Recall the expression for the windowed Fourier transform:

$$\tilde{f}(\omega, t) = \int_{-\infty}^{\infty} du\ e^{-2\pi\, i\omega u}\ \bar{g}(u-t)\, f(u) \equiv g^*_{\omega,t} f\,, \tag{5.9}$$

where $g \in L^2(\mathbf{R})$. The reconstruction formula obtained in Chapter 2 represents f as a continuous superposition of *all* the "notes" $g_{\omega,t}$:

$$f(u) = C^{-1} \iint d\omega\, dt\ g_{\omega,t}(u)\, \tilde{f}(\omega, t), \tag{5.10}$$

where $C \equiv \|g\|^2$. The vectors $g_{\omega,t}$ form a continuous frame, which we now call $\mathcal{G}_P$, parameterized by the time-frequency plane $P = \mathbf{R}^2$ equipped with the measure $ds\, dt$. This frame is tight, with frame bound C. (Hence the metric operator is $G = CI$ and the reciprocal frame vectors are $g^{\omega,t} \equiv C^{-1} g_{\omega,t}$.) As pointed out earlier, this frame is highly redundant. In this section we find some discrete subframes of $\mathcal{G}_P$, i.e., discrete subsets of $\mathcal{G}_P$ that themselves form a frame. This will be done explicitly for two types of windows:

(a) when g has compact support in the time domain, namely $g(u) = 0$ outside the time interval $a \le u \le b$ (such a function is called *time-limited*);

(b) when g is band-limited, with $\hat{g}(\omega) = 0$ outside the frequency band $\alpha \le \omega \le \beta$.

For other types of windows, it is usually more difficult to give an explicit construction. Some general theorems related to this question are stated in Section 5.4.

Note that g cannot be both time-limited and band-limited unless it vanishes identically, for if g is band-limited, then it extends to an entire-analytic function in the complex time plane, defined by (5.8). But an entire function cannot vanish on any interval (such as (b, ∞)), unless it vanishes identically. Hence g cannot also be time-limited. (The same argument applies when g is time-limited: Then $\hat{g}(\omega)$ extends to an entire function in the complex frequency plane, hence it cannot be band-limited unless it vanishes identically.) However, it is possible for g to be time-limited and for $\hat{g}(\omega)$ to be concentrated *mainly* in $[\alpha, \beta]$, or conversely for g to be band-limited and $g(t)$ to be concentrated *mainly* in $[a, b]$, or for g to be neither time-limited nor band-limited but $g(t)$ and $\hat{g}(\omega)$ to be small outside of two bounded intervals. Such windows are in fact desirable, since they provide good localization simultaneously in time and frequency (Section 2.2).

We now derive a time-frequency version of the Shannon sampling theorem, assuming that the window is time-limited but *not* assuming that the signal f is either time-limited or band-limited. Then the localized signal $f_t(u) \equiv$

$\bar{g}(u-t)\,f(u)$ has compact support in $[a+t, b+t]$; hence it can be expanded in that interval in a Fourier series with harmonic frequencies $\omega_m = m/(b-a)$:

$$f_t(u) = \nu \sum_{m \in \mathbf{Z}} e^{2\pi\, im\nu u}\, c_m(t), \qquad \nu \equiv (b-a)^{-1}, \tag{5.11}$$

where

$$\begin{aligned} c_m(t) &= \int_{a+t}^{b+t} du\; e^{-2\pi\, im\nu u}\, \bar{g}(u-t)\, f(u) \\ &= \int_{-\infty}^{\infty} du\; e^{-2\pi\, im\nu u}\, \bar{g}(u-t)\, f(u) = \tilde{f}(m\nu, t). \end{aligned} \tag{5.12}$$

That is, the Fourier coefficients of $f_t(u)$ are *samples* of $\tilde{f}(\omega, t)$ at the discrete frequencies $\omega_m = m\nu$. (Note the similarity with Shannon's theorem: In this case, we are sampling a time-limited function at discrete frequencies, and the sampling interval ν is the reciprocal of the support-width of the signal in the time domain.) Our next goal is to recover the whole signal $f(u)$ from (5.11) by using only discrete values of t, i.e., sampling in time as well as in frequency. The procedure will be analogous to the derivation of the continuous reconstruction formula (5.10), but the integral over t will be replaced by a sum over appropriate discrete values. Note that (5.11) cannot hold outside the interval $[a+t, b+t]$, since the left-hand side vanishes there while the right-hand side is periodic. To get a global equality, we multiply both sides of (5.11) by $g(u-t)$, using (5.12):

$$\begin{aligned} |g(u-t)|^2\, f(u) &= \nu \sum_{m \in \mathbf{Z}} e^{2\pi\, im\nu u}\, g(u-t)\, \tilde{f}(m\nu, t) \\ &= \nu \sum_{m \in \mathbf{Z}} g_{m\nu, t}(u)\, \tilde{f}(m\nu, t). \end{aligned} \tag{5.13}$$

We could now recover $f(u)$ from (5.13) by integrating both sides with respect to t and using $\int dt\, |g(u-t)|^2 = \|g\|^2$. But this would defeat our purpose of finding a discrete subframe of $\mathcal{G}_P$, since it would give an expression for $f(u)$ as a *semi-discrete* superposition of frame vectors, i.e., discrete in frequency but continuous in time. Instead, choose $\tau > 0$ and consider the function

$$H_\tau(u) \equiv \tau \sum_{n \in \mathbf{Z}} |g(u - n\tau)|^2, \tag{5.14}$$

which is a discrete approximation (Riemann sum) to $\|g\|^2$. Thus, for well-behaved (e.g., continuous) windows,

$$\lim_{\tau \to 0+} H_\tau(u) = \|g\|^2 \equiv C. \tag{5.15}$$

Note that for any $\tau > 0$, H_τ is periodic with period τ. Summing (5.13) over $t_n = n\tau$, we obtain

$$H_\tau(u)\, f(u) = \nu\tau \sum_{n \in \mathbf{Z}} \sum_{m \in \mathbf{Z}} g_{m\nu, n\tau}(u)\, \tilde{f}(m\nu, n\tau). \tag{5.16}$$

In order to reconstruct f, we must be able to divide by $H_\tau(u)$. We need to know that the sum (5.14) converges and that the result is (almost) everywhere positive. Convergence is no problem in this case, since (5.14) contains only a finite number of terms due to the compact support of g. Positivity is a bit more subtle. Clearly a *necessary* condition for $H_\tau > 0$ is that $0 < \tau \leq b - a$, since otherwise $H_\tau(u) = 0$ when $b < u < a + \tau$. That is, *we certainly cannot reconstruct* f *unless* $0 < \nu\tau \leq 1$. But we really need a bit more than positivity, as the next example shows.

Example 5.2: An Unstable Reconstruction. Let

$$g(u) = \begin{cases} \sqrt{u} & 0 < u \leq 1 \\ 0 & \text{otherwise,} \end{cases} \tag{5.17}$$

so $a = 0, b = 1$. Taking the maximal value $\tau = 1$, $H_1(u)$ is the "sawtooth" function with period 1 defined by $H_1(u) = u$, $0 < u \leq 1$. Although $H_1(u) > 0$ for all u, division by $H_1(u)$ in (5.16) will uncontrollably magnify any errors in the right-hand side, since

$$H_1(u)^{-1} g_{m\nu,n\tau}(u) = \begin{cases} e^{2\pi i m\nu u} / \sqrt{u-n} & n < u \leq n+1 \\ 0 & \text{elsewhere.} \end{cases} \tag{5.18}$$

Hence any small roundoff error in the computed sample $\tilde{f}(m\nu, n\tau)$ will result in a huge error in $f(u)$ near $u = n$.

Thus, to have a *numerically stable* reconstruction, we must have more than positivity: Assume that the *greatest lower bound* ("infimum") of $H_\tau(u)$ is positive. Later, when considering windows that are not time-limited, we will also need to assume that the sum in (5.14) converges to a bounded function, hence the *least upper bound* ("supremum") of H_τ is finite. Thus let†

$$A_\tau \equiv \inf_u H_\tau(u) > 0, \qquad B_\tau \equiv \sup_u H_\tau(u) < \infty. \tag{5.19}$$

In view of (5.15), we can also expect that for nice windows,

$$\lim_{\tau \to 0+} A_\tau = \lim_{\tau \to 0+} B_\tau = C. \tag{5.20}$$

Hence (5.19) should be satisfied for sufficiently small values of τ. In the above example, (5.19) holds if and only if $0 < \tau < 1$. Assuming that (5.19) holds, we have

$$0 < A_\tau \leq H_\tau(u) \leq B_\tau < \infty, \tag{5.21}$$

---

† Theoretically, since sets of zero measure do not count, the infimum and supremum in (5.19) can be replaced by the *essential infimum* and the *essential supremum* to give better bounds. But this is unnecessary for the windows used in practice; see Daubechies (1992), p. 102.

and we can solve (5.16) for $f(u)$. For convenience, let

$$\begin{aligned}\Gamma_{m,n}(t) &\equiv g_{m\nu,n\tau}(t)\\ \Gamma^{m,n}(t) &\equiv H_\tau(t)^{-1}\,\Gamma_{m,n}(t).\end{aligned} \tag{5.22}$$

Note that (5.21) implies $\|\Gamma^{m,n}\| \le A_\tau^{-1}\|\Gamma_{m,n}\| = C/A_\tau < \infty$, hence $\Gamma^{m,n} \in L^2(\mathbf{R})$. With this notation, we have

$$f(t) = \nu\tau \sum_{n\in\mathbf{Z}}\sum_{m\in\mathbf{Z}} \Gamma^{m,n}(t)\,\tilde{f}(m\nu, n\tau) \quad \text{a.e.} \tag{5.23}$$

We state our findings in the language of frames, as developed in Chapter 4.

Theorem 5.3. *Let $g \in L^2(\mathbf{R})$ be a compactly supported window of support width $b - a \equiv 1/\nu$, satisfying (5.21) for some $0 < \tau \le 1/\nu$. Let $P_{\nu,\tau}$ be the regular lattice $\{(m\nu, n\tau) : m, n \in \mathbf{Z}\}$ in the time-frequency plane, equipped with the measure $\mu_{\nu,\tau}$ assigning the value $\nu\tau$ to every point $(m\nu, n\tau)$. Then the family of vectors $g_{\omega,t} \in \mathcal{G}_P$ with $(\omega, t) \in P_{\nu,\tau}$, i.e.,*

$$\mathcal{G}_{\nu,\tau} = \{g_{m\nu,n\tau} \equiv \Gamma_{m,n} : m, n \in \mathbf{Z}\}, \tag{5.24}$$

is a frame parameterized by $P_{\nu,\tau}$. Its metric operator G_τ consists of multiplication by H_τ in the time domain:

$$(G_\tau f)(t) = H_\tau(t)\,f(t), \tag{5.25}$$

and the (best) frame bounds are A_τ and B_τ. They are related to the frame bound $C = \|g\|^2$ of the tight continuous frame $\mathcal{G}_P$ by

$$A_\tau \le C \le B_\tau. \tag{5.26}$$

A signal $f \in L^2(\mathbf{R})$ can be reconstructed from the samples of its WFT in $P_{\nu,\tau}$ by

$$f = \nu\tau \sum_{n\in\mathbf{Z}}\sum_{m\in\mathbf{Z}} \Gamma^{m,n}\,\tilde{f}(m\nu, n\tau), \tag{5.27}$$

and we have the corresponding resolution of unity,

$$\nu\tau \sum_{m,n\in\mathbf{Z}} \Gamma^{m,n}\,\Gamma^*_{m,n} = I, \tag{5.28}$$

where $\Gamma^{m,n} \equiv G_\tau^{-1}\Gamma_{m,n}$ are the vectors of the reciprocal frame $\mathcal{G}^{\nu,\tau}$.

Proof: Equations (5.25) and (5.16) give

$$G_\tau f = \nu\tau \sum_{m,n\in\mathbf{Z}} \Gamma_{m,n}\,\Gamma^*_{m,n}\,f = \int d\mu_{\nu,\tau}(m,n)\,\Gamma_{m,n}\,\Gamma^*_{m,n}\,f, \tag{5.29}$$

where the right-hand side is, by definition, the integral with respect to $d\mu_{\nu,\tau}$ (4.61). This shows that G_τ is indeed the metric operator of $\mathcal{G}_{\nu,\tau}$. For any $f \in L^2(\mathbf{R})$, the definition of G_τ gives

$$\langle f, G_\tau f \rangle = \int_{-\infty}^{\infty} dt\ \bar{f}(t)\, H_\tau(t)\, f(t) = \int_{-\infty}^{\infty} dt\ H_\tau(t)\, |f(t)|^2. \tag{5.30}$$

Hence (5.21) implies $A_\tau \|f\|^2 \leq \langle f, G_\tau f \rangle \leq B_\tau \|f\|^2$, which is the desired frame condition. To prove (5.26), note that

$$\int_0^\tau du\, H_\tau(u) = \tau \sum_{n \in \mathbf{Z}} \int_0^\tau du\, |g(u - n\tau)|^2 = \tau \|g\|^2 = \tau\, C. \tag{5.31}$$

Integrating (5.21) over $0 \leq u \leq \tau$ therefore gives (5.26). ■

Note: $P_{\nu,\tau}$ is usually given the *counting measure*, which assigns the value 1 to every sample point (m, n). In that case, the factor $\nu\tau$ in (5.23) and (5.27) must be absorbed into $\Gamma^{m,n}$. That means that G_τ must be replaced by $G'_\tau \equiv (\nu\tau)^{-1} G_\tau = ((b-a)/\tau) G_\tau$, and the frame bounds become $A'_\tau = (\nu\tau)^{-1} A_\tau$ and $B'_\tau = (\nu\tau)^{-1} B_\tau$. In addition, there is often an extra factor of 2π because of the convention of using $e^{\pm i\xi t}$ instead of $e^{\pm 2\pi i\omega t}$ in the Fourier transform, which means that frequencies are measured in radians (rather than cycles) per unit time, and we must replace ν with $\nu/2\pi$. For example, the frame bounds in Daubechies (1992) and Chui (1992a) corresponding to A_τ and B_τ are

$$A'_\tau = \frac{2\pi}{\nu\tau} A_\tau, \quad B'_\tau = \frac{2\pi}{\nu\tau} B_\tau. \tag{5.32}$$

We prefer the above convention because it is simpler and, with it, (5.27) is truly a discrete approximation (Riemann sum) for the continuous resolution of unity

$$\iint_{\mathbf{R}^2} d\omega\, dt\, g^{\omega,t}\, g^*_{\omega,t} = I, \tag{5.33}$$

with the measure $d\omega\, dt$ going over to $\nu\tau$, which is the area of the rectangular cell in the time-frequency plane labeled by the integers (m, n). A similar situation will be encountered in the constuction of discrete time-scale frames in Chapter 6.

$\mathcal{G}_{\nu,\tau}$ is a discrete subframe of $\mathcal{G}_P$, since every $\Gamma_{m,n}$ also belongs to $\mathcal{G}_P$. On the other hand, $\Gamma^{m,n}$ does *not* usually belong to $\mathcal{G}^P$ since $H_\tau(u)$ is not constant in general, so $\mathcal{G}^{\nu,\tau}$ is not a subframe of $\mathcal{G}^P$. Hence we do not have an *a priori* guarantee that all the vectors $\Gamma^{m,n}$ can be obtained by translating and modulating a single window. (For the continuous frame, this was assured by the fact that $g^{\omega,t} = C^{-1} g_{\omega,t}$.) This is clearly an important property, since otherwise each $\Gamma^{m,n}$ must be computed separately. Fortunately, it does hold, due to the periodicity of H_τ:

$$\begin{aligned}\Gamma^{m,n}(t) &= e^{2\pi\, im\nu t}\,\frac{g(t-n\tau)}{H_\tau(t)} = e^{2\pi\, im\nu t}\,\frac{g(t-n\tau)}{H_\tau(t-n\tau)} \\ &= e^{2\pi\, im\nu t}\,\Gamma^{0,0}(t-n\tau).\end{aligned} \tag{5.34}$$

Thus all the vectors $\Gamma^{m,n}$ are are obtained from $\Gamma^{0,0}$ by translation and modulation.

Let us step back in order to gain some intuitive understanding of the above construction. We started with the continuous frame $\mathcal{G}_P$, which is tight, and chose an appropriate subset $\mathcal{G}_{\nu,\tau}$. It is not surprising that $\mathcal{G}_{\nu,\tau}$ is still a frame for sufficiently small τ, since the sum representing G_τ in (5.25) must become $C\,I$ in the limit $\tau \to 0$. Of course, we cannot expect that sum to *equal* its limiting value in general, hence $\mathcal{G}_{\nu,\tau}$ cannot be expected to be tight. Fortunately, the formalism of frame theory can accomodate this possibility and still give us a resolution of unity and thus an exact way to do analysis and synthesis. The inequality (5.26) suggests that as τ increases, either $A_\tau \to 0$ or $B_\tau \to \infty$. In the case at hand, where g has compact support, we already know that $A_\tau = 0$ when $\tau > \nu^{-1}$.

Throughout the construction of $\mathcal{G}_{\nu,\tau}$, we regarded the window g as fixed, hence $\nu \equiv (b-a)^{-1}$ was also fixed. Implicitly, everything above depends on ν as well as on τ. Let us make this dependence explicit by writing $H_{\nu,\tau}$, $A_{\nu,\tau}$, and $B_{\nu,\tau}$ for H_τ, A_τ, and B_τ. Now consider a *scaled* version of g: $g'(t) = s^{-1/2}g(t/s)$ for some $s > 0$, so that $C' \equiv \|g'\|^2 = \|g\|^2 = C$. Then g' is supported in $[a', b'] = [sa, sb]$, hence $\nu' \equiv (b'-a')^{-1} = s^{-1}\nu$. Repeating the above construction with $\tau' \equiv s\tau$, we find

$$\begin{aligned}H_{\nu',\tau'}(t) &\equiv \tau' \sum_{n\in\mathbf{Z}} |g'(t-n\tau')|^2 \\ &= \tau \sum_{n\in\mathbf{Z}} \left| g\left(\frac{t}{s} - n\tau\right)\right|^2 = H_{\nu,\tau}(t/s).\end{aligned} \tag{5.35}$$

Hence

$$\begin{aligned}A_{\nu',\tau'} &\equiv \inf_t H_{\nu',\tau'}(t) = A_{\nu,\tau}, \\ B_{\nu',\tau'} &\equiv \sup_t H_{\nu',\tau'}(t) = B_{\nu,\tau}.\end{aligned} \tag{5.36}$$

This means that the set of vectors $\Gamma'_{m,n} \equiv g'_{m\nu',n\tau'}$ generated from g' by translations and modulations parameterized by the lattice $P_{\nu',\tau'}$ also forms a frame with frame bounds identical to those of $\mathcal{G}_{\nu,\tau}$. Note that the measure of a single sample point remains the same: $\nu'\tau' = \nu\tau$. In fact, by varying s, we obtain a whole *family* of frames with identical frame bounds. This shows that the above frame bounds $A_{\nu,\tau}$ and $B_{\nu,\tau}$ depend only on the product $\nu\tau$ as long as all the window functions are related by scaling. This is a special case of a method described in more detail at the end of Section 5.4; see also Baraniuk and Jones (1993) and Kaiser (1994a).

Note that

$$\rho(\nu,\tau) \equiv \frac{1}{\nu\tau} = \frac{b-a}{\tau} \tag{5.37}$$

is just the ratio of the width of the window to the time interval between samples. This measures the *redundancy* of our discrete frame. For example, if $\tau = (b-a)/2$, then the time axis is "covered twice" by H_τ and the redundancy is 2. As $\tau \to 0$ with ν fixed, $\mathcal{G}_{\nu,\tau}$ becomes infinitely redundant. Another interpretation of the redundancy is obtained by noting that $\rho(\nu,\tau)$ is our *sampling density* in the time-frequency plane, since we have exactly one sample in each cell with $\Delta t = \tau$ and $\Delta\omega = \nu$. The area of this cell, $\Delta A = \nu\tau$, is measured in *cycles* or repetitions. In fact, *the inequality $\nu\tau \le 1$, which is necessary for reconstruction, is really a manifestation of the uncertainty principle* (Section 2.2). It states, roughly, that *we need at least one sample for every "cycle" in order to reconstruct the signal.* The relation to the uncertainty principle is that the inequality $\nu\tau \le 1$ states, in essence, that not much "action" can take place in a cell of area $\Delta A = 1$ in the time-frequency plane, so one sample in each such cell is enough to determine f.

For the sake of efficiency, it may seem desirable to take τ as large as possible in order to minimize the number of time samples. If we choose the maximum value $\tau = b-a$, then $|g(u-t_n)|^2$ and $|g(u-t_{n+1})|^2$ do not overlap, which forces $g(t)$ to be discontinuous in order to satisfy (5.21). Such discontinuities in g lead to discontinuities in f_t, which means that $\hat{f}_t(\omega) \equiv \tilde{f}(\omega,t)$ must contain high-frequency components to produce them. That is, discontinuous windows intoduce unnecessary "noise" into the WFT. If we attempt to gain efficiency by taking $\tau \approx b-a$ while maintaining a constant value of the lower bound A_τ and keeping g continuous, then g is forced to change rapidly near the endpoints of its support in order to satisfy (5.21). This also introduces high-frequency components into the WFT, hence more coefficients $\tilde{f}(m\nu, n\tau)$ with high values of m are significantly large. Thus a reduction of the time sampling rate (within the constraint $\tau \le b-a$) inevitably leads to an increase in the number of frequency samples, *if* the frame bounds are held fixed. The only alternative, if g is to remain continuous, is to allow A_τ to become small as $\tau \to b-a$, since $H(t)$ must be small in the overlap of two consecutive windows. But then the reconstruction becomes unstable, as already discussed. For these reasons, it is better to let $g(t)$ be continuous and leave some margin between τ and $b-a$. This suggests that *some* oversampling pays off! This is exactly the kind of tradeoff that is characteristic of the uncertainty principle.

Equation (5.23) is similar to the Shannon formula (5.1), except that now the signal is being sampled in the joint time-frequency domain rather than merely in the time domain. Another important difference is that we did not need to assume that f is band-limited, since the formula works for any $f \in L^2(\mathbf{R})$.

Incidentally, the interpretation that we need at least one sample per "cycle" also applies to the Shannon sampling theorem. There, the sampling is done in time only, with intervals $\Delta t \le 1/2\Omega$, where Ω is the bandwidth of f. But then 2Ω is precisely the width of the support of $\hat{f}(\omega)$, so we still have $\Delta\omega \cdot \Delta t \le 1$. (This is actually a correct and simple way to understand the Nyquist sampling rate!) A more detailed comparison between these two sampling theorems will be made in Section 5.3. But first we indicate how a reconstruction similar to (5.23) is obtained when g has compact support in the frequency domain rather than the time domain. The derivation is entirely analogous to the above because of the symmetry of the WFT with respect to the Fourier transform (Section 2.2).

Theorem 5.4. *Let $g \in L^2(\mathbf{R})$ be a band-limited window, with $\hat{g}(\omega) = 0$ outside the interval $\alpha \le \omega \le \beta$. Let $\tau = (\beta - \alpha)^{-1}$, let $0 < \nu \le 1/\tau$, and define*

$$K_\nu(\omega) = \nu \sum_{m \in \mathbf{Z}} |\hat{g}(\omega - m\nu)|^2, \tag{5.38}$$

which is a discrete approximation to $\|\hat{g}\|^2 = C$. Suppose that

$$A'_\nu \equiv \inf_\omega K_\nu(\omega) > 0 \quad \text{and} \quad B'_\nu \equiv \sup_\omega K_\nu(\omega) < \infty. \tag{5.39}$$

Let $P_{\nu,\tau}$ and $\mathcal{G}_{\nu,\tau}$ be defined as in Theorem 5.3. Then $\mathcal{G}_{\nu,\tau}$ is a frame parameterized by $P_{\nu,\tau}$. Its metric operator G_ν consists of multiplication by K_ν in the frequency domain:

$$(G_\nu f)^\wedge(\omega) = K_\nu(\omega)\, \hat{f}(\omega), \tag{5.40}$$

and the (best) frame bounds are A'_ν and B'_ν. They are related to the frame bound $C = \|g\|^2$ of the tight continuous frame $\mathcal{G}_P$ by

$$A'_\nu \le C \le B'_\nu. \tag{5.41}$$

A signal $f \in L^2(\mathbf{R})$ can be reconstructed from the samples of its WFT in $P_{\nu,\tau}$ by

$$\hat{f}(\omega) = \nu\tau \sum_{m,n \in \mathbf{Z}} \hat{\Gamma}^{m,n}(\omega)\, \tilde{f}(m\nu, n\tau), \tag{5.42}$$

and we have the corresponding resolution of unity,

$$\nu\tau \sum_{m,n \in \mathbf{Z}} \Gamma^{m,n}\, \Gamma^*_{m,n} = I, \tag{5.43}$$

where $\Gamma^{m,n} \equiv G_\nu^{-1} \Gamma_{m,n}$ are the vectors of the reciprocal frame $\mathcal{G}^{\nu,\tau}$.

Proof: Exercise.

The time-frequency sampling formula (5.23) was stated in Kaiser (1978b) and proved in Kaiser (1984). These results remained unpublished and were discovered independently and studied in more detail and rigor by Daubechies, Grossmann, and Meyer (1986).

5.3 Time Sampling versus Time-Frequency Sampling

We now show that for band-limited signals, Theorem 5.4 implies the Shannon sampling theorem as a special case. Thus, suppose that $\hat{f}(\omega) = 0$ outside of the frequency band $|\omega| \leq \Omega$ for some $\Omega > 0$. For the window, choose $g(t)$ such that $\hat{g}(\omega) = P_\Omega(\omega)$, where

$$P_\Omega(\omega) \equiv \begin{cases} 1 & \text{if } |\,\omega| \leq \Omega \\ 0 & \text{if } |\,\omega| > \Omega. \end{cases} \tag{5.44}$$

P_Ω is called the *ideal low-pass filter* with bandwidth Ω (Papoulis [1962]), since multiplication by P_Ω in the frequency domain completely filters out all frequency components with $|\omega| > \Omega$ while leaving all other frequency components unaffected. Taking $\hat{g} = P_\Omega$ gives

$$g(t) = \check{P}_\Omega(t) = \int_{-\Omega}^{\Omega} d\omega\; e^{2\pi\, i\omega t} = \frac{\sin(2\pi\Omega t)}{\pi t}\,. \tag{5.45}$$

Since $\hat{g}$ has compact support, Theorem 5.4 applies with the time sampling interval $\tau = (2\Omega)^{-1}$. In this case, we can choose the maximum frequency sampling interval $\nu = 1/\tau = 2\Omega$, since this gives $K(\omega) = 2\Omega$, so (5.39) holds with $A'_\nu = B'_\nu = \nu$ (except when ω is an odd multiple of Ω, and this set is insignificant since it has zero measure). We claim that all terms with $m \neq 0$ in (5.42) vanish. That is because

$$\begin{aligned} \tilde{f}(2m\Omega, t) &= \int_{-\infty}^{\infty} d\omega\;\; e^{2\pi\, i(\omega-2m\Omega)t}\,\overline{\hat{g}}(\omega - 2m\Omega)\,\hat{f}(\omega) \\ &= \int_{-\Omega}^{\Omega} d\omega\, e^{2\pi\, i\omega t}\,\hat{f}(\omega + 2m\Omega), \end{aligned} \tag{5.46}$$

and the support of $\hat{f}(\omega+2m\Omega)$ is in $-(1+2m)\Omega \leq \omega \leq (1-2m)\Omega$. Furthermore, (5.46) also shows that $\tilde{f}(0,t) = f(t)$, so

$$\tilde{f}(m\nu, n\tau) = \delta^0_m\, f(n\tau) \tag{5.47}$$

in (5.42). Also

$$\Gamma^{\,0,n}(t) = \frac{1}{2\Omega}\,\Gamma_{0,n}(t) = \frac{1}{2\Omega}\,g(t - n\tau) = \frac{\sin\left[2\pi\Omega(t - n\tau)\right]}{2\pi\Omega(t - n\tau)}\,. \tag{5.48}$$

Substituting (5.47) and (5.48) into (5.42) and taking the inverse Fourier transform gives Shannon's sampling formula, as claimed.

Even for band-limited signals, Theorem 5.4 leads to an interesting generalization of the Shannon formula. Again, suppose that $\hat{f}(\omega) = 0$ for $|\omega| > \Omega$. Choose an integer $M \geq 0$, and let $L = 2M + 1$. For the window, choose $g(t) = \check{P}_{\Omega/L}(t)$. Then $\tau = L/2\Omega$. Choose the maximum frequency sampling interval $\nu = 1/\tau = 2\Omega/L$. Then $K(\omega) \equiv 2\Omega/L$ a.e., so (5.39) holds. Now the sum over m in (5.42) reduces to the *finite* sum over $|m| \leq M$, since $L = 2M + 1$

subbands of width $2\Omega/L$ are needed to cover the bandwidth of $\hat{f}$. Thus we have the following sampling theorem.

Theorem 5.5. *Let f be band-limited, with $\hat{f}(\omega) = 0$ for $|\omega| > \Omega$. Choose an integer $M = 0, 1, 2, \ldots$, and let*

$$\nu = \frac{2\Omega}{2M+1}, \qquad \tau = \frac{1}{\nu}. \tag{5.49}$$

Then

$$f(t) = \sum_{|m| \le M} \sum_{n \in \mathbf{Z}} e^{2\pi i m \nu t} \, \frac{\sin\left[2\pi\Omega\,(t - n\tau)\right]}{2\pi\Omega\,(t - n\tau)} \, \tilde{f}(m\nu, n\tau). \tag{5.50}$$

Hence the reconstruction of band-limited signals can be done "in parallel" in $2M+1$ different frequency "subbands" B_m $(m = -M, \ldots, M)$ of $[-\Omega, \Omega]$, where

$$B_m = \{\omega : |\omega - m\nu| \le \Omega/L\} = \left[\frac{(2m-1)}{2M+1}\,\Omega\,, \frac{(2m+1)}{2M+1}\,\Omega\right]. \tag{5.51}$$

Proof: Exercise.

One advantage of this scheme is that now the sampling rate for each subband is

$$R_M = \frac{2\Omega}{2M+1} = \frac{R_{\text{Nyquist}}}{2M+1}. \tag{5.52}$$

When $M = 0$, (5.50) reduces to the Shannon formula with the Nyquist sampling rate. For $M > 1$, a more relaxed sampling rate can be used in each subband. Note, however, that the sampling rate is the same in each subband. Ideally, one should use lower sampling rates on low-frequency subbands than on those with high frequencies, since low-frequency signals vary more slowly than high-frequency signals. This is precisely what happens in wavelet analysis, as we see in the next chapter, and it is one of the reasons why wavelet analysis is so efficient.

In practical computations, only a finite number of numerical operations can be performed. Hence the sums over m and n in (5.23) and (5.42), as well as the sum over n in (5.50), must be truncated. That is, the reconstruction must in every cases be approximated by a finite sum

$$f_{MN}(u) = \nu\tau \sum_{|m| \le M} \sum_{|n| \le N} \Gamma^{m,n}(u)\, \tilde{f}(m\nu, n\tau). \tag{5.53}$$

This approximation is good if the window is well localized in the frequency domain as well as the time domain and if M and N are taken sufficiently large. For example, we have seen above that for a band-limited signal, the sum over m automatically reduces to a finite sum if $\hat{g}(\omega)$ has compact support. Although a band-limited signal cannot also be time-limited, it either decays rapidly or it is only of interest in a finite time interval $|t| \le t_{\max}$. In that case, N can be chosen so that $|n\tau| \leq t_{\max} + \Delta t$ for $|n| < N$, where Δt is a "fudge factor" taking into account the width of the window. Then f_{MN} gives a good approximation for f only in the time interval of interest.

5.4 Other Discrete Time-Frequency Subframes

Suppose we begin with a window function $g \in L^2(\mathbf{R})$ such that neither g nor $\hat{g}$ have compact support. Let $\nu > 0$ and $\tau > 0$, and let $P_{\nu,\tau}$ and $\mathcal{G}_{\nu,\tau}$ be as in Theorems 5.3 and 5.4. It is natural to ask under what conditions $\mathcal{G}_{\nu,\tau}$ is a frame in $L^2(\mathbf{R})$. As expected (because of the uncertainty principle), it can be shown that $\mathcal{G}_{\nu,\tau}$ *cannot* be a frame if $\nu\tau > 1$, but all the known proofs of this in the general setting are difficult.

Although no single necessary and sufficient condition for $\mathcal{G}_{\nu,\tau}$ to be a frame when $\nu\tau \leq 1$ is known in general, it is possible to give two conditions, one necessary and the other sufficient, which are fairly close to one another under certain conditions. Once we know that $\mathcal{G}_{\nu,\tau}$ is a frame and have some estimates for its frame bounds, the recursive approximation scheme of Section 4.4 can be used to compute the reciprocal frame, reconstructing a signal from the values of its WFT sampled in $P_{\nu,\tau}$. The proofs of the necessary and the sufficient conditions are rather involved and will not be given here. We refer the reader to the literature. To simplify the statements, we combine the notations introduced in (5.14) and (5.38). Let

$$\begin{aligned} H_\tau(t) &= \tau \sum_{n\in\mathbf{Z}} |g(t-n\tau)|^2 \\ K_\nu(\omega) &= \nu \sum_{m\in\mathbf{Z}} |\hat{g}(\omega - m\nu)|^2 . \end{aligned} \tag{5.54}$$

These are discrete approximations to $\|g\|^2$ and $\|\hat{g}\|^2$ by Riemann sums, so that if $g(t)$ and $\hat{g}(\omega)$ are piecewise continuous,

$$\begin{aligned} \lim_{\tau\to 0+} H_\tau(t) &= \|g\|^2 = C \\ \lim_{\nu\to 0+} K_\nu(\omega) &= \|\hat{g}\|^2 = C. \end{aligned} \tag{5.55}$$

Let

$$\begin{aligned} A_\tau &= \inf_t H_\tau(t), \quad B_\tau = \sup_t H_\tau(t) \\ A'_\nu &= \inf_\omega K_\nu(\omega), \quad B'_\nu = \sup_\omega K_\nu(\omega). \end{aligned} \tag{5.56}$$

When both g and $\hat{g}$ are well behaved, (5.55) suggests that

$$\begin{aligned} \lim_{\tau\to 0+} A_\tau &= \lim_{\tau\to 0+} B_\tau = C \\ \lim_{\nu\to 0+} A'_\nu &= \lim_{\nu\to 0+} B'_\nu = C. \end{aligned} \tag{5.57}$$

Hence for sufficiently small ν and τ we expect that

$$\begin{aligned} 0 < A_\tau \le H_\tau(t) \le B_\tau < \infty, \\ 0 < A'_\nu \le K_\nu(\omega) \le B'_\nu < \infty. \end{aligned} \tag{5.58}$$

As in Theorems 5.3 and 5.4, these inequalities imply

$$0 < A_\tau \le C \le B_\tau < \infty, \qquad 0 < A'_\nu \le C \le B'_\nu. \tag{5.59}$$

When g was time-limited and $\mathcal{G}_{\nu,\tau}$ was a frame, then the *best* frame bounds for $\mathcal{G}_{\nu,\tau}$ were A_τ and B_τ. Similarly, when g was band-limited and $\mathcal{G}_{\nu,\tau}$ was a frame, then the best frame bounds for $\mathcal{G}_{\nu,\tau}$ were A'_ν and B'_ν. In the general case we are considering now, neither of these statements is necessarily true. The best frame bounds (when $\mathcal{G}_{\nu,\tau}$ is a frame) depend on both ν and τ, and they cannot be computed exactly in general. Instead, we have the following theorem.

Theorem 5.6: (Necessary condition). *Let $\mathcal{G}_{\nu,\tau}$ be a subframe of $\mathcal{G}_P$ parameterized by $P_{\nu,\tau}$, with frame bounds $0 < A_{\nu,\tau} \le B_{\nu,\tau} < \infty$. Then $H_\tau(t)$ and $K_\nu(\omega)$ satisfy the following inequalities:*

$$\begin{aligned} A_{\nu,\tau} \le H_\tau(t) \le B_{\nu,\tau} \quad a.e. \\ A_{\nu,\tau} \le K_\nu(\omega) \le B_{\nu,\tau} \quad a.e. \end{aligned} \tag{5.60}$$

Consequently, the frame bounds of $\mathcal{G}_{\nu,\tau}$ are related to the frame bound C of $\mathcal{G}_P$ by

$$A_{\nu,\tau} \le C \le B_{\nu,\tau}. \tag{5.61}$$

Equation (5.61) follows from (5.60) and (5.31). For the proof of (5.60), see Chui and Shi (1993). Note that our normalization is different from theirs because of our convention of placing the 2π factor in the exponent of the Fourier transform and assigning the measure $\nu\tau$ to each sample point. Thus, in Chui and Shi, as well as in Daubechies (1992), (5.60) and (5.61) are replaced by

$$\begin{aligned} A_{\nu,\tau} \le \frac{2\pi}{\nu\tau} H_\tau(t) \le B_{\nu,\tau} \quad \text{a.e.} \\ A_{\nu,\tau} \le \frac{2\pi}{\nu\tau} K_\nu(\omega) \le B_{\nu,\tau} \quad \text{a.e.} \\ A_{\nu,\tau} \le \frac{2\pi}{\nu\tau} C \le B_{\nu,\tau}. \end{aligned} \tag{5.62}$$

Theorem 5.7: (Sufficient condition). *Let $g(t)$ satisfy the regularity condition*

$$|g(t)| \le N(1+|t|)^{-b} \tag{5.63}$$

for some $b > 1$ *and* $N > 0$. *Let* $\tau > 0$, *and suppose that*

$$A_\tau \equiv \inf_t H_\tau > 0, \qquad B_\tau \equiv \sup_t H_\tau < \infty. \tag{5.64}$$

Then there exists a "threshold" frequency $\nu_o(\tau) > 0$ *and, for every* $0 < \nu < \nu_o(\tau)$, *a number* $\delta(\nu, \tau)$ *with* $0 \leq \delta(\nu, \tau) < A_\tau$, *such that*

$$\begin{aligned} A_{\nu,\tau} &\equiv A_\tau - \delta(\nu, \tau) \\ B_{\nu,\tau} &\equiv B_\tau + \delta(\nu, \tau) \end{aligned} \tag{5.65}$$

are frame bounds for $\mathcal{G}_{\nu,\tau}$, *i.e., the operator*

$$G_{\nu,\tau} \equiv \nu\tau \sum_{m,n \in \mathbf{Z}} \Gamma_{m,n}\, \Gamma^*_{m,n} \tag{5.66}$$

satisfies $A_{\nu,\tau}\, I \leq G_{\nu,\tau} \leq B_{\nu,\tau}\, I$.

For the proof, see Proposition 3.4.1 in Daubechies (1992), where expression for $\delta(\nu, \tau)$ in terms of g are also given. Note again that her normalization differs from ours in the way described in (5.62).

Early work on time-frequency localizations tended to use Gaussian windows, as originally proposed by Gabor (1946). For this reason, such expansions are now often called *Gabor expansions*, whether or not they use Gaussian windows. Figure 5.1 shows a Gaussian window and several of its reciprocal (or dual) windows, taken with decreasing values of $\nu\tau$. As expected, the reciprocal windows appear to converge to the original window as $\nu\tau \to 0$ (assuming $\|g\| = 1$).

Much work has been done in recent years on time-frequency expansions, especially in spaces other than $L^2(\mathbf{R})$. See Benedetto and Walnut (1993) and the review by Bastiaans (1993). Interesting results in even more general situations (Banach spaces, irregular sampling, and groups other than the Weyl–Heisenberg group) appear in Feichtinger and Gröchenig (1986, 1989a, 1989b, 1992, 1993) and Gröchenig (1991).

To illustrate the last theorem, suppose that g is continuous with compact support in $[a, b]$ and $g(t) \neq 0$ in (a, b). Then Theorem 5.3 shows that $\mathcal{G}_{\nu,\tau}$ is a frame if and only if $0 < \tau < b - a \equiv 1/\nu$, since $A_\tau > 0$ if $0 < \tau < 1/\nu$ but $A_\tau = 0$ if $\tau = 1/\nu$. Hence Theorem 5.7 holds with $\nu_o(\tau) = 1/\tau$.

Note that it is *not* claimed that $A_{\nu,\tau}$ and $B_{\nu,\tau}$, as given by (5.65), are the *best* frame bounds. Once we choose $0 < \nu < \nu_o(\tau)$ so that $\mathcal{G}_{\nu,\tau}$ is guaranteed to be a frame, we can conclude that its best frame bounds satisfy

$$0 < A_{\nu,\tau} \leq A^{\text{best}}_{\nu,\tau} \leq B^{\text{best}}_{\nu,\tau} \leq B_{\nu,\tau} < \infty. \tag{5.67}$$

An expression such as (5.65), together with an expression for $\delta(\nu, \tau)$, is called an *estimate* for the (best) frame bounds. It can be shown that $\nu_o(\tau) \leq 1/\tau$ for all $\tau > 0$, for if $\nu_o(\tau) = 1/\tau + \varepsilon$ for some $\tau > 0$ and $\varepsilon > 0$, then we could choose

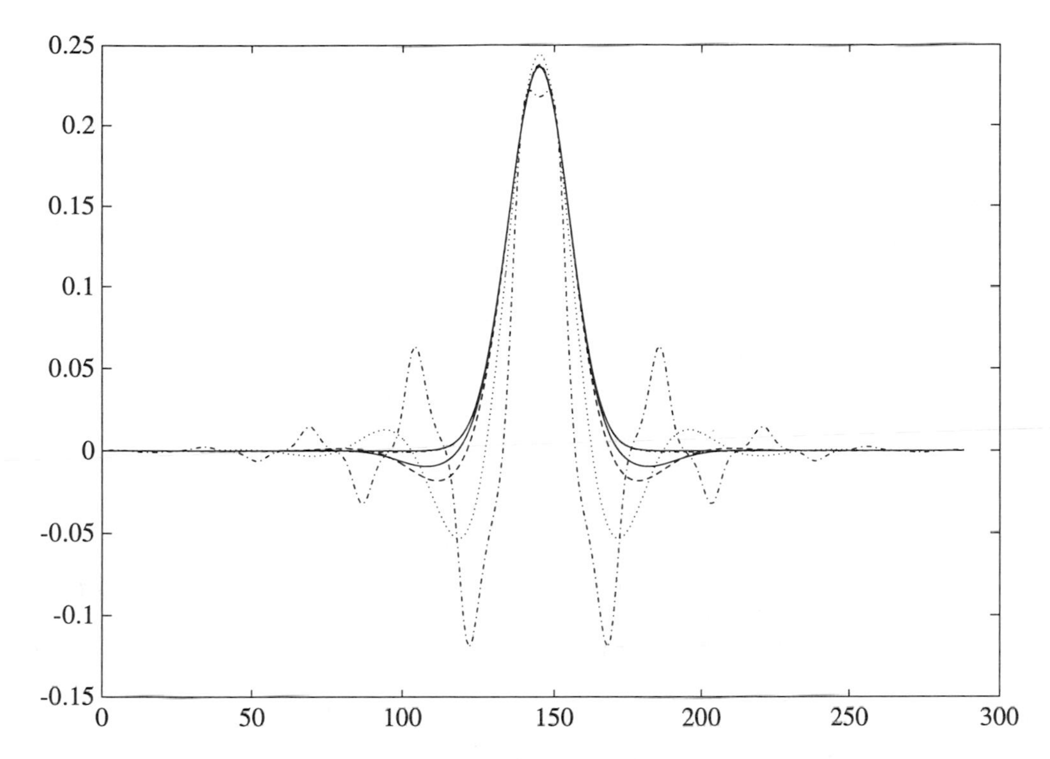

Figure 5.1. A Gaussian window and several of its reciprocal windows, with decreasing values of $\nu\tau$. This figure was produced by Hans Feichtinger with MATLAB using his irregular sampling toolbox (IRSATOL).

$\nu = 1/\tau + \varepsilon/2 < \nu_o(\tau)$. For this choice, $\tau\,\nu > 1$, in violation of the uncertainty principle.

To see how close the necessary and sufficient conditions of Theorems 5.6 and 5.7 are, note that if $\mathcal{G}_{\nu,\tau}$ is a frame whose existence is guaranteed by Theorem 5.7, then Theorem 5.6 implies that

$$\begin{aligned} A_{\nu,\tau} &\leq \inf_t H_\tau(t) = A_\tau, \\ B_{\nu,\tau} &\geq \sup_t H_\tau(t) = B_\tau\,. \end{aligned} \tag{5.68}$$

This is a somewhat weaker statement than (5.65) (if we have a way to compute $\delta(\nu,\tau)$). In fact, this shows that the "gap" between the necessary condition of Theorem 5.6 and the sufficient condition of Theorem 5.7 is precisely the "fudge factor" $\delta(\nu,\tau\,)$.

Because of the symmetry of the WFT with respect to the Fourier transform, Theorem 5.7 can be restated with the roles of time and frequency reversed. That is, begin with $\nu > 0$ and suppose that $A'_\nu \equiv \inf_\omega K_\nu(\omega) > 0$ and $B'_\nu \equiv \sup_\omega K_\nu(\omega) < \infty$. Let $\hat{g}(\omega)$ satisfy a regularity condition similar to (5.63). Then there exists a threshold time interval $\tau_o(\nu) > 0$ and, for every $0 < \tau < \tau_o(\nu)$, a number $0 \le \delta'(\nu, \tau) < A'_\nu$ such that

$$\begin{aligned} A'_{\nu,\tau} &\equiv A'_\nu - \delta'(\nu, \tau), \\ B'_{\nu,\tau} &\equiv B'_\nu + \delta'(\nu, \tau) \end{aligned} \tag{5.69}$$

are also frame bounds for $\mathcal{G}_{\nu,\tau}$. In general, the new bounds will differ from the old ones. A better set of bounds can then be obtained by choosing the larger of $A_{\nu,\tau}$ and $A'_{\nu,\tau}$ for the lower bound and the smaller of $B_{\nu,\tau}$ and $B'_{\nu,\tau}$ for the upper bound. Again, it can be shown that $\tau_o(\nu) \le 1/\nu$.

As in the case when g had compact support, $\mathcal{G}_{\nu,\tau}$ is a subframe of $\mathcal{G}_P$, but $\mathcal{G}^{\nu,\tau}$ is, in general, not a subframe of $\mathcal{G}^P$. It is therefore important to know whether the vectors $\Gamma^{m,n}$ can all be obtained by translating and modulating a single vector. When g had compact support, this was indeed shown to be the case, due to the periodicity of H_τ (5.34). It remains true in the general case at hand. We sketch the proof below, relegating the details to the Exercises.

Define the time translation operator T and the modulation operator M by

$$(Tf)(t) = f(t - \tau), \qquad (Mf)(t) = e^{2\pi i\nu t} f(t). \tag{5.70}$$

Then all the vectors in $\mathcal{G}_{\nu,\tau}$ are generated from g by powers of T and M according to

$$\Gamma_{m,n} = M^m T^n g, \tag{5.71}$$

and it can be shown that both of these operators commute with the metric operator defined by (5.66), i.e.,

$$T G_{\nu,\tau} = G_{\nu,\tau} T, \qquad M G_{\nu,\tau} = G_{\nu,\tau} M. \tag{5.72}$$

It follows that T and M also commute with $G_{\nu,\tau}^{-1}$. Consequently,

$$\Gamma^{m,n} \equiv G_{\nu,\tau}^{-1} \Gamma_{m,n} = G_{\nu,\tau}^{-1} M^m T^n g = M^m T^n G_{\nu,\tau}^{-1} g, \tag{5.73}$$

which states that $\Gamma^{m,n}$ is obtained from $G_{\nu,\tau}^{-1} g$ by a translation and modulation, exactly the way $\Gamma_{m,n}$ is obtained from g.

Note that Theorem 5.7 does not apply to the "critical" case $\nu\tau = 1$, since $\nu < \nu_o(\tau) \le \tau^{-1}$. Although discrete frames with $\nu\tau = 1$ do exist, our constructions in Section 5.2 suggest that they are problematic, since they seem to require bad behavior of either $g(t)$ of $\hat{g}(\omega)$. This objection is made precise in the general case (where neither g nor $\hat{g}$ have compact support) by the *Balian–Low theorem* (see Balian [1981], Low [1985], Battle [1988], Daubechies [1990]):

Theorem 5.8. *Let $\mathcal{G}_{\nu,\tau}$ be a frame with $\nu\tau = 1$. Then at least one of the following equalities holds:*

$$\int_{-\infty}^{\infty} dt \cdot t^2 \, |g(t)|^2 = \infty \quad \text{or} \quad \int_{-\infty}^{\infty} d\omega \cdot \omega^2 \, |\hat{g}(\omega)|^2 = \infty. \tag{5.74}$$

That is, either g decays very slowly in time or $\hat{g}$ decays very slowly in frequency.

The proof makes use of the so-called *Zak transform.* See Daubechies (1992).

Although no single necessary *and* sufficient condition for $\mathcal{G}_{\nu,\tau}$ to be a frame is known for a general window, a complete characterization of such frames with *Gaussian* windows now exists. Namely, let $g(t) = N\, e^{-\alpha t^2}$, for some $N, \alpha > 0$. As already mentioned, $\mathcal{G}_{\nu,\tau}$ cannot be a frame if $\nu\tau > 1$. Bacry, Grossmann, and Zak (1975) have shown that in the critical case $\nu\tau = 1$, $\mathcal{G}_{\nu,\tau}$ is *not* a frame even though the vectors $\Gamma_{m,n}$ span $L^2(\mathbf{R})$. (The prospective lower frame bound vanishes, and the reconstruction of f from the values of $\tilde{f}$ on $P_{\nu,\tau}$ is therefore unstable.) If $0 < \nu\tau < 1$, then Lyubarskii (1989) and Seip and Wallstén (1990) have shown independently that $\mathcal{G}_{\nu,\tau}$ *is* a frame. Thus, a necessary and sufficient condition for $\mathcal{G}_{\nu,\tau}$ to be a frame is that $0 < \nu\tau < 1$, when the window is Gaussian. The idea of using Gaussian windows to describe signals goes back to Gabor (1946). Bargmann et al. (1971) proved that with $\nu\tau \leq 1$ these vectors span $L^2(\mathbf{R})$, but they gave no reconstruction formula. (They used "value distribution theory" for entire functions of exponential type, which is made possible by the relation of the Gaussians to the so-called Bargmann transform.) Bastiaans (1980, 1981) found a family of dual vectors giving a "resolution of unity," but the convergence is very weak, in the sense of distributions; see Janssen (1981).

Discrete time-frequency frames $\mathcal{G}_{\nu,\tau}$ of the above form by no means exhaust the possibilities. For example, they include only *regular lattices* of samples (i.e., $\Delta\omega$ = constant and Δt = constant), and these lattices are parallel to the time and frequency axes. It has been discovered recently that various frames with very nice properties are possible by sampling *nonuniformly* in time and frequency. See Daubechies, Section 4.2.2.

Even for regular lattices, Theorems 5.3–5.7 do not exhaust the possibilities. There is a method of *deforming* a given frame to obtain a whole *family* of related frames. When applied to any of the above frames $\mathcal{G}_{\nu,\tau}$, this gives infinitely many frames *not* of the type $\mathcal{G}_{\nu,\tau}$ but just as good (with the same frame bounds). Some of these frames still use regular lattices in the time-frequency plane, but these lattices are now *rotated* rather than parallel to the axes (Kaiser [1994a]). Others use irregular lattices obtained from regular ones by a "stretching" or "shearing" operation (Baraniuk and Jones [1993]). To get an intuitive understanding of this method, consider the finite-dimensional Hilbert space $\mathbf{C}^N$. Let $R(\theta)$ be a *rotation* by the angle θ in some plane in $\mathbf{C}^N$. For example, in terms of an orthonormal basis $\{b_1, b_2, \ldots, b_N\}$ of $\mathcal{H} = \mathbf{C}^N$, we could let

$$\begin{aligned} R(\theta)b_1 &= b_1 \cos\theta + b_2 \sin\theta \\ R(\theta)b_2 &= b_2 \cos\theta - b_1 \sin\theta \\ R(\theta)b_k &= b_k, \quad 3 \leq k \leq N. \end{aligned} \tag{5.75}$$

That defines $R(\theta) : \mathbf{C}^N \to \mathbf{C}^N$ as a rotation in the plane spanned by $\{b_1, b_2\}$. In fact, $R(\theta)$ is *unitary*, i.e., $R(\theta)^* R(\theta) = I$. This means that $R(\theta)$ preserves inner products, and hence lengths and angles also (as a rotation should). Furthermore, the family $\{R(\theta) : \theta \in \mathbf{R}\}$ is a *one-parameter group* of unitary operators, which means that $R(\theta_1)R(\theta_2) = R(\theta_1 + \theta_2)$. (Actually, in this case, $R(\theta + 2\pi) = R(\theta)$, so only $0 \le \theta < 2\pi$ need to be considered.) Now given any frame $\mathcal{H}_M = \{h_m\} \subset \mathbf{C}^N$, let $h_m^\theta \equiv R(\theta)h_m$ and $\mathcal{H}_M^\theta \equiv \{h_m^\theta : m \in M\}$. Then it is easily seen that each $\mathcal{H}_M^\theta$ is also a frame, with the same frame bounds as $\mathcal{H}_M$, and $\mathcal{H}_M^0 = \mathcal{H}_M$. We call the family of frames $\{\mathcal{H}_M^\theta : \theta \in \mathbf{R}\}$ a *deformation* of $\mathcal{H}_M$. (Note that this is a *set* whose *elements* are themselves *sets* of vectors! This kind of mental "acrobatics" is common in mathematics, and, once absorbed, can make life much easier by organizing ideas and concepts. Of course, one can also get carried away with it into unnecessary abstraction.)

Returning to the general case, let $\mathcal{H}$ be an arbitrary Hilbert space, let M be a measure space, and let $\mathcal{H}_M$ be a generalized frame in $\mathcal{H}$. Any self-adjoint operator $H^* = H$ on $\mathcal{H}$ generates a one-parameter group of unitary operators on $\mathcal{H}$, defined by $R(\theta) = e^{i\theta H}$. (This operator can be defined by the so-called *spectral theorem* of Hilbert space theory.) If we define h^θ and $\mathcal{H}_M^\theta$ as above, it again follows that each $\mathcal{H}_M^\theta$ is a frame with the same frame bounds as $\mathcal{H}_M$, and $\mathcal{H}_M^0 = \mathcal{H}_M$. This is the general idea behind deformations of frames. There is an infinite number of distinct self-adjoint operators H on $\mathcal{H}$, and any two that are not proportional generate a distinct deformation. But most such deformations are not very interesting. For example, let $\mathcal{H} = L^2(\mathbf{R})$, and let $\mathcal{G}_{\nu,\tau}$ be one of the discrete time-frequency frames constucted in Sections 5.2 and 5.4. For a general choice of H, the vectors $\Gamma_{m,n}^\theta = e^{i\theta H}\Gamma_{m,n}$ cannot be interpreted as "notes" with a more or less well-defined time and frequency; hence $\mathcal{G}_{\nu,\tau}^\theta$ is generally not a time-frequency frame for $\theta \ne 0$. Furthermore, the vectors $\Gamma_{m,n}^\theta$ cannot, in general, be obtained from a single vector by applying simple operations such as translations and modulations. It is important to choose H so as to preserve the time-frequency character of the frame, and there is only a limited number of such choices. One choice is the so-called *harmonic oscillator Hamiltonian* of quantum mechanics, given by

$$(Hf)(t) \equiv \tfrac{1}{2}\left(t^2 f(t) - f''(t)\right), \tag{5.76}$$

which defines an *unbounded* self-adjoint operator on $L^2(\mathbf{R})$. It can be shown that $R(\pi/2)f = \hat{f}$ for any $f \in L^2(\mathbf{R})$, so $R(\pi/2)$ is just the Fourier transform. In fact, $R(\theta)$ looks like a *rotation* of the time-frequency plane. Specifically, we have, for any $(\omega, t) \in \mathbf{R}^2$,

$$g_{\omega,t}^\theta \equiv R(\theta)g_{\omega,t} = \alpha(\omega, t, \theta)\, g_{\omega(\theta),t(\theta)}, \tag{5.77}$$

where $|\alpha(\omega, t, \theta)| = 1$ and

$$\omega(\theta) = \omega \cos\theta - t \sin\theta, \qquad t(\theta) = t\cos\theta + \omega \sin\theta. \tag{5.78}$$

That is, up to an insignificant phase factor, $R(\theta)$ simply "rotates" the notes in the time-frequency plane! Consequently, if $\mathcal{G}_{\nu,\tau}$ is the discrete frame of Theorem 5.3 and we let $\Gamma^{\theta}_{m,n} \equiv R(\theta)\Gamma_{m,n}$ for $\Gamma_{m,n} \in \mathcal{G}_{\nu,\tau}$, then for each value of θ, the family

$$\mathcal{G}^{\theta}_{\nu,\tau} = \{\Gamma^{\theta}_{m,n} : m, n \in \mathbf{Z}\} \tag{5.79}$$

is also a discrete time-frequency frame, but parameterized by a *rotated* version $P^{\theta}_{\nu,\tau}$ of the lattice $P_{\nu,\tau}$. Furthermore, $\mathcal{G}^{0}_{\nu,\tau} = \mathcal{G}_{\nu,\tau}$ is the original frame, and $\mathcal{G}^{\pi/2}_{\nu,\tau}$ is the frame constructed in Theorem 5.4. Hence the family $\{\mathcal{G}^{\theta}_{\nu,\tau} : 0 \le \theta < 2\pi\}$ *interpolates* these two frames.

Another good choice of symmetry is given by

$$(Hf)(t) = -2itf'(t) - if(t). \tag{5.80}$$

That leads to reciprocal *dilations* in the time and frequency domains:

$$\begin{aligned} (e^{i\theta H} f)(t) &= e^{\theta} f(e^{2\theta} t) \\ (e^{i\theta H} f)^{\wedge}(\omega) &= e^{-\theta} \hat{f}(e^{-2\theta} \omega). \end{aligned} \tag{5.81}$$

Again, this gives a one-parameter group of unitary operators. The effect of these operators on a rectangular lattice of the type used in $\mathcal{G}_{\nu,\tau}$ is to compress the lattice by a factor $e^{2\theta}$ along the time axis while simultaneously stretching it by the same factor along the frequency axis. (Although this does not look like a rotation in the time-frequency plane, it is still a "rotation" in the infinite-dimensional space $L^2(\mathbf{R})$, in the sense that it preserves all inner products and hence all lengths and angles.) However, the deformed family of frames contains nothing new in this case, since $\mathcal{G}^{\theta}_{\nu,\tau}$ is just $\mathcal{G}_{\nu',\tau'}$ with $\nu' = e^{2\theta}\nu$ and $\tau' = e^{-2\theta}\tau$. In fact, this deformation was already used in Section 5.2 in the construction of $\mathcal{G}_{\nu',\tau'}$ from $\mathcal{G}_{\nu,\tau}$ (see (5.35) and (5.36)). There we had $\nu' = \sigma^{-1}\nu, \tau' = \sigma\tau$, so $\theta = -(1/2)\ln\sigma$. More interesting choices correspond to *quadratic* (rather than linear) distortions of the time-frequency plane, giving rise to a *shearing* action. This results in deformed time-frequency frames with *chirp-like* frame vectors (see Baraniuk and Jones [1993]).

5.5 The Question of Orthogonality

In this section we ask whether window functions $g(t)$ can be found such that $\mathcal{G}_{\nu,\tau}$ is not only a frame, but an *orthonormal basis.* We cannot pursue this question in all generality. Rather, we attempt to construct an orthonormal frame, or something close to it, by starting with a frame $\mathcal{G}_{\nu,\tau}$ generated by a time-limited window (Section 5.2) and defining

$$\Gamma^{\#}_{m,n} \equiv G^{-1/2}\Gamma_{m,n} \tag{5.82}$$

in the spirit of (4.77). Then

$$G^{\#}_{\nu,\tau} \equiv \nu\tau \sum_{m,n\in\mathbf{Z}} \Gamma^{\#}_{m,n}\,(\Gamma^{\#}_{m,n})^* = I, \tag{5.83}$$

and $\mathcal{G}^{\#}_{\nu,\tau}$ is a tight frame with frame bound $A = 1$. Since the time translation operator T and the modulation operator M commute with G, they also commute with $G^{-1/2}$. Hence

$$\Gamma^{\#}_{m,n} = G^{-1/2}M^mT^ng = M^nT^n\,G^{-1/2}g = M^nT^ng^{\#}, \tag{5.84}$$

where

$$g^{\#}(t) \equiv (G^{-1/2}g)(t) = \frac{g(t)}{\sqrt{H_\tau(t)}} \tag{5.85}$$

by (5.25). This proves that the frame vectors $\Gamma^{\#}_{m,n}$ are all obtained from $g^{\#}$ in the same way that $\Gamma_{m,n}$ are obtained from g.

Proposition 5.9. *Let $g \in L^2(\mathbf{R})$, let $0 < A_\tau \le H_\tau(t) \le B_\tau < \infty$, and let $g^{\#}$ be defined as above. Then $\|g^{\#}\| = 1$. If g is time-limited (so that the metric operator of $\mathcal{G}_{\nu,\tau}$ is given by (5.25)), then $\mathcal{G}^{\#}_{\nu,\tau}$ is normalized, i.e., $\|\Gamma^{\#}_{m,n}\| = 1$ for all $m, n \in \mathbf{Z}$.*

Proof: (5.85) implies that

$$H^{\#}_\tau(t) \equiv \tau \sum_{n\in\mathbf{Z}} |g^{\#}(t-n\tau)|^2 = 1 \quad \text{for all } t. \tag{5.86}$$

Integrating this over $0 \le t \le \tau$ and using (5.31), we obtain $\|g^{\#}\| = 1$. ■

From (5.83) it follows that for any $f \in L^2(\mathbf{R})$,

$$\|f\|^2 = \nu\tau \sum_{m,n\in\mathbf{Z}} |\langle\, \Gamma^{\#}_{m,n}\,, f\,\rangle|^2\,. \tag{5.87}$$

In particular, taking $f = \Gamma_{j,k}$,

$$\begin{aligned} 1 = \|\Gamma^{\#}_{j,k}\|^2 &= \nu\tau \sum_{m,n\in\mathbf{Z}} |\langle\, \Gamma^{\#}_{m,n}\,, \Gamma^{\#}_{j,k}\,\rangle|^2 \\ &= \nu\tau\left(1 + \sum_{(m,n)\ne(j,k)} |\langle\, \Gamma^{\#}_{m,n}\,, \Gamma^{\#}_{j,k}\,\rangle|^2\right) \\ &\equiv \nu\tau + \mathcal{O}^{\#}_{j,k}, \end{aligned} \tag{5.88}$$

where $\mathcal{O}_m$ was defined for arbitrary discrete frames in Section 4.5. Note that $\mathcal{G}^{\#}_{\nu,\tau}$ is "symmetric" in the sense that $\mathcal{O}^{\#}_{j,k}$ is independent of j and k, and its orthogonality index is $\mathcal{O}^{\#} \equiv \sup_{j,k} \mathcal{O}^{\#}_{j,k} = 1-\nu\tau$. Hence $\mathcal{G}^{\#}_{\nu,\tau}$ is not orthonormal if $\nu\tau < 1$. This is to be expected, since $\mathcal{G}^{\#}_{\nu,\tau}$ is then redundant and cannot be a basis. (Recall that $\rho(\nu,\tau) = (\nu\tau)^{-1}$ is a measure of the redundancy.) When $\nu\tau \approx 1$, then $\mathcal{O}^{\#} \approx 0$ and $\mathcal{G}^{\#}_{\nu,\tau}$ is "nearly" orthonormal. In the limit as $\nu\tau \to 1^-$, (5.88) shows that $\mathcal{G}^{\#}_{\nu,\tau}$ *becomes* orthonormal, provided the original window g is such that the lower bound A_τ of H_τ remains positive as $\nu\tau \to 1^-$. We have seen that this, in turn, requires that g (hence also $g^{\#}$) be discontinuous so that its Fourier transform has poor decay. Hence the above method can lead to orthonormal bases, but only bases with a poor frequency resolution. Similarly, when g is band-limited, this method can give orthonormal frames in the limit $\nu\tau \to 1^-$, but these have poor time resolution. The non-existence of orthonormal bases of notes with good time-frequency resolution seems to be a general feature of the WFT. See Daubechies (1992), where some related new developments are also discussed in Section 4.2.2.

Exercises

5.1. Prove Theorem 5.4.

5.2. Prove Theorem 5.5.

5.3. Define $G : L^2(\mathbf{R}) \to L^2(\mathbf{R})$ by (5.40). Show that

$$A'_\nu\, f^* f \le f^* G f \le B'_\tau\, f^* f.$$

Hence (5.39) implies that $\mathcal{G}_{\nu,\tau}$ is a frame.

5.4. (a) Show that the operators T and M defined in (5.70) are *unitary*, i.e., $T^* = T^{-1}$ and $M^* = M^{-1}$.

(b) Show that $\Gamma_{m,n} = M^m T^n g$. *Warning*: $(MTf)(t)$ is *not* $M(f(t-\tau))$ but $e^{2\pi\, i\nu t}\,(Tf)(t)$.

(c) Show that $TM = e^{-2\pi\, i\nu t}\, MT$.

(d) Show that $M\Gamma_{m,n} = \Gamma_{m+1,n}$, and hence $\Gamma^*_{m,n} M^{-1} = \Gamma^*_{m+1,n}$.

(e) Derive a similar set of relations for $T\,\Gamma_{m,n}$ and $\Gamma^*_{m,n} T^{-1}$.

(f) Use (d), (e), and the definition (5.66) of $G_{\nu,\tau}$ to prove (5.72).

(g) Show that (5.72) implies $G^{-1}_{\nu,\tau} T = T\, G^{-1}_{\nu,\tau}$ and $G^{-1}_{\nu,\tau} M = M\, G^{-1}_{\nu,\tau}$, which implies (5.73).

(h) Give an explicit formula for $\Gamma^{m,n}(t)$ in terms of the function $\breve{g}(t)$ defined by $\breve{g} \equiv G^{-1}_{\nu,\tau}\, g$. (You are *not* asked to compute $\breve{g}$!)

5.5. Show that the operators defined in (5.75) form a one-parameter group of unitary operators in $\mathbf{C}^N$, i.e.,

$$R(\theta)^* R(\theta) = I, \quad R(\theta_1)R(\theta_2) = R(\theta_1 + \theta_2). \tag{5.89}$$

5.6. Show that the operators $U(\theta) = e^{i\theta H}$ in (5.81) form a one-parameter group of unitary operators in $L^2(\mathbf{R})$.

Chapter 6

Discrete Time-Scale Analysis

Summary: In Chapter 3, we expressed a signal f as a continuous superposition of a family of wavelets $\psi_{s,t}$, with the CWT $\tilde{f}(s,t)$ as the coefficient function. In this chapter we discuss *discrete* constructions of this type. In each case the discrete wavelet family is a subframe of the continuous frame $\{\psi_{s,t}\}$, and the discrete coefficient function is a *sampled version* of $\tilde{f}(s,t)$. The salient feature of discrete wavelet analysis is that *the sampling rate is automatically adjusted to the scale.* That means that a given signal is sampled by first dividing its frequency spectrum into "bands," quite analogous to musical scales in that corresponding frequencies on adjacent bands are related by a constant *ratio* $\sigma > 1$ (rather than a constant *difference* $\nu > 0$, as is the case for the discrete WFT). Then the signal in each band is sampled at a rate proportional to the frequency scale of that band, so that high-frequency bands get sampled at a higher rate than that of low-frequency bands. Under favorable conditions the signal can be reconstructed from such samples of its CWT as a discrete superposition of reciprocal wavelets.

Prerequisites: Chapters 1, 3, and 4.

6.1 Sampling in the Time-Scale Domain

In Chapter 5, we derived some sampling theorems that enabled us to reconstruct a signal from the values of its WFT at the discrete set of frequencies and times given by $P_{\nu,\tau} = \{(m\nu, n\tau) : m, n \in \mathbf{Z}\}$. This is a *regular lattice* in the time-frequency plane, and under certain conditions the "notes" $\gamma_{m,n}$ labeled by $P_{\nu,\tau}$ formed a discrete subframe $\mathcal{G}_{\nu,\tau}$ of the continuous frame $\mathcal{G}_P$ of all notes. We now look for "regular" discrete subsets of the time-scale plane that are appropriate for the wavelet transform.

Regular sampling in scale means that the step from each scale to the next is the same. Since scaling acts by multiplication rather than by addition, we fix an arbitrary *scale factor* $\sigma > 1$ and consider only the scales $s_m = \sigma^m$, $m \in \mathbf{Z}$. Note that this gives only positive scales: Positive values of m give large (coarse) scales, and negative values give fine scales. If the mother wavelet ψ has both positive and negative frequency components as in Theorem 3.1, positive scales will suffice; if ψ has only positive frequencies as in Theorem 3.3, then we include the negative scales $-s_m = -\sigma^m$ as well in order to sample the negative-frequency part of f. To see how the sampling in time should be done, note first that the CWT has the following *scaling property*: If we scale the signal by σ just as we scaled the wavelet, i.e., define†

$$f_\sigma(t) \equiv \sigma^{-1/2} f(t/\sigma) \tag{6.1}$$

(this is a stretched version of f, since $\sigma > 1$), then its CWT is related to that of f by

$$\tilde{f}_\sigma(\sigma s, \sigma t) = \tilde{f}(s,t) \tag{6.2}$$

(exercise). This expresses the fact that *the CWT does not have any preference of one scale over another.* In order for a discrete subset of wavelets to inherit this "scale democracy," it is necessary to proceed as follows: In going from the scale $s_m = \sigma^m$ to the next larger one, $s_{m+1} = \sigma s_m$, we must increase the time-sampling interval Δt by a factor of σ. Thus we choose $\Delta t = \sigma^m \tau$, where $\tau > 0$ is the fixed time-sampling interval at the unit scale $s = 1$. The signal is sampled only at times $t_{m,n} = n\sigma^m \tau$ $(n \in \mathbf{Z})$ within the scale σ^m, which means that *the time-sampling rate is automatically adjusted to the scale.* The regular lattice $P_{\nu,\tau}$ in the time-frequency plane is therefore replaced by the set

$$S_{\sigma,\tau} = \{(\sigma^m, n\sigma^m\tau) : m, n \in \mathbf{Z}\}$$

in the time-scale plane. This is sometimes called a *hyperbolic lattice* since it looks regular to someone living in the "hyperbolic half-plane" $\{(s,t) \in \mathbf{R}^2 : s > 0\}$, where the geometry is such that the closer one gets to the t-axis $(s = 0)$, the more one shrinks.

6.2 Discrete Frames with Band-Limited Wavelets

In the language of Chapter 4, the set of all wavelets $\psi_{s,t}$ obtained from a given mother wavelet ψ by all possible translations and scalings forms a continuous frame labeled by the time-scale plane S, equipped with the measure $ds\,dt/s^2$. This frame will now be denoted by $\mathcal{W}_S$. In this section we construct a class of discrete subframes of $\mathcal{W}_S$ in the simplest case, namely when $\hat{\psi}(\omega)$ has compact support. The procedure will be similar to that employed in Section 5.2 when $\hat{g}(\omega)$ had compact support. To motivate the construction, we first build a class of *semi-discrete* frames, parameterized by discrete time but continuous scale. The discretization of scale will then be seen as a natural extension. These frames first appeared, in a somewhat different form, in Daubechies, Grossmann, and Meyer (1986).

Suppose we begin with a band-limited prospective mother wavelet: $\hat{\psi}(\omega) = 0$ outside the frequency band $\alpha \le \omega \le \beta$, for some $-\infty < \alpha < \beta < \infty$. Recall

† We now let $p = 1/2$ in (3.5), where we had defined $\psi_s(u) = |s|^{-p}\psi(u/s)$.

the frequency-domain expression for the CWT (setting $p = 1/2$ in (3.11)):

$$\tilde{f}(s,t) = \hat{\psi}_{s,t}^* \hat{f} = |s|^{1/2} \int_{-\infty}^{\infty} d\omega \;\; e^{2\pi\, i\omega t}\, \overline{\hat{\psi}}(s\,\omega)\, \hat{f}(\omega), \tag{6.3}$$

where $s \neq 0$. (At this point, we consider negative as well as positive scales.) In developing the *continuous* wavelet transform, we took the inverse Fourier transform of (6.3) with respect to t to recover $\overline{\hat{\psi}}(s\,\omega)\, \hat{f}(\omega)$. But this function is now band-limited, vanishing outside the interval $I_s = \{\omega : \alpha \leq s\,\omega \leq \beta\}$. Hence it can also be expanded in a Fourier series (rather than as a Fourier integral), provided we limit ω to the support interval. The width of the support is $(\beta - \alpha)/|s|$, which varies with the scale. Let

$$\tau = (\beta - \alpha)^{-1}. \tag{6.4}$$

This will be the time-sampling interval at unit scale ($s = 1$). The time-sampling interval at scale s is $|s|\tau$. Thus

$$|s|^{1/2}\, \overline{\hat{\psi}}(s\,\omega)\, \hat{f}(\omega) = |s|\tau \sum_{n\in\mathbf{Z}} e^{-2\pi\, in\omega s\tau}\, c_n(s), \quad \omega \in I_s, \tag{6.5}$$

where

$$c_n(s) = \int_{-\infty}^{\infty} d\omega \;\; e^{2\pi\, in\omega s\tau}\, |s|^{1/2}\, \overline{\hat{\psi}}(s\,\omega)\, \hat{f}(\omega) = \tilde{f}(s, ns\tau). \tag{6.6}$$

(The integral over $\omega \in I_s$ can be extended to $\omega \in \mathbf{R}$ because the integrand vanishes when $\omega \notin I_s$.) The Fourier series (6.5) is valid only for $\omega \in I_s$, since the left-hand side vanishes outside this interval while the right-hand side is periodic. To make it valid globally, multiply both sides by $|s|^{-1/2}\, \hat{\psi}(s\,\omega)$:

$$\begin{aligned} |\hat{\psi}(s\,\omega)|^2\, \hat{f}(\omega) &= \tau \sum_{n\in\mathbf{Z}} e^{-2\pi\, in\omega s\tau}\, |s|^{1/2}\, \hat{\psi}(s\,\omega)\, \tilde{f}(s, ns\tau) \\ &= \tau \sum_{n\in\mathbf{Z}} \hat{\psi}_{s,ns\tau}(\omega)\, \tilde{f}(s, ns\tau). \end{aligned} \tag{6.7}$$

We now attempt to recover $\hat{f}(\omega)$ by integrating both sides of (6.7) with respect to the measure ds/s, using *positive* scales only:

$$\int_0^{\infty} \frac{ds}{s}\, |\hat{\psi}(s\,\omega)|^2\, \hat{f}(\omega) = \tau \sum_{n\in\mathbf{Z}} \int_0^{\infty} \frac{ds}{s}\, \hat{\psi}_{s,ns\tau}(\omega)\, \tilde{f}(s, ns\tau). \tag{6.8}$$

Define the function

$$Y(\omega) \equiv \int_0^{\infty} \frac{ds}{s}\, |\hat{\psi}(s\,\omega)|^2, \tag{6.9}$$

already encountered in Chapter 3. Then (6.8) shows that $\hat{f}$ (and hence also f) can be fully recovered if division by $Y(\omega)$ makes sense for (almost) all ω. But if $\omega > 0$, then the change of variables $\xi = s\,\omega$ gives

$$Y(\omega) = \int_0^{\infty} \frac{d\xi}{\xi}\, |\hat{\psi}(\xi)|^2 \equiv C_+, \tag{6.10}$$

whereas if $\omega < 0$, the change $\xi = -s\,\omega$ gives

$$Y(\omega) = \int_0^\infty \frac{d\xi}{\xi}\,|\hat{\psi}(-\xi)|^2 \equiv C_-. \tag{6.11}$$

Thus, to recover f from the values of $\tilde{f}(s, ns\tau)$ with $s > 0$ and $n \in \mathbf{Z}$, a necessary and sufficient condition is that both of these constants are positive and finite:

$$0 < C_\pm < \infty. \tag{6.12}$$

This is, in fact, exactly the admissibility condition of Theorem 3.1. For (6.12) to hold, we must have $\hat{\psi}(\omega) \to 0$ as $\omega \to 0$ and the support of $\hat{\psi}$ must include $\omega = 0$ in its interior, i.e., $\alpha < 0 < \beta$. If this fails, e.g., if $\alpha \ge 0$, then negative as well as positive scales will be needed in order to recover f. We deal with this question later. For now, assume that (6.12) does hold. Then division by $Y(\omega)$ makes sense for all $\omega \ne 0$ (clearly, $Y(0) = 0$ since $\hat{\psi}(0) = 0$), and (6.8) gives

$$\begin{aligned}\hat{f}(\omega) &= \tau Y(\omega)^{-1} \sum_{n\in\mathbf{Z}} \int_0^\infty \frac{ds}{s}\,\hat{\psi}_{s,ns\tau}(\omega)\,\tilde{f}(s, ns\tau) \\ &\equiv \tau \sum_{n\in\mathbf{Z}} \int_0^\infty \frac{ds}{s}\,\hat{\psi}^{s,ns\tau}(\omega)\,\tilde{f}(s, ns\tau), \qquad \omega \ne 0,\end{aligned} \tag{6.13}$$

where $\psi^{s,ns\tau}$ is the reciprocal wavelet $\psi^{s,t} \in \mathcal{W}^S$ with $t = ns\tau$. Taking the inverse Fourier transform, we have the following vector equation in $L^2(\mathbf{R})$:

$$f = \tau \sum_{n\in\mathbf{Z}} \int_0^\infty \frac{ds}{s}\,\psi^{s,ns\tau}\,\tilde{f}(s, ns\tau). \tag{6.14}$$

This gives a *semi-discrete* subframe of $\mathcal{W}_S$, using only the vectors $\psi_{s,t}$ with $t = ns\tau$, $n \in \mathbf{Z}$, which we denote by $\mathcal{W}_\tau$. The frame bounds of $\mathcal{W}_\tau$ are the same as those of $\mathcal{W}_S$, namely

$$A = \min\{C_+, C_-\}, \qquad B = \max\{C_+, C_-\}. \tag{6.15}$$

In particular, if $C_- = C_+$, then $\mathcal{W}_\tau$ is tight. The reciprocal frame $\mathcal{W}^\tau$ is, similarly, a subframe of $\mathcal{W}^S$. Since $\tilde{f}(s, ns\tau) = (\psi_{s,ns\tau})^* f$, we also have the semi-discrete resolution of unity in $L^2(\mathbf{R})$:

$$\tau \sum_{n\in\mathbf{Z}} \int_0^\infty \frac{ds}{s}\,\psi^{s,ns\tau}(\psi_{s,ns\tau})^* = I. \tag{6.16}$$

As claimed, the sampling time interval $\Delta t = s\tau$ is automatically adjusted to the scale s. Note that the measure over the continuous scales in (6.16) is given (in differential form) by ds/s, rather than ds/s^2 as in the case of continuous time ((3.32) with $p = 1/2$). The two measures differ in that ds/s is *scaling-invariant*, while ds/s^2 is not, for if we change the scale by setting $s = \lambda s'$ for some fixed

$\lambda > 0$, then $ds'/s' = ds/s$ while $ds'/(s')^2 = \lambda ds/s^2$. The extra factor of s can be traced to the discretization of time. In the continuous case, the measure in the time-scale plane can be decomposed as

$$\frac{ds\,dt}{s^2} = \frac{ds}{s}\frac{dt}{s}, \tag{6.17}$$

which is the scaling-invariant measure in s, multiplied by the scaling-*covariant* measure in t. The latter takes into account the scaling of time that accompanies a change of scale. When time is discretized, dt/s is replaced by $s\tau/s = \tau$, which explains the factor τ in front of the sum in (6.14). Thus we interpret $\tau ds/s$ as a *semi-discrete measure* on the set of positive scales and discrete times, each n being assigned a measure τ. With this interpretation, the right-hand side of (6.14) is a Riemann sum (in the time variable) approximating the integral over the time-scale plane associated with the continuous frame.

We are now ready to discretize the scale. Instead of allowing all $s > 0$, consider only $s_m = \sigma^m$, with a fixed scale factor $\sigma > 1$ and $m \in \mathbf{Z}$. Then the increment in s_m is $\Delta s_m \equiv s_{m+1} - s_m = (\sigma - 1)s_m$. The discrete version of ds/s is therefore $\Delta s_m/s_m = \sigma - 1$. The fact that this is independent of m means that, like its continuous counterpart ds/s, the discrete measure $\Delta s_m/s_m$ treats all scales on equal footing. When $\hat{\psi}(\omega)$ is sufficiently well behaved (e.g., continuous), the integral representing $Y(\omega)$ can be written as

$$Y(\omega) = \lim_{\sigma \to 1+} \left[(\sigma - 1) \sum_{m \in \mathbf{Z}} |\hat{\psi}(\sigma^m \omega)|^2 \right]. \tag{6.18}$$

Hence, for $\sigma > 1$ we define

$$Y_\sigma(\omega) \equiv (\sigma - 1) \sum_{m \in \mathbf{Z}} |\hat{\psi}(\sigma^m \omega)|^2, \tag{6.19}$$

which is a Riemann sum approximating the integral. Y_σ will play a similar role in discretizing the CWT as that played by the functions H_τ and K_ν in discretizing the WFT in Chapter 5. By (6.10) and (6.11),

$$\lim_{\sigma \to 1+} Y_\sigma(\omega) = Y(\omega) = C_\pm \quad \text{for } \pm\omega > 0 \tag{6.20}$$

when $\hat{\psi}$ is well behaved. For somewhat more general ψ, (6.20) holds only for *almost* all $\pm\omega > 0$, being violated at most on a set of measure zero. Note that $Y_\sigma(\omega)$ is *scaling-periodic*, in the sense that

$$Y_\sigma(\sigma^k \omega) = Y_\sigma(\omega) \quad \text{for all } k \in \mathbf{Z}. \tag{6.21}$$

That implies that $Y_\sigma(\omega)$ *must be discontinuous at* $\omega = 0$. To see this, note first that $Y_\sigma(0) = 0$ since $\hat{\psi}(\omega) = 0$ and that $Y_\sigma(\omega)$ cannot vanish identically by (6.20), so $Y_\sigma(\omega_0) = \rho > 0$ for some $\omega_0 \neq 0$. Then (6.21) implies that $Y_\sigma(\sigma^k \omega_0) =$

ρ for all $k \in \mathbf{Z}$. Since $\sigma^k \omega_0 \to 0$ as $k \to -\infty$, $Y_\sigma(\omega)$ is discontinuous at $\omega = 0$. (In fact, the argument shows that Y_σ is both right- and left-discontinuous at $\omega = 0$.)

Another consequence of the dilation-periodicity is obtained by choosing an arbitrary frequency $\nu > 0$ and integrating $Y_\sigma(\pm\omega)$ over a single "period" $\nu \le \omega \le \nu\sigma$ with the scaling-invariant measure $d\omega/\omega$:

$$\begin{aligned}\int_\nu^{\sigma\nu} \frac{d\omega}{\omega} Y_\sigma(\pm\omega) &= (\sigma - 1) \sum_{m\in\mathbf{Z}} \int_\nu^{\nu\sigma} \frac{d\omega}{\omega} \, |\hat{\psi}(\pm\sigma^m\omega)|^2 \\ &= (\sigma - 1) \sum_{m\in\mathbf{Z}} \int_{\nu\sigma^m}^{\nu\sigma^{m+1}} \frac{d\xi}{\xi} \, |\hat{\psi}(\pm\xi)|^2 \\ &= (\sigma - 1) \int_0^\infty \frac{d\xi}{\xi} \, |\hat{\psi}(\pm\xi)|^2 = (\sigma - 1)\, C_\pm .\end{aligned} \tag{6.22}$$

The right-hand side of (6.22) is independent of ν because of the scaling-periodicity.

Let $A_{\pm,\sigma}$ and $B_{\pm,\sigma}$ be the greatest lower bound ("infimum") and least upper bound ("supremum") of $Y_\sigma(\pm\omega)$ over $\omega > 0$:[†]

$$A_{\pm,\sigma} = \inf_{\omega>0} Y_\sigma(\pm\omega), \qquad B_{\pm,\sigma} = \sup_{\omega>0} Y_\sigma(\pm\omega). \tag{6.23}$$

Because of the scaling-periodicity, it actually suffices to choose any $\nu > 0$ and to take the infimum and supremum over the single "period" $[\nu, \nu\sigma]$. In the continuous case, we only had to assume the admissibility condition $0 < C_\pm < \infty$ to obtain a reconstruction. This must now be replaced by the *stronger* set of conditions $A_{\pm,\sigma} > 0$, $B_{\pm,\sigma} < \infty$. Thus we have, by assumption,

$$0 < A_{\pm,\sigma} \le Y_\sigma(\pm\omega) \le B_{\pm,\sigma} < \infty \quad \text{for (almost) all } \omega > 0, \tag{6.24}$$

which is called the *stability condition* in Chui (1992a). When $\hat{\psi}$ behaves reasonably, (6.20) suggests that $A_{\pm,\sigma} \to C_\pm$ and $B_{\pm,\sigma} \to C_\pm$ as $\sigma \to 1^+$, and the stability condition reduces to the admissibility condition.

Applying (6.22) to (6.24), we get

$$0 < A_{\pm,\sigma} \ln\sigma \le (\sigma - 1)\, C_\pm \le B_{\pm,\sigma} \ln\sigma < \infty. \tag{6.25}$$

Hence the stability condition *implies* the admissibility condition, proving that it is stronger than the latter as claimed.

[†] Technically, to get better frame bounds, one should use the *essential* infimum and supremum, which ignore sets of measure zero. But, as in the time-frequency case discussed in Chapter 5, this is usually unnecessary in practice; see Daubechies (1992), p. 102. However, it *is* important to exclude $\omega = 0$ in (6.23), since otherwise $A_{\pm,\sigma}$ would vanish.

Note: In the definition (6.19) of $Y_\sigma(\omega)$, we could have chosen a factor of $\ln \sigma$ instead of $\sigma - 1$ to represent $\Delta s_m / s_m$, since $\sigma - 1 \approx \ln \sigma$ when $\sigma \approx 1$. (This can be arrived at by $\Delta \sigma^m = \Delta e^{m \ln \sigma} \approx \sigma^m \Delta(m \ln \sigma) = \sigma^m \ln \sigma$.) Equation (6.25) would then simplify to $0 < A_{\pm,\sigma} \le C_\pm \le B_{\pm,\sigma} < \infty$. A similar simplification occurs in (6.37), (6.52), and (6.58), which relate the frame bounds $C_\pm$ of the continuous frame $\mathcal{W}_S$ to the frame bounds of $\mathcal{W}_{\sigma,\tau}$.

Returning to (6.7), we now repeat the process that led to (6.14), but this time using only discrete scales. Introduce the notation

$$\Psi_{m,n}(t) = \psi_{\sigma^m, n\sigma^m\tau}(t) = \sigma^{-m/2}\, \psi(\sigma^{-m} t - n\tau). \tag{6.26}$$

Then $\tilde{f}(\sigma^m, n\sigma^m\tau) = \Psi_{m,n}^* f$. Multiplying (6.7) by $\sigma - 1$ and summing over $s_m = \sigma^m$, we get

$$Y_\sigma(\omega)\, \hat{f}(\omega) = (\sigma - 1)\, \tau \sum_{m,n \in \mathbf{Z}} \hat{\Psi}_{m,n}(\omega)\, \Psi_{m,n}^* f. \tag{6.27}$$

The factor $(\sigma - 1)\tau$ on the right-hand side will be regarded as the measure assigned to each sample point in the set $S_{\sigma,\tau} \equiv \{(\sigma^m, n\sigma^m\tau) : m, n \in \mathbf{Z}\}$. That is, the continuous measure is now replaced by the discrete measure as follows:

$$\frac{ds\, dt}{s^2} = \frac{ds}{s}\, \frac{dt}{s} \to \frac{\Delta s_m}{s_m}\, \frac{\Delta t_{m,n}}{s_m} = (\sigma - 1)\, \tau. \tag{6.28}$$

Equation (6.27) shows that the metric operator for the set of vectors $\{\Psi_{m,n}\}$ is the multiplication by $Y_\sigma(\omega)$ in the frequency domain (i.e., a *filter*):

$$(G_\sigma f)^\wedge(\omega) = Y_\sigma(\omega)\, \hat{f}(\omega). \tag{6.29}$$

Let

$$\begin{aligned} A_\sigma &\equiv \min\{A_{+,\sigma}, A_{-,\sigma}\} = \inf_{\omega \neq 0} Y_\sigma(\omega) \\ B_\sigma &\equiv \max\{B_{+,\sigma}, B_{-,\sigma}\} = \sup_{\omega \neq 0} Y_\sigma(\omega). \end{aligned} \tag{6.30}$$

Then (6.24) implies that

$$0 < A_\sigma \le Y_\sigma(\omega) \le B_\sigma < \infty \quad \text{a.e.} \tag{6.31}$$

For any $f \in L^2(\mathbf{R})$, Parseval's identity implies

$$\begin{aligned} \langle f, G_\sigma f \rangle = \langle \hat{f}, (G_\sigma f)^\wedge \rangle &= \int_{-\infty}^{\infty} d\omega\, \overline{\hat{f}(\omega)}\, Y_\sigma(\omega)\, \hat{f}(\omega) \\ &= \int_{-\infty}^{\infty} d\omega\, Y_\sigma(\omega)\, |\hat{f}(\omega)|^2. \end{aligned} \tag{6.32}$$

Hence by (6.31),

$$A_\sigma \|f\|^2 \le \langle f, G_\sigma f \rangle \le B_\sigma \|f\|^2. \tag{6.33}$$

This proves that $\{\Psi_{m,n}\}$ is a frame with G_σ as its metric operator and A_σ and B_σ as its frame bounds. Equation (6.27) gives

$$\hat{f}(\omega) = (\sigma - 1)\,\tau \sum_{m,n\in\mathbf{Z}} \hat{\Psi}^{m,n}(\omega)\,\Psi^*_{m,n}\,f, \tag{6.34}$$

where

$$\hat{\Psi}^{m,n}(\omega) = Y_\sigma(\omega)^{-1}\,\hat{\Psi}_{m,n}(\omega) \tag{6.35}$$

are the reciprocal frame vectors in the frequency domain. We summarize our findings, which amount to a discrete version of Theorem 3.1:

Theorem 6.1. *Let $\psi \in L^2(\mathbf{R})$ be band-limited, with $\hat{\psi}(\omega) = 0$ outside of the interval $\alpha \le \omega \le \beta$. Let $\tau \equiv (\beta - \alpha)^{-1}$, and let $Y_\sigma(\omega)$ satisfy (6.24) for some $\sigma > 1$. Let $S_{\sigma,\tau} = \{(\sigma^m, n\sigma^m\tau) : m, n \in \mathbf{Z}\}$, equipped with the measure $\mu_{\sigma,\tau}$ assigning the value $(\sigma - 1)\tau$ to every point $(\sigma^m, n\sigma^m\tau)$. Then the family of wavelets $\psi_{s,t} \in \mathcal{W}_S$ with $(s,t) \in S_{\sigma,\tau}$, i.e.,*

$$\mathcal{W}_{\sigma,\tau} = \{\psi_{\sigma^m,\, n\sigma^m\tau} \equiv \Psi_{m,n} \,:\, m, n \in \mathbf{Z}\}, \tag{6.36}$$

is a frame parameterized by $S_{\sigma,\tau}$. Its metric operator G_σ is given by (6.29), and the (best) frame bounds are A_σ and B_σ. They are related to the frame bounds of the continuous frame $\mathcal{W}_S$ by

$$A_\sigma \le \frac{\sigma - 1}{\ln\sigma}\,A \le \frac{\sigma - 1}{\ln\sigma}\,B \le B_\sigma. \tag{6.37}$$

A signal $f \in L^2(\mathbf{R})$ can be reconstructed from its samples in $S_{\sigma,\tau}$ by

$$f = (\sigma - 1)\tau \sum_{m,n\in\mathbf{Z}} \Psi^{m,n}\,\tilde{f}(\sigma^m, n\sigma^m\tau), \tag{6.38}$$

and we have the corresponding resolution of unity

$$(\sigma - 1)\tau \sum_{m,n\in\mathbf{Z}} \Psi^{m,n}\,\Psi^*_{m,n} = I. \quad \blacksquare \tag{6.39}$$

Note: As in the time-frequency case, the *counting measure* is usually taken as the measure on $S_{\sigma,\tau}$. In that case, the factor $(\sigma - 1)\,\tau$ must be absorbed into $\Psi^{m,n}$, which means that G_σ is redefined by $G_\sigma \to G'_\sigma \equiv [(\sigma - 1)\,\tau]^{-1}G_\sigma$. Then the frame bounds must be redefined in the same way. In addition, there can be an extra factor of 2π due to a different definition of the Fourier transform, as discussed in Chapter 5. For example, according to the conventions in Daubechies (1992) and Chui (1992a), (6.37) becomes

$$0 < A'_\sigma \le \frac{2\pi}{\tau\ln\sigma}\,A \le \frac{2\pi}{\tau\ln\sigma}\,B \le B'_\sigma < \infty. \tag{6.40}$$

(The dependence of the bounds on $\tau = (\beta - \alpha)^{-1}$ is, of course, implicit throughout, since τ depends on the mother wavelet. Later, when discussing frames whose wavelets are not band-limited, we make this dependence explicit.) We prefer the above form with $(\sigma - 1)\tau$ as the measure per sample point, because it seems to be simpler and more intuitive. The sums in (6.38) and (6.39) are approximations to the continuous case, and the relation (6.37) between the bounds of the discrete and continuous frames is simpler. The factor $2\pi/(\tau \ln \sigma)$ in (6.40) diverges as $\sigma \to 1^+$, whereas our factor $(\sigma - 1)/\ln \sigma \to 1$. In fact, when $\hat{\psi}$ is a reasonable function, then $A_\sigma \to A$ and $B_\sigma \to B$ as $\sigma \to 1^+$, and we have complete continuity between $\mathcal{W}_{\sigma,\tau}$ and $\mathcal{W}_S$.

The stability condition for Theorem 6.1 implies that $C_\pm > 0$, so $\hat{\psi}(\omega)$ must have support in negative as well as positive frequencies. Sometimes it is useful to work with wavelets that have only positive (or only negative) frequency components. Thus suppose we begin with a mother wavelet ψ_+ such that $\hat{\psi}_+(\omega) = 0$ outside the interval $\alpha \le \omega \le \beta$, where $0 \le \alpha < \beta < \infty$. Then $C_- = 0$, and the above construction must be modified. A glance at (6.3) suggests two solutions to the problem:

(a) Include negative as well as positive scales in the reconstruction. This works because for $s < 0$, $\hat{\psi}(s\,\omega)$ "probes" the negative-frequency part of the signal.

(b) Choose another band-limited wavelet ψ_- with $\hat{\psi}_-(\omega) = 0$ for $\omega \ge 0$, and use ψ_+ to recover the positive-frequency part f_+ of f and ψ_- to recover the negative-frequency part f_-, using only *positive* scales in both cases. Thus *two* mother wavelets are used instead of one.

In option (a), we double the number of scales, and in (b) we double the number of mother wavelets. In fact, (a) is really a special case of (b), since the choice $\psi_-(t) \equiv \psi_+(-t)$ gives $\hat{\psi}_-(\omega) = \hat{\psi}_+(-\omega)$ and

$$\begin{aligned} \tilde{f}_-(s,t) &\equiv |s|^{1/2} \int_{-\infty}^{\infty} d\omega \; e^{2\pi\, i\omega t}\, \overline{\hat{\psi}_-(s\,\omega)}\, \hat{f}(\omega) \\ &= |s|^{1/2} \int_{-\infty}^{\infty} d\omega \; e^{2\pi\, i\omega t}\, \overline{\hat{\psi}_+(-s\,\omega)}\, \hat{f}(\omega) \equiv \tilde{f}_+(-s,t). \end{aligned} \tag{6.41}$$

Thus we get greater generality at no extra cost by taking option (b). In particular, notice that since $\hat{\psi}_+(-\omega) \ne \overline{\hat{\psi}_+(\omega)}$, ψ_+ is necessarily complex-valued. By taking $\psi_-(t) \equiv \bar{\psi}_+(t)$, we get $\hat{\psi}_-(\omega) = \overline{\hat{\psi}_+(-\omega)}$ (supported in $\omega < 0$ as desired), and

$$\psi_+(t) + \psi_-(t) = 2\,\mathrm{Re}\,\psi_+(t), \quad \psi_+(t) - \psi_-(t) = 2i\,\mathrm{Im}\,\psi_+(t). \tag{6.42}$$

This means that instead of using ψ_+ and ψ_- as the two mother wavelets, we can also use the real and imaginary parts of ψ_+. Although two mother wavelets are used, they are actually the real and imaginary parts of a function analytic in the upper-half complex time plane (Section 3.3). This is analogous to the situation in Fourier analysis, where either complex exponentials or cosines and sines can be used.

We now give a version of Theorem 6.1 appropriate to the case of option (b). For simplicity, we assume that $\hat{\psi}_+(\omega)$ and $\hat{\psi}_-(-\omega)$ are both supported in $\alpha \le \omega \le \beta$, with $0 \le \alpha < \beta < \infty$. Let $\tau = (\beta - \alpha)^{-1}$ as before. Choose $\sigma > 1$ and denote the discrete sets of wavelets generated by ψ_+ and ψ_- by

$$\Psi_{\pm,m,n}(t) = \sigma^{-m/2}\psi_\pm(\sigma^{-m}t - n\tau). \tag{6.43}$$

We then get the following counterparts of (6.27):

$$Y_{\pm,\sigma}(\omega)\,\hat{f}(\omega) = (\sigma - 1)\,\tau \sum_{m,n\in\mathbf{Z}} \hat{\Psi}_{\pm,m,n}(\omega)\,\Psi_{\pm,m,n}{}^* f, \tag{6.44}$$

where

$$Y_{\pm,\sigma}(\omega) \equiv (\sigma - 1)\sum_{m\in\mathbf{Z}} |\hat{\psi}_\pm(\sigma^m\omega)|^2. \tag{6.45}$$

If $\hat{\psi}_+(\omega)$ and $\hat{\psi}_-(\omega)$ are well behaved, then

$$\lim_{\sigma\to 1+} Y_{\pm,\sigma}(\omega) = \int_0^\infty \frac{ds}{s}\,|\hat{\psi}_\pm(s\,\omega)|^2 \equiv C_\pm. \tag{6.46}$$

Clearly $Y_{\pm,\sigma}(\omega) = 0$ when $\pm\omega \le 0$. Let

$$\begin{aligned} A_{\pm,\sigma} &= \inf_{\pm\omega>0} Y_{\pm,s,t}(\omega), \qquad & B_{\pm,\sigma} &= \sup_{\pm\omega>0} Y_{\pm,s,t}(\omega) \\ A_\sigma &= \min\{A_{+,\sigma}, A_{-,\sigma}\}, \qquad & B_\sigma &= \max\{B_{+,\sigma}, B_{-,\sigma}\}. \end{aligned} \tag{6.47}$$

To obtain a stable reconstruction of f from the samples

$$\tilde{f}_\pm(\sigma^m, n\sigma^m\tau) \equiv (\Psi_{\pm,m,n})^* f, \tag{6.48}$$

we must assume that $A_{\pm,\sigma} > 0$ and $B_{\pm,\sigma} < \infty$, i.e.,

$$0 < A_{\pm,\sigma} \le Y_{\pm,\sigma}(\omega) \le B_{\pm,\sigma} < \infty \quad \text{for (almost) all } \pm\omega > 0. \tag{6.49}$$

Theorem 6.1 has the following straightforward extension.

Theorem 6.2. *Let $\psi_\pm \in L^2(\mathbf{R})$ be band-limited, with $\hat{\psi}_\pm(\omega) = 0$ outside of the intervals $\alpha \le \pm\omega \le \beta$, where $0 \le \alpha < \beta$. Let $\tau \equiv (\beta - \alpha)^{-1}$, and let $Y_{\pm,\sigma}(\omega)$ satisfy (6.49) for some $\sigma > 1$. Then the family of wavelets*

$$\mathcal{W}_{\sigma,\tau} = \{\Psi_{+,m,n}, \Psi_{-,m,n} : m, n \in \mathbf{Z}\} \tag{6.50}$$

is a frame. Its metric operator G_σ is given by

$$(G_\sigma f)^\wedge(\omega) = \left[Y_{+,\sigma}(\omega) + Y_{-,\sigma}(\omega)\right]\hat{f}(\omega), \tag{6.51}$$

and the (best) frame bounds are A_σ *and* B_σ. *They are related to the frame bounds of the continuous frame* $\mathcal{W}_S$ *by*

$$A_\sigma \le \frac{\sigma - 1}{\ln \sigma} A \le \frac{\sigma - 1}{\ln \sigma} B \le B_\sigma. \tag{6.52}$$

A signal $f \in L^2(\mathbf{R})$ *can be reconstructed from its samples* $\tilde{f}_\pm(\sigma^m, n\sigma^m\tau)$ *by*

$$f = (\sigma - 1)\tau \sum_{\substack{m,n \in \mathbf{Z} \\ P = \pm}} \Psi^{P,m,n}\, \tilde{f}_P(\sigma^m, n\sigma^m\tau), \tag{6.53}$$

and we have the corresponding resolution of unity

$$(\sigma - 1)\tau \sum_{\substack{m,n \in \mathbf{Z} \\ P = \pm}} \Psi^{P,m,n}\, (\Psi_{P,m,n})^* = I. \tag{6.54}$$

Proof: Except for (6.51), the proof is identical to that of Theorem 6.1. By definition, the metric operator for $\mathcal{W}_{\sigma,\tau}$ is

$$G_\sigma = (\sigma - 1)\tau \sum_{\substack{m,n \in \mathbf{Z} \\ P = \pm}} \Psi_{P,m,n}\, (\Psi_{P,m,n})^*. \tag{6.55}$$

Hence by (6.44),

$$\begin{aligned} (G_\sigma f)^\wedge(\omega) &= (\sigma - 1)\tau \sum_{\substack{m,n \in \mathbf{Z} \\ P = \pm}} \hat{\Psi}_{P,m,n}(\omega)\, (\Psi_{P,m,n})^* f \\ &= Y_{+,\sigma}(\omega)\hat{f}(\omega) + Y_{-,\sigma}(\omega)\hat{f}(\omega). \quad \blacksquare \end{aligned} \tag{6.56}$$

6.3 Other Discrete Wavelet Frames

The above constructions of discrete wavelet frames relied on the assumption that ψ is band-limited. When this is not the case, the method employed in the last section fails. Unlike the WFT, the CWT is not symmetric with respect to the Fourier transform (i.e., it cannot be viewed in terms of translations and scalings in the frequency domain), hence we cannot apply the same method to build frames with wavelets that are compactly supported in time. Usually it is more difficult to construct frames from wavelets that are not band-limited. In this section we state some general theorems about discrete wavelet frames of the type $\mathcal{W}_{\sigma,\tau} = \{\Psi_{m,n}\}$, parameterized by $S_{\sigma,\tau}$ with $\sigma > 1$ and $\tau > 0$, *without* assuming that the mother wavelet is band-limited. As in Section 5.4, there is no known condition that is both necessary and sufficient, but separate necessary and sufficient conditions can be stated.

Theorem 6.3: (Necessary condition). *Let $\mathcal{W}_{\sigma,\tau}$ be a subframe of $\mathcal{W}_S$ parameterized by $S_{\sigma,\tau}$, with frame bounds $A_{\sigma,\tau}$, $B_{\sigma,\tau}$. Then $Y_\sigma(\omega)$ satisfies the following inequality:*

$$A_{\sigma,\tau} \le Y_\sigma(\omega) \le B_{\sigma,\tau} \quad a.e. \tag{6.57}$$

Consequently, the frame bounds of $\mathcal{W}_{\sigma,\tau}$ are related to those of $\mathcal{W}_S$ by

$$A_{\sigma,\tau} \le \frac{(\sigma-1)}{\ln\sigma} A \le \frac{(\sigma-1)}{\ln\sigma} B \le B_{\sigma,\tau}. \tag{6.58}$$

As before, (6.22) is used to obtain (6.58) from (6.57). The difficult part is to prove (6.57). For this, see Chui and Shi (1993). Again note that, as explained in (6.40), our normalization differs from theirs and from that in Daubechies (1992). In terms of their conventions, (6.58) becomes

$$A'_{\sigma,\tau} \le \frac{2\pi}{\tau\ln\sigma} A \le \frac{2\pi}{\tau\ln\sigma} B \le B_{\sigma,\tau}. \tag{6.59}$$

Theorem 6.4: (Sufficient condition). *Let ψ satisfy the regularity condition*

$$|\hat{\psi}(\omega)| \le N\,|\omega|^a\,(1+|\omega|)^{-b} \tag{6.60}$$

for some $0 < a < b-1$ and $N > 0$. Let $\sigma > 1$, and suppose that

$$\begin{aligned} A_\sigma &\equiv \inf_{\omega\neq 0} Y_\sigma(\omega) > 0, \\ B_\sigma &\equiv \sup_{\omega\neq 0} Y_\sigma(\omega) < \infty. \end{aligned} \tag{6.61}$$

Then there exists a "threshold" time-sampling interval $\tau_o(\sigma) > 0$ and, for every $0 < \tau < \tau_o(\sigma)$, a number $\delta(\sigma,\tau)$ with $0 \le \delta(\sigma,\tau) < A_\sigma$, such that

$$\begin{aligned} A_{\sigma,\tau} &\equiv A_\sigma - \delta(\sigma,\tau), \\ B_{\sigma,\tau} &\equiv B_\sigma + \delta(\sigma,\tau) \end{aligned} \tag{6.62}$$

are frame bounds for $\mathcal{W}_{\sigma,\tau}$, i.e., the operator

$$G_{\sigma,\tau} \equiv (\sigma-1)\tau \sum_{m,n\in\mathbf{Z}} \Psi_{m,n}\,\Psi^*_{m,n} \tag{6.63}$$

satisfies $A_{\sigma,\tau}\, I \le G_{\sigma,\tau} \le B_{\sigma,\tau}\, I$.

For the proof, see Proposition 3.4.1 in Daubechies (1992), where an expression for $\delta(\sigma,\tau)$ in terms of $\hat{\psi}$ are also given. Note again the difference in normalizations. Thus, in Daubechies, (6.62) becomes

$$\begin{aligned} A'_{\sigma,\tau} &= \frac{2\pi}{\tau\ln\sigma}\left[A_\sigma - \delta(\sigma,\tau)\right], \\ B'_{\sigma,\tau} &= \frac{2\pi}{\tau\ln\sigma}\left[B_\sigma + \delta(\sigma,\tau)\right]. \end{aligned} \tag{6.64}$$

Note that it is *not* claimed that $A_{\sigma,\tau}$ and $B_{\sigma,\tau}$, as given by (6.62), are the *best* frame bounds. Once we choose $0 < \tau < \tau_o(\sigma)$, so that $\mathcal{W}_{\sigma,\tau}$ is guaranteed to be a frame, we can conclude that its best frame bounds satisfy

$$0 < A_{\sigma,\tau} \leq A^{\text{best}}_{\sigma,\tau} \leq B^{\text{best}}_{\sigma,\tau} \leq B_{\sigma,\tau} < \infty. \tag{6.65}$$

Thus (6.62), together with an expression for $\delta(\sigma,\tau)$, gives an *estimate* for the best frame bounds.

To get some insight into this theorem, suppose we were dealing with the situation in Theorem 6.2, where $\hat{\psi}_\pm(\pm\omega)$ both have support in $[\alpha,\beta]$, with $0 < \alpha < \beta$. Then the *maximum* time-sampling interval is $\tau_{\text{max}} = (\beta-\alpha)^{-1}$. Of course, we can always oversample, choosing $\alpha' \leq \alpha$ and $\beta' \geq \beta$, so that $0 < \tau \leq \tau_{\text{max}}$. Suppose, furthermore, that $\hat{\psi}$ is continuous, with $\hat{\psi}(\omega) \neq 0$ for $\alpha < |\omega| < \beta$. Then it is clear that the stability condition (6.61) is satisfied if and only if $\alpha < \sigma\alpha < \beta$. Thus $\mathcal{W}_{\sigma,\tau}$ is a frame if and only if

$$0 < \tau \leq \frac{1}{(\beta-\alpha)} < \frac{1}{(\sigma-1)\alpha} \equiv \tau_o(\sigma). \tag{6.66}$$

In the general case under consideration in Theorem 6.4, no support parameters such as α and β are given. But the regularity condition (6.60) implies that $\hat{\psi}(\omega)$ vanishes as $|\omega|^a$ near $\omega = 0$ and decays faster than $|\omega|^{-1}$ as $\omega \to \pm\infty$. Consequently, $\hat{\psi}(\omega)$ is concentrated *mainly* in a double interval $\alpha_{\text{effective}} \leq |\omega| \leq \beta_{\text{effective}}$, for some $0 < \alpha_{\text{effective}} < \beta_{\text{effective}} < \infty$. Hence we expect $\tau_o(\sigma) \sim [(\sigma-1)\alpha_{\text{effective}}]^{-1}$ in Theorem 6.4.

Let us define the *sampling density* $\Re(\sigma,\tau)$, *in samples per unit measure*, as the reciprocal of the measure $\mu_{\sigma,\tau} = (\sigma-1)\tau$ per sample. Then (6.66) shows that for band-limited wavelets,

$$\Re(\sigma,\tau) \equiv \frac{1}{(\sigma-1)\tau} \geq \alpha. \tag{6.67}$$

In the general context, (6.67) can be expected to hold with α replaced by $\alpha_{\text{effective}}$ on the right-hand side. In the time-frequency case, a similar role to (6.67) was played by the inequality $\rho(\nu,\tau) \geq 1$, where $\rho(\nu,\tau)$ was the sampling density in samples per unit *area* of the time-frequency plane. But now we no longer measure area, since $\mu_{\sigma,\tau}$ is a discrete version of $d\mu(s,t) = ds\,dt/s^2$, which is not Lebesgue measure in the time-scale plane. Still, the notion of sampling density survives in the above form. But since α can be chosen as small as one likes, *there is no absolute lower bound on the sampling frequency for wavelet frames.* (This can also be seen from a "dimensional" argument: ν and τ determine a dimensionless number $\nu\tau$, which then serves as an upper limit for the measure per sample; since s is a pure number, there is no way of making a dimensionless number from s and τ.)

To see how close the necessary and sufficient conditions of Theorems 6.3 and 6.4 are, note that if $\mathcal{W}_{\sigma,\tau}$ is a frame whose existence is guaranteed by Theorem 6.4, then Theorem 6.3 implies that

$$\begin{aligned} A_{\sigma,\tau} &\le \inf_{\omega\neq 0} Y_\sigma(\omega) \equiv A_\sigma, \\ B_{\sigma,\tau} &\ge \sup_{\omega\neq 0} Y_\sigma(\omega) \equiv B_\sigma . \end{aligned} \tag{6.68}$$

This is a somewhat weaker statement than (6.62), if we know how to compute $\delta(\sigma,\tau)$. This shows that the gap between Theorems 6.3 and 6.4 is the "fudge factor" $\delta(\sigma,\tau)$.

In the case of the WFT, we saw that quite generally the whole reciprocal frame $\mathcal{G}^{\nu,\tau}$ could be generated from the single basic window $G_{\nu,\tau}^{-1}g$, which makes it unnecesary to compute the vectors $\Gamma^{m,n}$ individually. Let us now see if the vectors $\Psi^{m,n}$ also have this important property. Define the time translation and dilation operators $T, D : L^2(\mathbf{R}) \to L^2(\mathbf{R})$ by

$$\begin{aligned} (Tf)(t) &\equiv f(t-\tau), \\ (Df)(t) &\equiv \sigma^{-1/2} f(\sigma^{-1}t). \end{aligned} \tag{6.69}$$

They are both *unitary*, i.e., $T^* = T^{-1}$ and $D^* = D^{-1}$. For any $m, n \in \mathbf{Z}$, we have

$$\begin{aligned} (D^m\, T^n\, \psi)(t) &= \sigma^{-m/2} (T^n\, \psi)(\sigma^{-m}\, t) \\ &= \sigma^{-m/2}\psi(\sigma^{-m}\, t - n\tau) = \Psi_{m,n}(t), \end{aligned} \tag{6.70}$$

hence $\Psi_{m,n} = D^m\, T^n\, \psi$. The reciprocal frame vectors are therefore

$$\Psi^{m,n} = G_{\sigma,\tau}^{-1}\, D^m\, T^n \psi. \tag{6.71}$$

If $\mathcal{W}_{\sigma,\tau}$ is tight, then $G_{\sigma,\tau} = A_{\sigma,\tau}\, I$ and $\Psi^{m,n} = D^m\, T^n\,(A_{\sigma,\tau}^{-1}\psi)$, which shows that $\mathcal{W}^{\sigma,\tau}$ is generated from the single vector $A_{\sigma,\tau}^{-1}\psi$. When $\mathcal{W}_{\sigma,\tau}$ is *not* tight, the situation is more difficult. A *sufficient* condition is that D and T *commute* with $G_{\sigma,\tau}$, i.e., $DG_{\sigma,\tau} = G_{\sigma,\tau}D$ and $TG_{\sigma,\tau} = G_{\sigma,\tau}T$, since this would imply (multiplying by $G_{\sigma,\tau}^{-1}$ from both sides) that they also commute with $G_{\sigma,\tau}^{-1}$, hence

$$\Psi^{m,n} = D^m\, T^n\, G_{\sigma,\tau}^{-1}\psi, \tag{6.72}$$

showing that $\mathcal{W}^{\sigma,\tau}$ is generated from $G_{\sigma,\tau}^{-1}\psi$. Thus let us check to see if D and T commute with $G_{\sigma,\tau}$. By (6.70) and the unitarity of D,

$$D\Psi_{m,n} = \Psi_{m+1,n}, \quad \text{hence} \quad \Psi_{m,n}^*\, D^{-1} = \Psi_{m+1,n}^*. \tag{6.73}$$

Therefore

$$\begin{aligned} DG_{\sigma,\tau}D^{-1} &= (\sigma-1)\tau \sum_{m,n\in\mathbf{Z}} D\Psi_{m,n}\,\Psi_{m,n}^*\,D^{-1} \\ &= (\sigma-1)\tau \sum_{m,n\in\mathbf{Z}} \Psi_{m+1,n}\Psi_{m+1,n}^* = G_{\sigma,\tau}\,, \end{aligned} \tag{6.74}$$

which proves that D commutes with $G_{\sigma,\tau}$. It follows that $\Psi^{m,n} = D^m \Psi^{0,n}$, and hence only the vectors $\Psi^{0,n}$ need to be computed for all $n \in \mathbf{Z}$. Unfortunately, T does *not* commute with $G_{\sigma,\tau}$ in general, so the vectors $\Psi^{0,n}$ cannot be written as $T^n \Psi^{0,0}$. To see this, note that

$$(T\Psi_{m,n})(t) = \Psi_{m,n}(t-\tau) = \sigma^{-m/2}\psi(\sigma^{-m}(t-\tau) - n\tau). \tag{6.75}$$

If $m > 0$, then $T\Psi_{m,n}$ is not a frame vector in $\mathcal{W}_{\sigma,\tau}$ and we don't have the counterpart of (6.73), which was responsible for (6.74). In other words, if $m > 0$, then we must translate in time steps that are multiples of $\sigma^m \tau > \tau$ in order to remain in the time-scale sampling lattice $T_{\sigma,\tau}$.

As a result of the above analysis, we conclude that *when $\mathcal{W}_{\sigma,\tau}$ is not tight, the reciprocal frame $\mathcal{W}^{\sigma,\tau}$ may not be generated by a single wavelet.* We may need to compute an infinite number of vectors $\Psi^{0,n}$ in order to get an exact reconstruction. When $G_{\sigma,\tau}$ does not commute with T, then neither does $G_{\sigma,\tau}^{-1/2}$. Hence the self-reciprocal frame $\breve{\mathcal{W}}_{\sigma,\tau}$ derived from $\mathcal{W}_{\sigma,\tau}$ (Section 4.5) is also not generated from a single mother wavelet.

This shows that it is especially important to have tight frames when dealing with wavelets (as opposed to windowed Fourier transforms). In practice, it may suffice for $\mathcal{W}_{\sigma,\tau}$ to be *snug*, i.e., "almost tight," with $\delta \equiv (B_{\sigma,\tau} - A_{\sigma,\tau})/(B_{\sigma,\tau} + A_{\sigma,\tau}) \ll 1$, since then the first terms of the recursive procedures in Section 4.4 and 4.5 give reasonable approximations: $G_{\sigma,\tau}^{-1} \approx \alpha_{\sigma,\tau}$ and $G_{\sigma,\tau}^{-1/2} \approx \sqrt{\alpha_{\sigma,\tau}}$, where $\alpha_{\sigma,\tau} \equiv 2/(B_{\sigma,\tau} + A_{\sigma,\tau})$. Hence

$$\Psi^{m,n} \approx \alpha_{\sigma,\tau}\Psi_{m,n}, \qquad \breve{\Psi}_{m,n} \approx \sqrt{\alpha_{\sigma,\tau}}\,\Psi_{m,n}. \tag{6.76}$$

Exercises

6.1. Prove (6.2).

6.2. Given $\psi_+ \in L^2(\mathbf{R})$ such that $\hat{\psi}_+(\omega) = 0$ outside the interval $[\alpha, \beta]$, where $0 \le \alpha < \beta < \infty$, let $\psi_-(t) \equiv \overline{\psi_+(t)}$ and

$$\begin{aligned} \psi_R(t) &\equiv 2\,\mathrm{Re}\,\psi_+(t) = \psi_+(t) + \psi_-(t) \\ \psi_I(t) &\equiv 2i\,\mathrm{Im}\,\psi_+(t) = \psi_+(t) - \psi_-(t). \end{aligned} \tag{6.77}$$

The frame vectors $\Psi_{\pm,m,n}$ defined in (6.43) decompose as

$$\Psi_{R,m,n} = \Psi_{+,m,n} + \Psi_{-,m,n}, \qquad \Psi_{I,m,n} = \Psi_{+,m,n} - \Psi_{-,m,n}. \tag{6.78}$$

(a) By Theorem 6.2, the vectors $\mathcal{W}_{\sigma,\tau} \equiv \{\Psi_{+,m,n}, \Psi_{-,m,n}\}$ form a frame. Show that $\mathcal{W}'_{\sigma,\tau} \equiv \{\Psi_{R,m,n}, \Psi_{I,m,n}\}$ is also a frame by computing its metric operator G'_σ in terms of the metric operator G_σ of $\mathcal{W}_{\sigma,\tau}$. (*Hint*: Use (6.55).) Use G'_σ to find the reciprocal vectors $\Psi^{R,m,n}$ and $\Psi^{I,m,n}$ in terms of $\Psi^{+,m,n}$ and $\Psi^{-,m,n}$.

(b) If we had started with $\psi(t) \equiv \psi_R(t)$ and followed the procedure in Theorem 6.1, we would have arrived at a frame $\mathcal{W}''_{\sigma,\tau} \equiv \{\Psi_{m,n}\}$, which *seems* to be the same as the family $\{\Psi_{R,m,n}\}$. Use (6.55) to prove that $\{\Psi_{R,m,n}\}$ is *not* a frame. Explain this apparent contradiction. *Hint*: In spite of appearances, $\mathcal{W}'_{\sigma,\tau}$ is *not* more redundant than $\mathcal{W}''_{\sigma,\tau}$! Carefully review the construction of $\mathcal{W}''_{\sigma,\tau}$.

Chapter 7

Multiresolution Analysis

Summary: In all of the frames studied so far, the analysis (computation of $\tilde{f}(\omega, t)$ or $\tilde{f}(s, t)$ or their discrete samples) must be made directly by computing the relevant integrals for all the necessary values of the time-frequency or time-scale parameters. Around 1986, a radically new method for performing discrete wavelet analysis and synthesis was born, known as *multiresolution analysis.* This method is completely *recursive* and therefore ideal for computations. One begins with a version $f^0 = \{f_n^0\}_{n \in Z}$ of the signal *sampled* at regular time intervals $\Delta t = \tau > 0$. f^0 is split into a "blurred" version f^1 at the coarser scale $\Delta t = 2\tau$ and "detail" d^1 at scale $\Delta t = \tau$. This process is repeated, giving a sequence $f^0, f^1, f^2, \ldots$ of more and more blurred versions together with the details $d^1, d^2, d^3, \ldots$ removed at every scale ($\Delta t = 2^m \tau$ in f^m and d^{m-1}). As a result of some wonderful magic, each d^m can be written as a superposition of wavelets of the type $\Psi_{m,n}$ encountered in Chapter 6, with a very special mother wavelet that is determined recursively by the "filter coefficients" associated with the blurring process. After N iterations, the original signal can be reconstructed as $f^0 = f^N + d^1 + d^2 + \cdots + d^N$. The wavelets constructed in this way not only form a frame: They form an *orthonormal basis* with as much locality and smoothness as desired (subject to the uncertainty principle, of course). The unexpected existence of such bases is one of the reasons why wavelet analysis has gained such widespread popularity.

Prerequisites: Chapters 1, 3, and 6.

7.1 The Idea

In Chapter 6, we saw that continuous wavelet frames can be discretized if the mother wavelet is reasonable, the scale factor σ is chosen to be sufficiently close to unity, and the time sampling interval τ is sufficiently small. The continuous frames are infinitely redundant; hence the resulting discrete frames with $\sigma \approx 1$ and $\tau \approx 0$ can be expected to be highly redundant. Since a continuous frame exists for any wavelet satisfying the admissibility condition, the existence of such discrete frames with $\sigma \approx 1$ is not unexpected. In this chapter we discuss a rather surprising development: The existence of wavelets that form *orthonormal bases* rather than merely frames. Being bases, these frames have *no* redundancy, hence they cannot be expected to be obtained by discretizing a "generic" continuous frame. In fact, they use the scale factor 2, which cannot be regarded as being close to unity, and satisfy a strict set of constraints beyond the admissibility condition. The construction of the new wavelets is based on a new method called *multiresolution analysis* (MRA) that, to a large extent, explains the "miracle" of their existence.

We now fix the basic time sampling interval at $\tau = 1$ and the scale factor at $\sigma = 2$. Whereas the choice $\tau = 1$ is merely one of convenience (τ can be rescaled by shifting scales), the choice $\sigma = 2$ is fundamental in that many of the constructions given below fail when $\sigma \neq 2$. The time translation and dilation operators $T, D : L^2(\mathbf{R}) \to L^2(\mathbf{R})$ are defined, respectively, by

$$\begin{aligned} (Tf)(t) &= f(t-1), \;\Rightarrow\; (T^n f)(t) = f(t-n), && n \in \mathbf{Z} \\ (Df)(t) &= 2^{-1/2} f(t/2), \;\Rightarrow\; (D^m f)(t) = 2^{-m/2} f(2^{-m} t), && m \in \mathbf{Z}. \end{aligned} \tag{7.1}$$

They are invertible and satisfy $\langle Tf, Tg \rangle = \langle f, g \rangle = \langle Df, Dg \rangle$, hence they are unitary:

$$T^* = T^{-1}, \qquad D^* = D^{-1}. \tag{7.2}$$

Df is a stretched version of f, and $D^{-1} f$ is a compressed version. It can be easily checked that $DTf = T^2 Df$ for all $f \in L^2(\mathbf{R})$, and so we have the operator equations

$$DT = T^2 D, \qquad D^{-1} T = T^{1/2} D^{-1}. \tag{7.3}$$

The first equation states that translating a signal by $\Delta t = 1$ and then stretching it by a factor of 2 amounts to the same thing as first stretching it, then translating by $\Delta t = 2$. The second equation has a similar interpretation. These equations will be used to "square" T and take its "square root," respectively.

Rather than beginning with a mother wavelet, multiresolution analysis starts with a basic function $\phi(t)$ called the *scaling function.* ϕ serves as a kind of "potential" for generating ψ. The translated and dilated versions of ϕ, defined by

$$\begin{aligned} \phi_{m,n}(t) &\equiv (D^m T^n \phi)(t) = 2^{-m/2} (T^n \phi)(2^{-m} t) \\ &= 2^{-m/2} \phi(2^{-m} t - n), \end{aligned} \tag{7.4}$$

are used to *sample* signals at various times and scales. Generally, ϕ will be a "bump function" of some width W centered near $t = 0$. Then (7.4) shows that $\phi_{m,n}$ is a bump function of width $2^m W$ centered near $t = 2^m n$. Unlike wavelet samples, which only give "details" of the signal, the samples $\langle \phi_{m,n}, f \rangle \equiv \phi^*_{m,n} f$ are supposed to represent the values of the signal itself, averaged over a neighborhood of width $2^m W$ around $t = 2^m n$. For notational convenience we also write

$$\phi_n(t) \equiv \phi_{0,n}(t) = \phi(t-n).$$

In order for ϕ to determine a multiresolution analysis, it will be required to satisfy certain conditions. The first of these is *orthonormality* within the scale $m = 0$:

$$\langle \phi_n, \phi_k \rangle = \delta_n^k. \tag{7.5}$$

Now $T^* = T^{-1}$ implies that $\langle T^n\phi, T^k\phi \rangle = \langle T^{n-k}\phi, \phi \rangle$; hence (7.5) is equivalent to the simpler condition

$$\langle \phi_n , \phi \rangle \equiv \phi_n^* \phi = \delta_n^0, \qquad n \in \mathbf{Z}. \tag{7.6}$$

Furthermore, it follows from $D^* = D^{-1}$ that

$$\langle \phi_{m,n} , \phi_{m,k} \rangle = \langle D^m \phi_n , D^m \phi_k \rangle = \langle \phi_n , \phi_k \rangle. \tag{7.7}$$

Hence (7.6) implies orthonormality at every scale. Note, however, that $\phi_{m,n}$'s at *different* scales need not be orthogonal.

Now let f be the constant function $f(t) \equiv 1$. If ϕ is integrable, then the inner products $\phi_n^* f$ are well defined. If their interpretation as samples of f is to make sense, then we should have

$$\phi_n^* f = \int_{-\infty}^{\infty} dt\ \bar{\phi}(t-n) = \int_{-\infty}^{\infty} dt\ \bar{\phi}(t) = \overline{\hat{\phi}}(0) = 1. \tag{7.8}$$

We therefore assume that $\hat{\phi}(0) = 1$ as our second condition, to which we will refer as the *averaging property.* Strictly speaking, this condition, as well as the orthonormality (7.6), is unnecessary for the general notion of multiresolution analysis; see Daubechies (1992), Propositions 5.3.1 and 5.3.2. However, the combination of (7.6) and (7.8) will be seen to be very natural from the point of view of filtering and sampling (Section 7.3).

For any fixed $m \in \mathbf{Z}$, let V_m be the (closed) subspace of $L^2(\mathbf{R})$ spanned by $\{\phi_{m,n} : n \in \mathbf{Z}\}$, i.e.,

$$V_m = \Big\{ f = \sum_n \phi_{m,n}\, u_n : \|f\|^2 = \sum_n |u_n|^2 < \infty \Big\}. \tag{7.9}$$

($\|f\|^2 = \sum_n |u_n|^2$, which follows from (7.5), shows that we may identify V_0 with the sequence space $\ell^2(\mathbf{Z})$ by $u(T)\phi \leftrightarrow \{u_n\}$. This is useful in computations.) Clearly $\{\phi_{m,n} : n \in \mathbf{Z}\}$ is an orthonormal basis for V_m. Since

$$\sum_n \phi_{m,n}\, u_n = D^m \sum_n \phi_n\, u_n \tag{7.10}$$

and $\sum_n \phi_n\, u_n \in V_0$, it follows that

$$V_m = \Big\{ D^m f : f \in V_0 \Big\} \equiv D^m V_0, \qquad m \in \mathbf{Z}. \tag{7.11}$$

(That is, V_m is the image of V_0 under D^m.) The orthogonal projection to V_m is the operator $P_m : L^2(\mathbf{R}) \to L^2(\mathbf{R})$ defined by

$$P_m = \sum_n \phi_{m,n}\, \phi_{m,n}^* = D^m \sum_n \phi_n\, \phi_n^*\, D^{-m} = D^m P_0\, D^{-m}, \tag{7.12}$$

where we have used $\phi^*_{m,n} = (D^m \phi_n)^* = \phi^*_n D^{-m}$. Note that $P^*_m = P_m = P^2_m$ (these are the algebraic properties required of all orthogonal projections). P_m gives a partial "reconstruction" of f from its samples $\phi^*_{m,n} f$ at the scale m:

$$f_m(t) \equiv (P_m f)(t) = \sum_n \phi_{m,n}(t)\, \phi^*_{m,n} f. \tag{7.13}$$

Here $\phi_{m,n}(t)$ acts like a scaled "pixel function" giving the shape of a "dot" at $t = n2^m$ and scale m.

When a signal is sampled at $\Delta t = 2^m$, we expect that the detail at scales less than 2^m will be lost. This suggests that V_m *should be regarded as containing signal information only down to the time scale* $\Delta t = 2^m$. That idea is made precise by requiring that

$$V_{m+1} \subset V_m \quad \text{for all } m \in \mathbf{Z}. \tag{7.14}$$

Equation (7.14) does not follow from any of the above, so it is really a new requirement. It can be reduced to a simple condition on ϕ. Namely, since $D\phi \in V_1$ and $V_1 \subset V_0$, we must have $D\phi \in V_0$. Thus

$$D\phi = \sum_n h_n \phi_n = \sum_n h_n T^n \phi \equiv h(T)\phi \tag{7.15}$$

for some coefficients h_n. Equation (7.15) is called a *dilation equation* or *two-scale relation* for ϕ. It will actually be seen to *determine* ϕ, up to normalization. $\{h_n\}$ is known as the *two-scale sequence* or the set of *filter coefficients* for ϕ. We have formally introduced the operator $h(T) = \sum_n h_n T^n$ on V_0 (or $L^2(\mathbf{R})$). $h(T)$ will be interpreted as an *averaging operator* that, according to (7.15), represents the stretched "pixel" $D\phi$ as a superposition of unstretched pixels ϕ_n. To make the formal expression $h(T)$ less abstract, the reader is advised to think of $h(T)$ as the operation that must be performed on ϕ in order to produce $D\phi$: translating ϕ to different locations, multiplying by the filter coefficients and summing. The dilation equation can also be written as $D^{-1}h(T)\phi = \phi$, which means that when ϕ is first averaged and then the average is compressed, this gives back ϕ. (In mathematical language, ϕ is a *fixed point* of the operator $D^{-1}h(T)$.) This can be used to *determine* ϕ from $h(T)$, up to normalization. Explicitly, (7.15) states that

$$\begin{aligned}\phi(t) &= (D^{-1}h(T)\phi)(t) = \sqrt{2}\big(h(T)\phi\big)(2t) \\ &= \sqrt{2}\sum_n h_n \phi(2t-n).\end{aligned} \tag{7.16}$$

It will be convenient to use a notation similar to that in (7.15) for all vectors in V_0:

$$\sum_n u_n \phi_n = \sum_n u_n T^n \phi \equiv u(T)\phi. \tag{7.17}$$

As with $h(T)$, it is useful to think of $u(T)$ as the *operation* that must be performed on ϕ in order to produce the vector $u(T)\phi$ in V_0. Vectors in V_m can likewise be written as $D^m u(T)\phi$. When the above sums contain an infinite number of terms, care must be taken in defining the operators $h(T)$ and $u(T)$, which may be unbounded. A precise definition can be made in the frequency domain, where T becomes multiplication by $e^{-2\pi i\omega}$ (since $(Tf)^\wedge(\omega) = e^{-2\pi i\omega}\hat{f}(\omega)$), and hence $u(T)$ acts on $f \in L^2(\mathbf{R})$ by

$$\big(u(T)f\big)^\wedge(\omega) = \sum_n u_n\, e^{-2\pi in\omega}\,\hat{f}(\omega) \equiv u\big(e^{-2\pi i\omega}\big)\hat{f}(\omega). \tag{7.18}$$

Since $\sum_n |u_n|^2 < \infty$, the function $u(z)$ in (7.18) is square-integrable on the unit circle $|z| = 1$, with

$$\|u\|^2 \equiv \int_0^1 d\omega\, |u\big(e^{-2\pi i\omega}\big)|^2 = \sum_n |u_n|^2. \tag{7.19}$$

For a general sequence $\{u_n\} \in \ell^2(\mathbf{Z})$, (7.18) defines $u(T)$ as an *unbounded* operator on $L^2(\mathbf{R})$. However, we can approximate vectors in V_0 by *finite* linear combinations of ϕ_n's. Then $u(T)$ is a "Laurent polynomial" in T, i.e., a polynomial in T and T^{-1}, and $u(z)$ is a *trigonometric polynomial*; hence $u(T)$ is a bounded operator on $L^2(\mathbf{R})$. As for $h(T)$, we are interested mainly in the case when ϕ has compact support. Then so does $D\phi$, hence the sum in (7.15) must be finite and $h(T)$ is also a Laurent polynomial. The set of all Laurent polynomials in T will be denoted by $\mathcal{P}$. It is an *algebra*, meaning that the sums and products of such polynomials again belong to $\mathcal{P}$. To simplify the analysis, we will generally assume that ϕ has compact support, although many of the results generalize to noncompactly supported ϕ.

The dilation equation (7.15) is not only necessary in order to satisfy (7.14), but it is also sufficient. To see this, note that (7.3) implies $Du(T) = u(T^2)D$. Hence a typical vector in $V_1 = DV_0$ can be written as

$$Du(T)\phi = u(T^2)D\phi = u(T^2)h(T)\phi = h(T)u(T^2)\phi. \tag{7.20}$$

The right-hand side belongs to V_0, which establishes $V_1 \subset V_0$. Applying D^m to both sides then gives $V_{m+1} \subset V_m$ as claimed.

To give a formal definition of multiresolution analysis, we first collect the essential properties arrived at so far. We have a sequence of closed subspaces V_m of $L^2(\mathbf{R})$ such that

(a) $\phi_n(t) \equiv \phi(t-n)$ forms an orthonormal basis for V_0,

(b) $V_m \equiv D^m V_0$, and

(c) $V_{m+1} \subset V_m$.

This nested sequence of spaces is said to form a *multiresolution analysis* if, in addition to the above, we have for every $f \in L^2(\mathbf{R})$:

$$\begin{aligned} &\text{(d)} \quad \lim_{m\to\infty} P_m f = 0 \text{ in } L^2(\mathbf{R}), \text{ i.e.,} \quad \lim_{m\to\infty} \|P_m f\| = 0; \\ &\text{(e)} \quad \lim_{m\to-\infty} P_m f = f \text{ in } L^2(\mathbf{R}), \text{ i.e.,} \quad \lim_{m\to-\infty} \|f - P_m f\| = 0. \end{aligned} \tag{7.21}$$

Since V_0 is generated by ϕ and V_m is determined by V_0, the additional properties (d) and (e) must ultimately reduce to requirements on ϕ. $P_m f$ will be interpreted as a version of f *blurred* to the scale $\Delta t = 2^m$. Hence the limit of $P_m f$ as $m \to \infty$ should be a constant function. Since the only constant in $L^2(\mathbf{R})$ is 0, we expect (d) to hold. On the other hand, if ϕ is reasonable, we expect $P_m f$ to approximate f more and more closely as $m \to -\infty$, hence (e) should hold. A condition making ϕ sufficiently reasonable is $\int dt\, |\phi(t)| < \infty$, i.e., that ϕ be *integrable.*

Proposition 7.1. *Let ϕ be an integrable function satisfying the orthonormality and averaging conditions (7.6) and (7.8). Then (7.21d) and (7.21e) hold.*

This is a special case of Propositions 5.3.1 and 5.3.2 in Daubechies (1992), where ϕ is neither required to satisfy (7.6) nor (7.8). In that case, it is assumed that the translates ϕ_n form a "Riesz basis" for V_0 (their linear span) and $\hat{\phi}(\omega)$ is bounded for all ω and continuous near $\omega = 0$, with $\hat{\phi}(0) \neq 0$. These are close to being the minimal conditions that ϕ has to meet in order to generate a multiresolution analysis. When $\phi(t)$ is integrable, then $\hat{\phi}(\omega)$ is continuous for all ω. Furthermore, if ϕ satisfies (7.6), then $\hat{\phi}(\omega)$ is also square-integrable and hence necessarily bounded. In the more general case when ϕ is *not* required to be orthogonal to its translates, it is necessary to use the so-called "orthogonalization trick" to generate a new scaling function $\phi^{\#}$ that does satisfy (7.6) (see Daubechies (1992), pp. 139-140). In general, $\phi^{\#}$ and its associated mother wavelet do not have compact support (even if ϕ does). But they can still have good decay. For example, the wavelets of Battle (1987) and Lemarié (1988) have exponential decay as well as being C^k (k times continuously differentiable) for any chosen positive integer k. (Since they are smooth, they also decay well in the frequency domain; hence they give rise to good time-frequency localization.)

The construction of wavelets from a multiresolution analysis begins by considering the orthogonal complements of V_{m+1} in V_m:

$$W_{m+1} \equiv \{ f \in V_m : \langle f, g \rangle = 0 \text{ for all } g \in V_{m+1} \}. \tag{7.22}$$

We write $W_{m+1} = V_m \ominus V_{m+1}$ and $V_m = V_{m+1} \oplus W_{m+1}$. Thus, every $f^m \in V_m$ has a unique decomposition $f^m = f^{m+1} + d^{m+1}$ with $f^{m+1} \in V_{m+1}$ and $d^{m+1} \in W_{m+1}$. Since $W_{m+1} \subset V_m$ and W_m is orthogonal to V_m, it follows that W_{m+1} is also orthogonal to W_m. Similarly, we find that *all the spaces* W_m *(unlike the spaces* V_m*) are mutually orthogonal.* Furthermore, since D^m preserves orthogonality, it follows that

$$W_m = \{D^m f : f \in W_0\} \equiv D^m W_0. \tag{7.23}$$

Let $Q_m : L^2(\mathbf{R}) \to L^2(\mathbf{R})$ denote the orthogonal projection to W_m. Then

$$\begin{aligned} W_m = D^m W_0 &\Rightarrow Q_m = D^m Q_0 D^{-m}, \\ V_m \perp W_m &\Rightarrow P_m Q_m = Q_m P_m = 0, \\ V_m = V_{m+1} \oplus W_{m+1} &\Rightarrow P_m = P_{m+1} + Q_{m+1}. \end{aligned} \tag{7.24}$$

Note also that when $k > m$, then $V_k \subset V_m$ and $W_k \subset V_m$, hence

$$P_k P_m = P_m P_k = P_k\,, \quad Q_k P_m = P_m Q_k = Q_k\,, \quad k > m. \tag{7.25}$$

The last equation of (7.24) can be iterated by repeatedly replacing the P's with their decompositions:

$$P_m = P_M + \sum_{k=m+1}^{M} Q_k, \quad M > m. \tag{7.26}$$

Any $f \in L^2(\mathbf{R})$ can now be decomposed as follows: Starting with the projection $f^m \equiv P_m f \in V_m$ for some $m \in \mathbf{Z}$, we have

$$f^m = P_M f + \sum_{k=m+1}^{M} Q_k f = P_M f^m + \sum_{k=m+1}^{M} Q_k f^m\,, \tag{7.27}$$

by (7.25). In practice, any signal can only be approximated by a sampled version, which may then be identified with f^m for some m. Since the sampling interval for f^m is $\Delta t = 2^m$, we may call 2^{-m} the *resolution* of the signal. Although the "finite" form (7.27) is all that is really ever used, it is important to ask what happens when the resolution of the initial signal becomes higher and higher. The answer is given by examining the limit of (7.27) as $m \to -\infty$. Since $P_m f \to f$ by (7.21e), (7.27) becomes

$$f = P_M f + \sum_{k=-\infty}^{M} Q_k f, \qquad f \in L^2(\mathbf{R}). \tag{7.28}$$

This gives f as a sum of a blurred version $f^M \equiv P_M f$ and successively finer detail $d^k = Q_k f$, $k \le M$. If the given signal f actually belongs to some V_m, then $Q_k f = 0$ for $k \le m$ and (7.28) reduces to (7.27). Equation (7.28) gives an orthogonal decomposition of $L^2(\mathbf{R})$:

$$L^2(\mathbf{R}) = V_M \oplus \bigoplus_{k=-\infty}^{M} W_k. \tag{7.29}$$

If we now let $M \to \infty$ and use (7.21d), we obtain

$$f = \sum_{k=-\infty}^{\infty} Q_k f, \qquad f \in L^2(\mathbf{R}). \tag{7.30}$$

This gives the orthogonal decomposition

$$L^2(\mathbf{R}) = \bigoplus_{k=-\infty}^{\infty} W_k. \tag{7.31}$$

We will see later that the form (7.28) is preferable to (7.30) because the latter only makes sense in $L^2(\mathbf{R})$, whereas the former makes sense in most reasonable spaces of functions. It will be shown that $Q_k f$ is a superposition of wavelets, each of which has zero integral due to the admissibility condition; hence the right-hand side of (7.30) formally integrates to zero, while the left-hand side may not. This apparent paradox is resolved by noting that (7.30) does *not* converge in the sense of $L^1(\mathbf{R})$. No such "paradox" arises from (7.28).

To make the preceeding more concrete, we now give the simplest example, which happens to be very old. More interesting examples will be seen later in this chapter and especially in Chapter 8.

Example 7.2: The Haar Multiresolution Analysis. The simplest function satisfying the orthonormality condition (7.6) is the characteristic function of the interval $I \equiv [0, 1)$:

$$\phi(t) = \chi_{[0,1)}(t) \equiv \begin{cases} 1 & \text{if } 0 \le t < 1 \\ 0 & \text{otherwise.} \end{cases}$$

Its Fourier transform is

$$\hat{\phi}(\omega) = e^{-\pi i \omega} \frac{\sin(\pi\omega)}{\pi\omega}, \tag{7.32}$$

hence $\hat{\phi}(0) = 1$ as required in Proposition 7.1. The basis for V_m is $\phi_{m,n}(t) = 2^{-m/2}\chi_{m,n}(t)$, where $\chi_{m,n}$ is the charactersitic function of $I_{m,n} \equiv \big[n/2^m, (n+1)/2^m\big)$. The projection of $f \in L^2(\mathbf{R})$ to V_m is therefore

$$P_m f = \sum_n \phi_{m,n}(t)\, \phi^*_{m,n} f = \sum_n \chi_{m,n}(t) f_{m,n}, \tag{7.33}$$

where

$$f_{m,n} \equiv 2^{-m/2}\phi^*_{m,n} f = 2^{-m} \int_{n/2^m}^{(n+1)/2^m} f(t)\, dt \tag{7.34}$$

is the average of f in $I_{m,n}$. Equation (7.33) gives an approximation of f by functions constant on all the intervals $I_{m,n}$. To show that ϕ indeed generates a multiresolution analysis, the properties (c)–(e) must be verified. Recall that (c) hinges on the dilation equation (7.15). But

$$D\phi = \frac{1}{\sqrt{2}}\chi_{[0,2)} = \frac{1}{\sqrt{2}}\Big(\chi_{[0,1)} + \chi_{[1,2)}\Big) = \frac{1}{\sqrt{2}}(\phi + T\phi), \tag{7.35}$$

so (7.15) holds with

$$h(T) = \frac{1}{\sqrt{2}}(1+T). \tag{7.36}$$

7.2 Operational Calculus for Subband Filtering Schemes

We are now ready to use the multiresolution analysis developed above to construct an orthonormal basis of wavelets for $L^2(\mathbf{R})$. Define the operators $H : L^2(\mathbf{R}) \to L^2(\mathbf{R})$ and $G : L^2(\mathbf{R}) \to L^2(\mathbf{R})$ by

$$H = P_0 D^{-1} = D^{-1} P_1 \,, \qquad G = Q_0 D^{-1} = D^{-1} Q_1 . \tag{7.37}$$

Then

$$H^* = DP_0 = P_1 D, \qquad G^* = DQ_0 = Q_1 D \tag{7.38}$$

and

$$\begin{aligned} HH^* &= P_0^2 = P_0 & GG^* &= Q_0^2 = Q_0 \\ HG^* &= P_0 Q_0 = 0 & GH^* &= Q_0 P_0 = 0 \\ H^*H &= P_1^2 = P_1 & G^*G &= Q_1^2 = Q_1 \\ H^*G &= P_1 Q_1 = 0 & G^*H &= Q_1 P_1 = 0 \end{aligned} \tag{7.39}$$

$$HH^* + GG^* = P_0 + Q_0 = P_{-1}$$

$$H^*H + G^*G = P_1 + Q_1 = P_0 .$$

Although we have defined H and G on all of $L^2(\mathbf{R})$, we will be most interested in their actions on $V_0 = V_1 \oplus W_1$. Since

$$\begin{aligned} P_1 V_1 &= V_1 \,, & P_1 W_1 &= \{0\}, \\ Q_1 V_1 &= \{0\} \,, & Q_1 W_1 &= W_1 \,, \end{aligned}$$

(7.37) implies that the images of V_1 and W_1 under H and G are

$$\begin{aligned} HV_1 &= D^{-1} V_1 = V_0 \,, & HW_1 &= \{0\}, \\ GW_1 &= D^{-1} W_1 = W_0 \,, & GV_1 &= \{0\}. \end{aligned} \tag{7.40}$$

In fact, since D^{-1} is unitary, H maps V_1 *isometrically* onto V_0 (i.e., $\|Hf\| = \|f\|$ for $f \in V_1$), and G maps W_1 isometrically onto W_0. Similarly, (7.38) implies that

$$H^* V_0 = DV_0 = V_1 \,, \qquad G^* W_0 = DW_0 = W_1 \,, \tag{7.41}$$

both actions being isometric. If $f^0 = f^1 + d^1 \in V_0$ with $f^1 \in V_1$ and $d^1 \in W_1$, then $Hf^0 = D^{-1} f^1$ and $Gf^0 = D^{-1} d^1$. Since f^1 and d^1 are interpreted as the "average" (low-frequency part) and "detail" (high-frequency part) in f_0, H is called a *low-pass filter* and G a *high-pass filter*. (Actually, Hf^0 and Gf^0 are *compressed* versions of the low-frequency and high-frequency parts of f^0 because of the action of D^{-1}. This does not alter their informational contents and is convenient for computations, since it keeps $\Delta t = 1$ for Hf^0 so that when the decomposition is iterated, the *density* of information remains constant and there is no need to keep track of the scale. Thus H acts on the sampled signal f^0 by *zooming out* with a factor of 2, losing half the detail – the high-frequency part – in the process.) H and G are also known as *quadrature mirror filters.* They are usually defined on V_0 only or, equivalently, on the sequence space $\ell^2(\mathbf{Z}) \approx V_0$; see Daubechies (1988). I believe the present definitions, starting on $L^2(\mathbf{R})$, are somewhat simpler and more transparent. Their restrictions to V_0 will be seen to coincide with the usual operators.

The point of defining H and G is that they can be used to characterize the subspaces V_1 and W_1 explicitly, as well as to generate the wavelets associated with the MRA in a conceptually clear fashion. To characterize V_1, note that it is the range of H^*. The action of H^* on $u(T)\phi \in V_0$ is made explicit by (7.20),

$$H^* v(T)\phi = Dv(T)\phi = h(T)v(T^2)\phi, \tag{7.42}$$

which follows from $DT = T^2 D$ and $D\phi = h(T)\phi$. Thus

$$V_1 = \text{closure of } \{h(T)v(T^2)\phi :\ v(T) \in \mathcal{P}\}, \tag{7.43}$$

where the closure needs to be taken since $v(T) \in \mathcal{P}$ has only a finite number of nonvanishing terms. (This means that we include all limits of vectors in the above set with finite norms.) Equation (7.43) is the desired characterization of V_1.

To characterize W_1, note that according to (7.40) it is the set of all vectors in V_0 annihilated by H (i.e., the *null space* of H). To find H explicitly, we first cast H^* in a slightly different form. Define the *up-sampling operator* $S_{\Uparrow} : V_0 \to V_0$ by

$$S_{\Uparrow} u(T)\phi \equiv u(T^2)\phi = \sum_n u_n T^{2n}\phi = \sum_n u_n \phi_{2n}\,. \tag{7.44}$$

$S_{\Uparrow}$ "doubles" the number of samples by separating them to $\Delta t = 2$ and inserting a zero between each neighboring pair (which restores $\Delta t = 1$). Thus $S_{\Uparrow} u(T)\phi$ is twice as wide as $u(T)\phi$ and differs from $Du(T)\phi$ only in that the "pixels" ϕ_n are unstretched. It looks like a picket fence with every other picket missing. Equation (7.41) shows that $H^* V_0 = V_1 \subset V_0$, hence H^* defines an operator on V_0, which we denote by the same symbol. (This amounts to a mild abuse of notation, but saves us from introducing yet another set of symbols.) According to (7.42) and (7.44), $H^* : V_0 \to V_0$ is given by

$$H^* u(T)\phi = h(T) S_{\Uparrow} u(T)\phi\,, \quad \text{or} \quad H^* = h(T) S_{\Uparrow} \text{ on } V_0. \tag{7.45}$$

Equation (7.45) has a simple and elegant interpretation: H^* acts on $u(T)\phi \in V_0$ by first up-sampling with $S_{\Uparrow}$, then *averaging* with $h(T)$. This averaging process fills in the missing "pickets," and the result, according to (7.42), is precisely the stretched version $Du(T)\phi$ of the original signal! For this reason, H^* is called an *interpolation operator.* It *zooms in* on the sampled signal, just as H zooms out.

The advantage of the form (7.45) of H^* is that it immediately gives the adjoint operator $H : V_0 \to V_0$ as

$$H = S_{\Uparrow}^* h(T)^* = S_{\Downarrow} h(T)^* \quad \text{on } V_0, \tag{7.46}$$

where $S_{\Downarrow} \equiv S_{\Uparrow}^*$ is called the *down-sampling operator*. To find H, we must find $S_{\Downarrow}$ explicitly. Note that every $v(T) \in \mathcal{P}$ can be written uniquely as a sum of even and odd polynomials:

$$v(T) = v_+(T^2) + Tv_-(T^2), \tag{7.47}$$

where

$$\begin{aligned} v_+(T^2) &= \tfrac{1}{2}[v(T) + v(-T)] = \sum_n v_{2n} T^{2n} \\ v_-(T^2) &= \tfrac{1}{2} T^{-1}[v(T) - v(-T)] = \sum_n v_{2n+1} T^{2n}. \end{aligned} \tag{7.48}$$

Equation (7.3) can be used to give $v_+(T)$ and $v_-(T)$ directly in terms of $v(T)$ and $v(-T)$:

$$\begin{aligned} v_+(T) &= \sum_n v_{2n} T^n = \tfrac{1}{2} D^{-1}[v(T) + v(-T)]D \\ v_-(T) &= \sum_n v_{2n+1} T^n = \tfrac{1}{2} D^{-1} T^{-1}[v(T) - v(-T)]D. \end{aligned} \tag{7.49}$$

Proposition 7.3. *The up-sampling and down-sampling operators have the following properties:*

$$\begin{aligned} &\text{(a)} \quad S_{\Downarrow} S_{\Uparrow} = I_{V_0} \\ &\text{(b)} \quad S_{\Downarrow} v(T)\phi = v_+(T)\phi = \sum_n v_{2n}\phi_n \\ &\text{(c)} \quad S_{\Downarrow} v(T) S_{\Uparrow} = v_+(T), \qquad S_{\Downarrow} T^{-1} v(T) S_{\Uparrow} = v_-(T) \\ &\text{(d)} \quad S_{\Uparrow} S_{\Downarrow} + T S_{\Uparrow} S_{\Downarrow} T^{-1} = I_{V_0}. \end{aligned} \tag{7.50}$$

Proof: Since $S_{\Downarrow} = S_{\Uparrow}^*$, we have

$$\begin{aligned} \langle u(T)\phi, S_{\Downarrow} S_{\Uparrow} v(T)\phi \rangle &= \langle S_{\Uparrow} u(T)\phi, S_{\Uparrow} v(T)\phi \rangle = \langle u(T^2)\phi, v(T^2)\phi \rangle \\ &= \langle u(T)\phi, v(T)\phi \rangle, \end{aligned} \tag{7.51}$$

where the last equality follows from the orthonormality condition, since $\langle \phi_{2n}, \phi_{2k} \rangle = \delta_{2k}^{2n} = \delta_k^n = \langle \phi_n, \phi_k \rangle$. That proves (a). To prove (b), note that for any $u(T) \in \mathcal{P}$, we have

$$\begin{aligned} \langle u(T)\phi, S_{\Downarrow} v(T)\phi \rangle &= \langle S_{\Uparrow} u(T)\phi, v(T)\phi \rangle \\ &= \langle u(T^2)\phi, v_+(T^2)\phi + Tv_-(T^2)\phi \rangle. \end{aligned} \tag{7.52}$$

The orthogonality of the ϕ_n's implies that odd polynomials are orthogonal to even ones. Hence (7.52) gives

$$\langle\, u(T)\phi, S_{\Downarrow}v(T)\phi\,\rangle = \langle\, u(T^2)\phi, v_+(T^2)\phi\,\rangle = \langle\, u(T)\phi, v_+(T)\phi\,\rangle. \tag{7.53}$$

Since this holds for all $u(T)\phi \in V_0$, it proves (b). To show (c), note that $S_{\Uparrow}u(T)v(T)\phi = u(T^2)v(T^2)\phi = u(T^2)S_{\Uparrow}v(T)\phi$; hence we have an operator commutation relation similar to (7.3):

$$S_{\Uparrow}u(T) = u(T^2)S_{\Uparrow}\,, \qquad u(T) \in \mathcal{P}. \tag{7.54}$$

Thus

$$\begin{aligned} S_{\Downarrow}v(T)S_{\Uparrow} &= S_{\Downarrow}[v_+(T^2) + Tv_-(T^2)]S_{\Uparrow} \\ &= S_{\Downarrow}S_{\Uparrow}v_+(T) + S_{\Downarrow}TS_{\Uparrow}v_-(T). \end{aligned} \tag{7.55}$$

But $S_{\Downarrow}TS_{\Uparrow}u(T)\phi = S_{\Downarrow}Tu(T^2)\phi = 0$ by (b), since $Tu(T^2)$ is odd; hence $S_{\Downarrow}TS_{\Uparrow} = 0$. Furthermore, $S_{\Downarrow}S_{\Uparrow} = I_{V_0}$ by (a), thus (7.55) reduces to $S_{\Downarrow}v(T)S_{\Uparrow} = v_+(T)$. The second equality in (c) follows from the first since the even part of $T^{-1}v(T)$ is $v_-(T^2)$. (d) follows from

$$\begin{aligned} S_{\Uparrow}S_{\Downarrow}v(T)\phi + TS_{\Uparrow}S_{\Downarrow}T^{-1}v(T)\phi &= S_{\Uparrow}v_+(T)\phi + TS_{\Uparrow}v_-(T)\phi \\ &= v_+(T^2)\phi + Tv_-(T^2)\phi = v(T)\phi. \quad \blacksquare \end{aligned}$$

Equation (7.50b) gives an interpretation of the down-sampling operator: $S_{\Downarrow}v(T)\phi$ is obtained by deleting all the odd samples in $v(T)\phi$, so that the resulting signal has *half* the width of the original one. Clearly, information is lost in the process, by contrast with up-sampling.

To characterize W_1, note that

$$\begin{aligned} Hu(T)\phi &= S_{\Downarrow}h(T)^*u(T)\phi \\ &= \tfrac{1}{2}D^{-1}\big[h(T)^*u(T) + h(-T)^*u(-T)\big]D\phi. \end{aligned} \tag{7.56}$$

Hence $Hu(T)\phi = 0$ if and only if $h(T)^*u(T)$ is odd. This can be arranged by choosing

$$u(T) = h(-T)^*v(T) \quad \text{with } v(T) \text{ odd}, \tag{7.57}$$

since $h(T)^*h(-T)^*$ is necessarily even. In fact, this gives a characterization of W_1 that complements that of V_1 given by (7.43), as shown next.

Proposition 7.4. *Fix an arbitrary odd integer λ and let*

$$g(T) = -T^{\lambda}h(-T)^*. \tag{7.58}$$

Then every $u(T) \in \mathcal{P}$ can be written uniquely as

$$u(T) = h(T)v(T^2) + g(T)w(T^2) \quad \textit{for some} \quad v(T), w(T) \in \mathcal{P}, \tag{7.59}$$

and

$$W_1 = \text{closure of } \{g(T)w(T^2)\phi : w(T) \in \mathcal{P}\}. \tag{7.60}$$

Proof: Equation (7.59) will be proved below. Assuming it is true, note that $h(T)v(T^2)\phi \in V_1$ and $g(T)w(T^2)\phi \in W_1$ (by (7.57), since $-T^\lambda w(T^2)$ is odd). Hence

$$h(T)v(T^2)\phi = P_1 u(T)\phi, \qquad g(T)w(T^2)\phi = Q_1 u(T)\phi. \tag{7.61}$$

Since $\{Q_1 u(T)\phi : u(T) \in \mathcal{P}\}$ is dense in W_1, that proves (7.60). We prove (7.59) by using Lemma 7.5 (which follows immediately below) to solve for $v(T)$ and $w(T)$. If (7.59) holds, then by (7.65) of the Lemma,

$$\begin{aligned} S_\Downarrow h(T)^* u(T) S_\Uparrow &= S_\Downarrow h(T)^* h(T) v(T^2) S_\Uparrow + S_\Downarrow h(T)^* g(T) w(T^2) S_\Uparrow \\ &= S_\Downarrow h(T)^* h(T) S_\Uparrow v(T) + S_\Downarrow h(T)^* g(T) S_\Uparrow w(T) \\ &= v(T). \end{aligned} \tag{7.62}$$

Similarly,

$$S_\Downarrow g(T)^* u(T) S_\Uparrow = w(T). \tag{7.63}$$

Thus $v(T)$ and $w(T)$ are uniquely determined, *if* (7.59) holds. It remains to show that given any $u(T) \in \mathcal{P}$, the substitution of (7.62) and (7.63) into the right-hand side of (7.59) indeed gives back $u(T)$. By (7.49) and (7.50)(c),

$$\begin{aligned} v(T^2) &= Dv(T)D^{-1} = DS_\Downarrow h(T)^* u(T) S_\Uparrow D^{-1} \\ &= \tfrac{1}{2}\left[h(T)^* u(T) + h(-T)^* u(-T)\right] \\ w(T^2) &= Dw(T)D^{-1} = DS_\Downarrow g(T)^* u(T) S_\Uparrow D^{-1} \\ &= \tfrac{1}{2}\left[g(T)^* u(T) + g(-T)^* u(-T)\right]. \end{aligned}$$

Hence

$$\begin{aligned} h(T)v(T^2) + g(T)w(T^2) &= \tfrac{1}{2}\left[h(T)h(T)^* + g(T)g(T)^*\right] u(T) \\ &\quad + \tfrac{1}{2}\left[h(T)h(-T)^* + g(T)g(-T)^*\right] u(-T). \end{aligned}$$

But by (7.66) of the Lemma,

$$\begin{aligned} h(T)h(T)^* + g(T)g(T)^* &= h(T)h(T)^* + h(-T)^* h(-T) = 2I_{V_0} \\ h(T)h(-T)^* + g(T)g(-T)^* &= h(T)h(-T)^* - h(-T)^* h(T) = 0. \end{aligned} \tag{7.64}$$

Thus (7.59) holds as claimed. ■

Lemma 7.5. *$h(T)$ and $g(T)$ satisfy the following "orthogonality" relations:*

$$\begin{aligned} S_\Downarrow h(T)^* h(T) S_\Uparrow &= S_\Downarrow g(T)^* g(T) S_\Uparrow = I_{V_0} \\ S_\Downarrow g(T)^* h(T) S_\Uparrow &= S_\Downarrow h(T)^* g(T) S_\Uparrow = 0. \end{aligned} \tag{7.65}$$

Equivalently,

$$\begin{aligned} h(T)^* h(T) + h(-T)^* h(-T) &= g(T)^* g(T) + g(-T)^* g(-T) = 2I_{V_0} \\ g(T)^* h(T) + g(-T)^* h(-T) &= h(T)^* g(T) + h(-T)^* g(-T) = 0. \end{aligned} \tag{7.66}$$

Proof: Recall from (7.39) that $HH^* = P_0$ on $L^2(\mathbf{R})$. For the restrictions to V_0, this means that $HH^* = I_{V_0}$. Thus by (7.45) and (7.46),

$$HH^* = S_{\Downarrow} h(T)^* h(T) S_{\Uparrow} = I_{V_0}. \tag{7.67}$$

Furthermore, since $g(T)^* g(T) = h(-T)h(-T)^* = h(-T)^* h(-T)$, we also have $S_{\Downarrow} g(T)^* g(T) S_{\Uparrow} = I_{V_0}$. Finally,

$$S_{\Downarrow} h(T)^* g(T) S_{\Uparrow} = -S_{\Downarrow} h(T)^* h(-T)^* T^{\lambda} S_{\Downarrow} = 0 \tag{7.68}$$

since $h(T)^* h(-T)^* T^{\lambda}$ is odd, and similarly $S_{\Downarrow} g(T)^* h(T) S_{\Uparrow} = 0$. That proves (7.65). Equation (7.66) follows from (7.65) and (7.49). ■

The intuitive meaning of $g(T)$ is that it is a *differencing operator* on V_0, just as $h(T)$ is an averaging operator. Equations (7.64) can be interpreted as stating that these two operators *complement* each other. The significance of the integer λ will be discussed later, once the wavelets have been defined. Clearly, a change in λ merely amounts to a change in the polynomial $w(T)$ representing $g(T)w(T^2)\phi \in W_1$.

We are at last able to define the wavelets associated with the multiresolution analysis. The mother wavelet belongs to W_0, just as the scaling function belongs to V_0. Since $W_0 = D^{-1}W_1$, a dense set of vectors in W_0 is given by

$$D^{-1}g(T)w(T^2)\phi = D^{-1}w(T^2)g(T)\phi = w(T)D^{-1}g(T)\phi, \tag{7.69}$$

with $w(T) \in \mathcal{P}$. We therefore define the mother wavelet as

$$\psi \equiv D^{-1}g(T)\phi \in W_0\,, \quad \text{i.e.,} \quad \psi(t) = \sqrt{2}\sum_n g_n \phi(2t-n), \tag{7.70}$$

so that

$$W_0 = \text{closure of } \{w(T)\psi : w(T) \in \mathcal{P}\}. \tag{7.71}$$

The shifted wavelets

$$\psi_n \equiv T^n \psi = T^n D^{-1} g(T)\phi = D^{-1}T^{2n} g(T)\phi \tag{7.72}$$

therefore span W_0. It follows that for any fixed $m \in \mathbf{Z}$, the vectors

$$\psi_{m,n} \equiv D^m T^n \psi\,, \quad \text{i.e.,} \quad \psi_{m,n}(t) = 2^{-m/2}\psi(2^{-m}t - n), \tag{7.73}$$

span W_m.

Proposition 7.6. *The wavlets $\{\psi_{m,n} : m, n \in \mathbf{Z}\}$ form an orthonormal basis for $L^2(\mathbf{R})$.*

Proof: We show that the wavelets ψ_n at scale $m = 0$ are orthonormal, and hence they form an orthonormal basis for their span W_0. It then follows from the unitarity of D that for any fixed m, $\{\psi_{m,n} : n \in \mathbf{Z}\}$ is an orthonormal basis for W_m. Since $L^2(\mathbf{R})$ is the orthogonal sum of the W_m's, the proposition follows.

$$\begin{aligned}\langle \psi_n, \psi_k \rangle &= \langle T^{2n} g(T)\phi, T^{2k} g(T)\phi \rangle \\ &= \langle T^{2n-2k}\phi, g(T)^* g(T)\phi \rangle.\end{aligned} \tag{7.74}$$

But $\phi = S_\Uparrow \phi$ and $T^{2n-2k}\phi = S_\Uparrow T^{n-k}\phi$, so by Lemma 7.5 and the orthonormality of the ϕ_n's,

$$\begin{aligned}\langle \psi_n, \psi_k \rangle &= \langle S_\Uparrow T^{n-k}\phi, g(T)^* g(T) S_\Uparrow \phi \rangle \\ &= \langle T^{n-k}\phi, S_\Downarrow g(T)^* g(T) S_\Uparrow \phi \rangle \\ &= \langle T^{n-k}\phi, \phi \rangle = \langle \phi_n, \phi_k \rangle = \delta_k^n . \quad \blacksquare\end{aligned} \tag{7.75}$$

We can now express the orthogonal projections Q_m to W_m in terms of the wavelet basis, just as the projections P_m to V_m were given in (7.12):

$$Q_m = \sum_n \psi_{m,n} \psi_{m,n}^* . \tag{7.76}$$

Thus (7.28) and (7.30) become

$$f = \sum_{n \in \mathbf{Z}} \phi_{M,n} \phi_{M,n}^* f + \sum_{k=-\infty}^{M} \sum_{n \in \mathbf{Z}} \psi_{k,n} \psi_{k,n}^* f \tag{7.77}$$

and

$$f = \sum_{k \in \mathbf{Z}} \sum_{n \in \mathbf{Z}} \psi_{k,n} \psi_{k,n}^* f, \tag{7.78}$$

respectively.

The operator $H^* = DP_0$, originally defined on $L^2(\mathbf{R})$, was restricted to $H^* : V_0 \to V_0$. Together with its adjoint, this operator was used to define the wavelets. It is useful to restrict $G^* = DQ_0$ similarly, since the four restrictions H, H^*, G, G^* collectively give rise to "subband filtering," a scheme for wavelet decompositions that is completely recursive and therefore useful in computations. According to (7.41), the image of W_0 under G^* is $W_1 \subset V_0$. Hence G^* restricts to an operator from W_0 to V_0, which we denote by the same symbol to avoid introducing more notation. This restriction $G^* : W_0 \to V_0$ acts by

$$\begin{aligned}G^* w(T)\psi &= DQ_0 w(T)\psi = Dw(T)\psi = w(T^2)D\psi \\ &= w(T^2)g(T)\phi = g(T)S_\Uparrow w(T)\phi.\end{aligned} \tag{7.79}$$

It can be shown that $G : V_0 \to W_0$ is given by

$$Gu(T)\phi = S_\Downarrow g(T)^* u(T)\psi, \tag{7.80}$$

where $S_\Downarrow$ and $S_\Uparrow$ act on W_0 exactly as they do on V_0:

$$S_\Uparrow w(T)\psi \equiv w(T^2)\psi, \quad \Rightarrow \quad S_\Downarrow w(T)\psi = w_+(T)\psi. \tag{7.81}$$

Together with

$$H^*u(T)\phi = h(T)S_{\Uparrow}u(T)\phi, \qquad Hu(T)\phi = S_{\Downarrow}h(T)^*u(T)\phi, \tag{7.82}$$

(7.79) and (7.80) give a complete description of the QMFs H, G, and their adjoints. In actual computations, these operators act on the sequences $\{u_n\}$ corresponding to $u(T)\phi \in V_0$ or $u(T)\psi \in W_0$. We have given the above "abstract" description because it seems to be easier to visualize. (Compare the actions of H^* given by (7.82) and (7.88) below, for example.) We now find the corresponding actions on sequences. Note, first of all, that

$$\begin{aligned} g(T) &= -T^\lambda h(-T)^* = -\sum_n (-1)^n \bar{h}_n T^{\lambda-n} \\ &= \sum_n (-1)^n \bar{h}_{\lambda-n} T^n. \end{aligned} \tag{7.83}$$

Hence

$$g_n = (-1)^n \bar{h}_{\lambda-n}\,. \tag{7.84}$$

$\{g_n\}$ is the set of filter coefficients for ψ, or the *high-pass filter sequence.* Daubechies wavelets will have $2N$ nonzero coefficients $h_0, h_1, \ldots, h_{2N-1}$, with $N = 1, 2, 3, \ldots$. ($N = 1$ gives the Haar wavelet.) Then the nonvanishing coefficients of $g(T)$ are in the same range (i.e., $g_n \neq 0$ only for $n = 0, 1, \ldots, 2N-1$) if and only if we choose $\lambda = 2N - 1$. This, in turn, results in ψ having the same support as ϕ. A change of λ to $\lambda' = \lambda + 2\mu$ gives $g'(T) = T^{2\mu}g(T)$, resulting in a new wavelet

$$\psi' = D^{-1}T^{2\mu}g(T)\phi = T^\mu \psi, \tag{7.85}$$

which is merely a translated version of ψ. This, then, is the significance of λ.

The action of H^* on sequences is obtained by first considering its action on the basis elements ϕ_k:

$$H^*\phi_k = h(T)\phi_{2k} = \sum_n h_n T^{n+2k}\phi = \sum_n h_{n-2k}\phi_n. \tag{7.86}$$

Hence

$$H^* \sum_k u_k \phi_k = \sum_k \sum_n u_k h_{n-2k} \phi_n = \sum_n \Big(\sum_k h_{n-2k}\, u_k \Big) \phi_n, \tag{7.87}$$

which gives

$$H^*\{u_k\}_{k\in\mathbf{Z}} = \left\{ \sum_k h_{n-2k}\, u_k \right\}_{n\in\mathbf{Z}}. \tag{7.88}$$

Similarly,

$$G^*\psi_k = \sum_n g_{n-2k}\phi_n\,, \quad H\phi_n = \sum_k \bar{h}_{n-2k}\phi_k\,, \quad G\phi_n = \sum_k \bar{g}_{n-2k}\phi_k. \tag{7.89}$$

These result in

$$
\begin{aligned}
G^*\{u_k\}_{k\in\mathbf{Z}} &= \left\{\sum_k g_{n-2k}\, u_k\right\}_{n\in\mathbf{Z}} \\
H\{u_n\}_{n\in\mathbf{Z}} &= \left\{\sum_n \bar{h}_{n-2k}\, u_n\right\}_{k\in\mathbf{Z}} \\
G\{u_n\}_{n\in\mathbf{Z}} &= \left\{\sum_n \bar{g}_{n-2k}\, u_n\right\}_{k\in\mathbf{Z}}
\end{aligned} \tag{7.90}
$$

In the case of compactly supported ϕ and ψ, all the above sums are finite; hence these equations are easy to implement numerically. H and G are then said to be *finite impulse response (FIR) filters.* The above gives a *recursive* scheme for the wavelet decomposition and reconstruction of signals, as follows: Note that for the restrictions $H : V_0 \to V_0$ and $G : V_0 \to W_0$ and their adjoints, the relations (7.39) imply

$$
\begin{gathered}
HH^* = I_{V_0}, \quad GG^* = I_{W_0}, \quad HG^* = GH^* = 0 \\
H^*H + G^*G = I_{V_0}.
\end{gathered} \tag{7.91}
$$

This can be applied to a sequence $u^0 = \{u_n^0\} \in \ell^2(\mathbf{Z}) \approx V_0$ to give the decomposition

$$
u^0 = H^*u^1 + G^*d^1, \quad \text{where} \quad u^1 = Hu^0, \quad d^1 = Gu^0. \tag{7.92}
$$

Since $u^1 \in \ell^2(\mathbf{Z}) \approx V_0$, we may similarly decompose u^1. Repeating this, we obtain the recursive formulas for synthesis and analysis,

$$
u^m = H^*u^{m+1} + G^*d^{m+1}, \quad \text{where} \quad u^{m+1} = Hu^m, \; d^{m+1} = Gu^m, \tag{7.93}
$$

where every step involves just a finite number of computations. In the electrical engineering literature, this scheme is known as *subband filtering* (Smith and Barnwell [1986], Vetterli [1986]). (It is also called *subband coding*; the cofficients $\{u^m, d^m\}$ are interpreted as a "coded" version of the signal, from which the original can be reconstructed by "decoding.") In view of the expressions (7.79), (7.80), and (7.82) for the filters, (7.93) can be written in a diagrammatic form as in Figure 7.1.

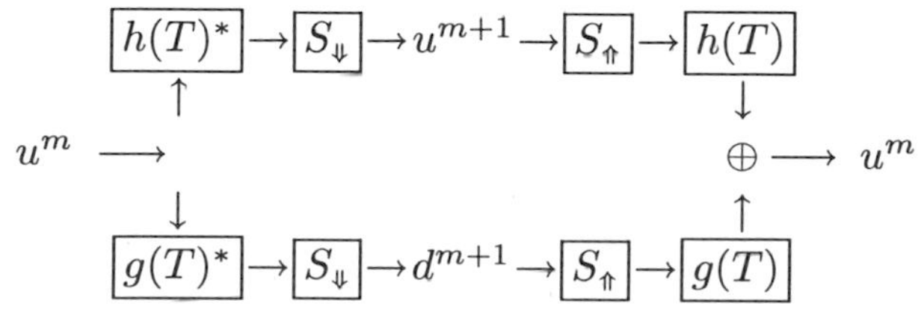

$$
\begin{array}{ccccccc}
 & \boxed{h(T)^*} \to \boxed{S_\Downarrow} \to u^{m+1} \to \boxed{S_\Uparrow} \to \boxed{h(T)} & \\
 & \uparrow \qquad\qquad\qquad\qquad\qquad\qquad \downarrow & \\
u^m \longrightarrow & \qquad\qquad\qquad\qquad\qquad\qquad\qquad \oplus & \longrightarrow u^m \\
 & \downarrow \qquad\qquad\qquad\qquad\qquad\qquad \uparrow & \\
 & \boxed{g(T)^*} \to \boxed{S_\Downarrow} \to d^{m+1} \to \boxed{S_\Uparrow} \to \boxed{g(T)} &
\end{array}
$$

Figure 7.1. Subband filtering scheme.

Repeated decomposition and reconstruction are represented by the pair of "dual" diagrams in Figures 7.2 and 7.3.

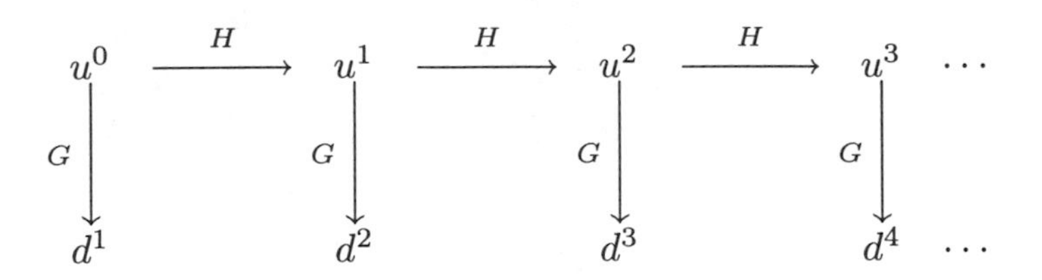

Figure 7.2. Wavelet decomposition with QMFs.

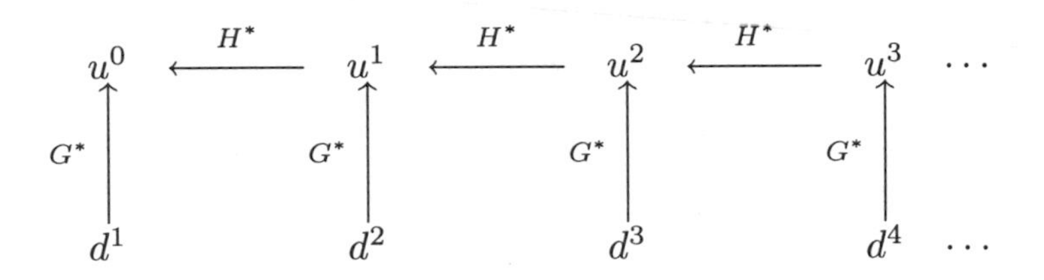

Figure 7.3. Wavelet composition with QMFs.

Let us now review the various requirement that have been imposed on the scaling function so far, in order to translate them into constraints on $h(T)$ and $g(T)$ or, equivalently, on the filter coefficients h_n and g_n. The requirements will be reviewed in an order that makes it possible to build on earlier ones to extract the constraints on $h(T)$ and $g(T)$.

(a) *The averaging property* $\hat{\phi}(0) \equiv \int dt\, \phi(t) = 1$ imposes no constraints on the filter coefficients.

(b) *The dilation equation* $D\phi = h(T)\phi$: Integrating over t and using (a), we obtain

$$\phi(t) = \sqrt{2} \sum_n h_n \phi(2t - n) \quad \Rightarrow \quad \sum_n h_n = \sqrt{2}. \tag{7.94}$$

(c) *The orthogonality condition*:

$$\begin{aligned} \delta_k^0 &= \langle\, \phi_k, \phi \,\rangle = \langle\, D\phi_k, D\phi \,\rangle = \langle\, h(T)\phi_{2k} \,, h(T)\phi \,\rangle \\ &= \sum_{n,l} \bar{h}_n h_l \langle\, \phi_{2k+n} \,, \phi_l \,\rangle \\ &= \sum_n \bar{h}_n h_{n+2k} \,. \end{aligned} \tag{7.95}$$

Note that both constraints (7.94) and (7.95) come from the interaction between the dilations and translations. These are all the requirements imposed so far;

more will be added later to obtain scaling functions and wavelets with some regularity properties. Since we have not set a limit on the number of nonvanishing coefficients h_n, *any* finite number of conditions can be imposed as long as they are mutually consistent. The relation between the linear condition (7.94) and the quadratic one (7.95) must be examined to ensure they do not conflict. An easy way to do this is to note that by (7.66),

$$\begin{aligned} 2I_{V_0} &= h(T)^* h(T) + h(-T)^* h(-T) \\ &= \sum_{n,l} \left[1 + (-1)^l\right] \bar{h}_n h_{n+l} T^l, \end{aligned} \tag{7.96}$$

and hence

$$\left[1 + (-1)^l\right] \sum_n \bar{h}_n h_{n+l} = 2\delta_l^0. \tag{7.97}$$

For odd l, (7.97) is an identity, whereas for $l = 2k$, (7.97) reduces to (7.95). The advantage of the form (7.96) of the orthogonality condition over (7.95) is that its relation to (7.94) is completely transparent. According to (7.18), taking the Fourier transform results in the substitution $T \to e^{-2\pi i\omega} \equiv z$. Hence $h(T)$ becomes $h(z)$, an ordinary Laurent polynomial in the complex variable z with $|z| = 1$. Similarly, $h(T)^* = \sum_n \bar{h}_n T^{-n}$ becomes $\sum_n \bar{h}_n z^{-n} = \sum_n \bar{h}_n \bar{z}^n = \overline{h(z)}$. Thus in the frequency domain (7.96) becomes

$$|h(z)|^2 + |h(-z)|^2 = 2. \tag{7.98}$$

Similarly, (7.94) translates to

$$h(1) = \sqrt{2}\,. \tag{7.99}$$

Now the relation between the two constraints is clear: Together they imply that

$$h(-1) = 0, \quad \text{i.e.,} \quad \sum_n (-1)^n h_n = 0. \tag{7.100}$$

Hence the sum of the even coefficients h_{2k} must equal the sum of the odd ones h_{2k+1}. Since the total sum equals $\sqrt{2}$, it follows that

$$\sum_k h_{2k} = \sum_k h_{2k+1} = \frac{1}{\sqrt{2}}\,. \tag{7.101}$$

Recall that $g(T) = -T^\lambda h(-T)^*$. Hence $|g(z)| = |h(-z)|$ for $|z| = 1$ and

$$|g(z)|^2 + |g(-z)|^2 = 2, \quad g(1) = 0, \quad g(-1) = \sqrt{2}\,. \tag{7.102}$$

Thus

$$\sum_k g_{2k} = -\sum_k g_{2k+1} = \frac{1}{\sqrt{2}}\,. \tag{7.103}$$

Equation (7.103) actually implies that *the mother wavelet ψ satisfies the admissibility condition*:

$$\psi(t) = \sqrt{2} \sum_n g_n \phi(2t - n) \Rightarrow \int_{-\infty}^{\infty} dt\ \psi(t) = \frac{1}{\sqrt{2}} \sum_n g_n = 0. \tag{7.104}$$

The filter coefficients h_n and g_n thus represent all the properties of $\phi(t)$ and $\psi(t)$, respectively, with the advantage that the continuous variable t is replaced by the discrete variable n that, for compactly supported ϕ, assumes only a finite number of values. Equations (7.95) and (7.98) are discrete representations (for compactly supported ϕ, *finite* representations) of the orthogonality condition; (7.103) is a discrete/finite representation of the admissibility condition; and (7.94) is a discrete/finite representation of $\int dt\, \phi(t) \neq 0$. (Recall that the dilation equation cannot fix the normalization of ϕ.) We will see later that further conditions can be imposed to assure some degree of regulariy for ϕ and ψ. These new conditions will also turn out to have discrete/finite representations in terms of the filter coefficients.

Let us examine the expansions (7.77) and (7.78) in light of the admissibility condition (7.104).

$$\begin{aligned} &\text{(a)} \qquad f(t) = \sum_{n \in \mathbf{Z}} \phi_{M,n}(t) \phi_{M,n}^* f + \sum_{k=-\infty}^{M} \sum_{n \in \mathbf{Z}} \psi_{k,n}(t) \psi_{k,n}^* f. \\ &\text{(b)} \qquad f(t) = \sum_{k \in \mathbf{Z}} \sum_{n \in \mathbf{Z}} \psi_{k,n}(t) \psi_{k,n}^* f. \end{aligned} \tag{7.105}$$

Since $\int dt\, \phi(t) = 1$ and $\int \psi_{m,n}(t) = 0$ implies

$$\int_{-\infty}^{\infty} dt\ \phi_{M,n}(t) = 2^{M/2}, \qquad \int_{-\infty}^{\infty} dt\ \psi_{m,n}(t) = 0, \tag{7.106}$$

a formal term-by-term integration of (7.105) gives

$$\begin{aligned} &\text{(a)} \qquad \int_{-\infty}^{\infty} dt\ f(t) = 2^{M/2} \sum_{n \in \mathbf{Z}} \phi_{M,n}^* f. \\ &\text{(b)} \qquad \int_{-\infty}^{\infty} dt\ f(t) = 0. \end{aligned} \tag{7.107}$$

(7.107b) cannot hold in general, since the left-hand side need not vanish for $f \in L^2(\mathbf{R})$. This "paradox" is a result of the term-by-term integration performed on (7.105b). The point is that the expansions (7.105b) converge in the sense of

$L^2(\mathbf{R})$ but *not necessarily* in the sense of L^1. That means that the partial sums

$$\begin{aligned}
&\text{(a)}\quad f_N(t) = \sum_{n=-N}^{N} \phi_{M,n}(t)\phi^*_{M,n}\, f + \sum_{k=-N}^{M}\sum_{n=-N}^{N} \psi_{k,n}(t)\psi^*_{k,n}\, f \\
&\text{(b)}\quad f_N(t) = \sum_{k=-N}^{N}\sum_{n=-N}^{N} \psi_{k,n}(t)\psi^*_{k,n}\, f
\end{aligned} \tag{7.108}$$

approach f in the L^2 norm $\|\cdot\|_2$ (which we have been denoting simply by $\|\cdot\|$ for convenience),

$$\|f - f_N\|_2^2 \equiv \int_{-\infty}^{\infty} dt\ |f(t) - f_N(t)|^2 \to 0 \quad \text{as } N \to \infty, \tag{7.109}$$

but not necessarily in the L^1 norm $\|\cdot\|_1$:

$$\|f - f_N\|_1 \equiv \int_{-\infty}^{\infty} dt\ |f(t) - f_N(t)| \not\to 0 \quad \text{as } N \to \infty, \tag{7.110}$$

in general. For example, let

$$g_N(t) = \begin{cases} N^{-3/4} & \text{if } |t| < N \\ 0 & \text{otherwise.} \end{cases} \tag{7.111}$$

Then as $N \to \infty$,

$$\|g_N\|_2^2 = 2N^{-1/2} \to 0 \quad \text{and} \quad \|g_N\|_1 = 2N^{1/4} \to \infty. \tag{7.112}$$

The function $f(t) - f_N(t)$ in (7.108b) behaves much like g_N: It becomes long and flat for large N, since more and more of the detail has been removed. As $N \to \infty$, its L^2 norm vanishes, but its L^1 norm does not vanish unless $\int dt f(t) = 0$, causing the "paradox" in (7.107b). As seen from (7.108a), no such paradox arises from (7.105)(a) because it contains the "remainder" term $P_M f$. Hence the expansion (a) is preferable to (b). (In fact, (a) holds in a great variety of function and distribution spaces, including L^p, Sobolev, etc.; see Meyer (1993a).)

Example 7.7: The Haar Wavelet. As a simple illustration, we apply the above construction to the Haar MRA of Example 7.2, where $\phi = \chi_{[0,1)}$. Equations (7.36) and (7.58) give

$$g(T) = -T^{\lambda}\, h(-T)^* = \frac{1}{\sqrt{2}}\,(T^{\lambda-1} - T^{\lambda}). \tag{7.113}$$

Hence

$$\begin{aligned}
\psi = D^{-1} g(T)\phi &= \frac{1}{\sqrt{2}}\,D^{-1}\Big(\chi_{[\lambda-1,\,\lambda)} - \chi_{[\lambda,\,\lambda+1)}\Big) \\
&= \chi_{[\frac{\lambda-1}{2},\,\frac{\lambda}{2})} - \chi_{[\frac{\lambda}{2},\,\frac{\lambda+1}{2})}.
\end{aligned} \tag{7.114}$$

Note that the support of ϕ is $[0,1]$, while that of ψ is $[\frac{\lambda-1}{2}, \frac{\lambda+1}{2}]$. The two supports therefore coincide only when $\lambda = 1$ is chosen in $g(T)$. This illustrates the point made earlier. For $\lambda = 1$, (7.114) is the Haar wavelet.

7.3 The View in the Frequency Domain

Although it is sometimes claimed that wavelet analysis replaces Fourier analysis (which may or may not be the case in practice), there can be no doubt that Fourier analysis plays an important role in the *construction* of wavelets. Furthermore, it gives wavelet decompositions like (7.28) and (7.30) a vital interpretation by explaining how the frequency spectrum of a signal $f \in V_m$ is divided up between the spaces V_{m+1} and W_{m+1}, which enhances our understanding of what is meant by "average" and "detail." Thus, although the natural domains for wavelet analysis are time and the time-scale plane, it is also useful to study it in the frequency domain.

In Section 7.2, we have referred to the inner products $\langle \phi_n, f \rangle$ as "samples" in V_0 (i.e., with $\Delta t = 1$) of $f \in L^2(\mathbf{R})$. However, these samples are not exactly the *values* $f(n)$, since ϕ is not the delta function. In fact,

$$\langle \phi_n, f \rangle = \langle \hat{\phi}_n, \hat{f} \rangle = \int_{-\infty}^{\infty} d\omega \; e^{2\pi\, in\omega}\, \overline{\hat{\phi}}(\omega)\hat{f}(\omega) = F(n), \tag{7.115}$$

where the function $F(t)$ is defined by its Fourier transform:

$$\hat{F}(\omega) \equiv \overline{\hat{\phi}}(\omega)\hat{f}(\omega). \tag{7.116}$$

Note that $\int d\omega\, |\hat{F}(\omega)| \le \|\hat{\phi}\|\|\hat{f}\| = \|\phi\|\|f\|$ by the Schwarz inequality and Plancherel's theorem; hence $\hat{F}(\omega)$ is integrable. $F(t)$ is therefore continuous, and its values $F(n)$ are well defined. Thus $\phi_n^* f$ *are actually the samples of the function* $F(t)$ *defined by (7.116) rather than samples of* $f(t)$. In fact, a general function $f \in L^2(\mathbf{R})$ cannot be sampled as is, since its value at any given point may be undefined (Section 1.3). In Section 5.1, we sampled band-limited functions, which are necessarily smooth and therefore have well-defined values. A general $f \in L^2(\mathbf{R})$ must be *smoothed* (e.g., by being passed through a low-pass filter) before it can be sampled. Equation (7.115) shows that ϕ performs just such a smoothing operation, the result being the function $F(t)$. The operator $f \mapsto F$ is a *filter* in the standard engineering sense, since it is performed through a multiplication (by $\overline{\hat{\phi}}$) in the frequency domain. Note that

$$\begin{aligned} F(n) &= \int_{-\infty}^{\infty} d\omega \; e^{2\pi\, in\omega}\, \hat{F}(\omega) = \sum_k \int_0^1 d\omega \; e^{2\pi\, in(\omega+k)}\, \hat{F}(\omega+k) \\ &= \int_0^1 d\omega \; e^{2\pi\, in\omega} \sum_k \hat{F}(\omega+k) \equiv \int_0^1 d\omega \; e^{2\pi\, in\omega}\, A(\omega). \end{aligned} \tag{7.117}$$

The samples $F(n)$ are therefore the Fourier coefficients of the 1-periodic function $A(\omega) = \sum_k \hat{F}(\omega + k)$. The copies $\hat{F}(\omega + k)$ with $k \neq 0$ are *aliases* of $\hat{F}(\omega)$, and if the support of $\hat{F}(\omega)$ is wider than 1, these aliases may intefere, making it impossible to recover $\hat{F}(\omega)$ from the samples. In other words, knowing $F(t)$ only at $t = n \in \mathbf{Z}$ gives information only about the *periodized* version of $\hat{F}$ obtained by aliasing.

So much for the sampling of $f \in L^2(\mathbf{R})$ in V_0. Now let us look at the partial reconstruction of f by $P_0 f = \sum_n u_n \phi_n = u(T)\phi$, where $u_n \equiv \phi_n^* f$ are the above samples. Recall that the time translation operator T acts in the frequency domain by (7.18):

$$(Tg)^\wedge(\omega) = z\,\hat{g}(\omega), \text{ where } z = z(\omega) \equiv e^{-2\pi\, i\omega}\,, \tag{7.118}$$

for any $g \in L^2(\mathbf{R})$. It is convenient to use the notation $\hat{g}(\omega) = e_\omega^* g$, even though $e_\omega(t) \equiv e^{2\pi\, i\omega t}$ does not belong to $L^2(\mathbf{R})$ (Section 1.4), for then we can write (7.118) as

$$e_\omega^* T g = z\, e_\omega^* g, \quad \text{or} \quad e_\omega^* T = z\, e_\omega^*. \tag{7.119}$$

The last equation is just the adjoint of $T^{-1} e_\omega(t) = e^{2\pi\, i\omega(t+1)} = \bar{z}\, e_\omega(t)$. Equation (7.119) states that e_ω^* is a "left eigenvector" of T with eigenvalue $z(\omega)$. (This can be made rigorous in terms of distribution theory.) Thus

$$(P_0 f)^\wedge(\omega) = e_\omega^* P_0 f = e_\omega^* u(T)\phi = u(z) e_\omega^* \phi = u(z)\hat{\phi}(\omega). \tag{7.120}$$

But $u_n = F(n)$, hence by (7.117),

$$u(z) \equiv \sum_n e^{-2\pi\, in\omega}\, u_n = A(\omega) \equiv \sum_k \hat{F}(\omega + k), \tag{7.121}$$

the "aliased" version of $\hat{F}(\omega)$. We therefore recognize our $u(T)$ notation of Section 7.2 as a symbolic form of Fourier series, to which it reverts under the Fourier transform. In fact, (7.120) *characterizes* V_0. That is, V_0 is the closure (in $L^2(\mathbf{R})$) of the set of all functions whose Fourier transform can be written as the product of an *arbitrary* trigonometric polynomial $u(z)$ with the *fixed* function $\hat{\phi}(\omega)$. (Taking the closure enlarges the set by allowing $u(z)$ to be any 1-periodic function square-integrable on $0 \leq \omega \leq 1$, as in (7.19).)

Let us digress a little in order to get an overview that ties together (7.115) through (7.121). The sampling process $f \mapsto \{u_n\}_{n \in \mathbf{Z}}$ defines an operator $S_0 : L^2(\mathbf{R}) \to \ell^2(\mathbf{Z})$, i.e., $S_0 f \equiv \{u_n\}_{n \in \mathbf{Z}}$. We call S_0 the *V_0-sampling operator.* (The V_m-sampling operator S_m can be defined similarly, and everything below can be extended to arbitrary $m \in \mathbf{Z}$. For simplicity we consider only $m = 0$.) If $\{u_n\} \in \ell^2(\mathbf{Z})$, then the corresponding Fourier series $u(z) \equiv \sum_n u_n z^n$ is a square-integrable function on the unit circle $\mathbf{T} \equiv \{z \in \mathbf{C} : |z| = 1\}$. Let us denote the space of all such functions by $L^2(\mathbf{T})$. The map $\{u_n\}_{n \in \mathbf{Z}} \mapsto u(z)$

then defines an operator $\wedge : \ell^2(\mathbf{Z}) \to L^2(\mathbf{T})$, which is a discrete analogue of the Fourier transform.[†]

We write $\{u_n\}^\wedge(z) = u(z)$. We define the operator $\hat{S}_0 : L^2(\mathbf{R}) \to L^2(\mathbf{T})$ as the "Fourier dual" of S_0, i.e., as the operator corresponding to S_0 in the frequency domain:

$$(\hat{S}_0 \hat{f})(z) \equiv (S_0 f)^\wedge(z) = \sum_n z^n \phi_n^* f. \tag{7.122}$$

That is, $\hat{S}_0$ is defined by the "commutative diagram" in Figure 7.4.

$$\begin{array}{ccc} f(t) \in L^2(\mathbf{R}) & \xrightarrow{S_0} & \{\phi_n^* f \equiv u_n\} \in \ell^2(\mathbf{Z}) \\ \wedge \big\downarrow & & \wedge \big\downarrow \\ \hat{f}(\omega) \in L^2(\mathbf{R}) & \xrightarrow{\hat{S}_0} & u(z) \in L^2(\mathbf{T}) \end{array}$$

Figure 7.4. The V_0-sampling operator S_0 and its Fourier dual, the aliasing operator $\hat{S}_0$.

Equations (7.121) and (7.116) give an explicit form for $\hat{S}_0$:

$$(\hat{S}_0 \hat{f})(z) = \sum_k \overline{\hat{\phi}(\omega + k)} \hat{f}(\omega + k). \tag{7.123}$$

(The right-hand side is 1-periodic; hence it depends only on $z = e^{-2\pi i\omega}$.) Thus $\hat{S}_0$ can be called the *aliasing operator* associated with V_0. Figure 7.4 gives the precise relation between sampling and aliasing: *Aliasing is the frequency-domain equivalent of sampling in the time domain.*

From the definition of S_0 it easily follows that $S_0^* : \ell^2(\mathbf{Z}) \to L^2(\mathbf{R})$ is given by $S_0^*\{u_n\} = u(T)\phi$. S_0^* is an interpolation operator, since it intepolates the samples $\{u_n\}$ to give the function $\sum_n u_n \phi(t - n)$. The relation between the adjoint operators S_0^* and $\hat{S}_0^*$ of S_0 and $\hat{S}_0$ is summarized by the diagram in Figure 7.5, where (7.120) has been used.

[†] The proper context for this is the theory of *Fourier analysis on locally compact Abelian groups* (Katznelson [1976]), where $\mathbf{T}$ (which is a group under multiplication) is seen to be the *dual group* of $\mathbf{Z}$, just as the "frequency domain" $\mathbf{R}_n$ is the dual group of the "time domain" $\mathbf{R}^n$; cf. Section 1.4. (The case $n = 1$ is deceptive since $\mathbf{R}$ is self-dual.)

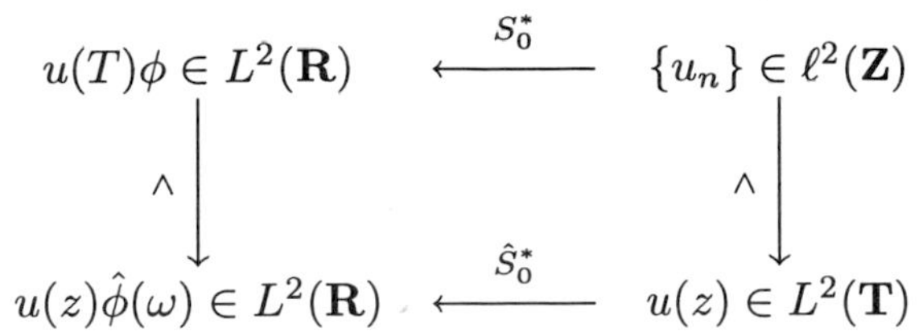

$$\begin{array}{ccc} u(T)\phi \in L^2(\mathbf{R}) & \xleftarrow{S_0^*} & \{u_n\} \in \ell^2(\mathbf{Z}) \\ \big\downarrow \wedge & & \big\downarrow \wedge \\ u(z)\hat{\phi}(\omega) \in L^2(\mathbf{R}) & \xleftarrow{\hat{S}_0^*} & u(z) \in L^2(\mathbf{T}) \end{array}$$

Figure 7.5. The V_0-interpolation operator S_0^* and its Fourier dual, the "de-aliasing" operator $\hat{S}_0^*$.

The operator $\hat{S}_0^*$ attempts to undo the effect of aliasing by producing the non-periodic function $u(z)\hat{\phi}(\omega)$ from $u(z)$. Of course, for a general $f \in L^2(\mathbf{R})$, the aliasing cannot be undone. In fact, by (7.123),

$$(\hat{S}_0^* \hat{S}_0 \hat{f})(\omega) = \sum_k \overline{\hat{\phi}(\omega + k)} \hat{f}(\omega + k) \hat{\phi}(\omega), \tag{7.124}$$

which does not coincide with $\hat{f}(\omega)$ in general. When *is* $\hat{S}_0^* \hat{S}_0 \hat{f} = \hat{f}$? Note that $S_0^* S_0 = P_0$, hence $\hat{S}_0^* \hat{S}_0 \hat{f} = \hat{f}$ if and only if $f \in V_0$. If $f = u(T)\phi \in V_0$, then $\hat{f}(\omega) = u(z(\omega))\hat{\phi}(\omega)$, and for $k \in \mathbf{Z}$,

$$\hat{f}(\omega + k) = u(z(\omega + k))\hat{\phi}(\omega + k) = u(z(\omega))\hat{\phi}(\omega + k). \tag{7.125}$$

Hence (7.124) becomes

$$(\hat{S}_0^* \hat{S}_0 \hat{f})(\omega) = \Big[\sum_k |\hat{\phi}(\omega + k)|^2\Big] u(z)\hat{\phi}(\omega). \tag{7.126}$$

The next proposition confirms that the right-hand side is, in fact, just $u(z)\hat{\phi}(\omega) = \hat{f}(\omega)$.

Proposition 7.8. *The orthonormality property* $\langle \phi_n, \phi \rangle = \delta_n^0$ *is equivalent to*

$$K(\omega) \equiv \sum_k |\hat{\phi}(\omega + k)|^2 = 1 \quad a.e. \tag{7.127}$$

Proof: Note that $K(\omega)$ is 1-periodic and $\int_0^1 d\omega\, K(\omega) = \|\hat{\phi}\|^2 = 1$; hence K is integrable and therefore possesses a Fourier series. Equations (7.115) and (7.117), with $f = \phi$, give

$$\langle \phi_n, \phi \rangle = \int_0^1 d\omega\; e^{2\pi i n \omega}\, K(\omega). \tag{7.128}$$

Hence the orthonormality condition states that the Fourier coefficients of $K(\omega)$ are $c_n = \delta_n^0$, which means that $K(\omega) = 1$ a.e. ∎

Note: (a) When $\phi(t)$ has compact support, then $\hat{\phi}(\omega)$ is analytic and therefore continuous, and (7.127) holds for all ω, not just "almost everywhere." (b) When $e^{\pm i\omega t}$ are used in the Fourier transform instead of $e^{\pm 2\pi i\omega t}$, then (7.127) becomes

$$\sum_k |\hat{\phi}(\omega + 2\pi k)|^2 = \frac{1}{2\pi} \quad \text{a.e.} \quad (\text{if } e^{\pm i\omega t} \text{ are used}). \tag{7.129}$$

This is the convention used in Chui (1992a), Daubechies (1992), and Meyer (1993a).

Proposition 7.8 gives the following intuitively appealing intepretation of $\hat{\phi}$: According to (7.123), the k-th copy $\hat{f}(\omega+k)$ of $\hat{f}$ gets modulated with the (complex) amplitude function $\overline{\hat{\phi}(\omega + k)}$ in the aliasing process. In the (partial) reconstruction $u(z) \mapsto u(z(\omega))\hat{\phi}(\omega)$, $\hat{\phi}(\omega)$ acts as a single amplitude function modulating all the different copies. If we restrict ω to $[-.5, .5)$ and regard $|\hat{\phi}(\omega + k)|^2$ as a *weight function* assigned to the k-th copy, then Proposition 7.8 shows that the total weight given to all copies is 1. Thus we may say that ϕ is a "scattered" de-aliasing filter: If $\hat{\phi}(\omega)$ were the characteristic function of a single period $[-.5 - k, .5 - k)$, then $u(z)\hat{\phi}(\omega)$ would filter out only that copy, thus de-aliasing $u(z)$ (see Example 7.10). But even when the support of $\hat{\phi}$ is "scattered" among many different periods, Proposition 7.8 guarantees that $\hat{S}_0^*$ performs an exact de-aliasing of $u(z)$ when $f \in V_0$. Note that Proposition 7.8 gives no preference to any copy over any other, since it is translation-invariant: For any α, $\hat{\phi}^\alpha(\omega) \equiv \hat{\phi}(\omega + \alpha)$ equally satisfies (7.127). The next proposition shows that when ϕ is required to satisfy the averaging property in addition to orthonormality, then it gives the strongest possible preference to the *original* $\hat{f}(\omega)$ over all the aliases $\hat{f}(\omega + k)$, $k \neq 0$; thus ϕ acts as a *low-pass de-aliasing filter.* The next proposition shows another important consequence of the averaging and orthonormality properties: *If any signal $f \in V_0$ is aliased with period 1 in the time domain, the result is a constant function.*

Proposition 7.9. *Let ϕ be integrable and satisfy $\langle \phi_n, \phi \rangle = \delta_n^0$ and $\hat{\phi}(0) = 1$. Then*

$$\begin{aligned} &(a) \quad \hat{\phi}(k) = \delta_k^0, \qquad k \in \mathbf{Z} \\ &(b) \quad \sum_n f(t - n) = \int_{-\infty}^{\infty} du\, f(u) \quad \textit{a.e. for all integrable } f \in V_0 \\ &(c) \quad \{\phi_n\} \textit{ form a partition of unity, i.e., } \sum_n \phi_n(t) = 1 \quad \textit{a.e.} \end{aligned} \tag{7.130}$$

Proof: Since ϕ is integrable, $\hat{\phi}$ is continuous; hence its values at the individual points $\omega = k$ are well defined. Together with (7.127), $\hat{\phi}(0) = 1$ implies that

$\hat{\phi}(k) = 0$ for all integers $k \neq 0$, which proves (a). Now let $f = u(T)\phi \in V_0$ and consider the time-aliased signal $P(t) \equiv \sum_n f(t-n)$, which is 1-periodic and integrable (since $u(T)$ is a polynomial) and hence possesses a Fourier series. Its Fourier coefficients are

$$\begin{aligned} c_k &= \int_0^1 dt\ e^{-2\pi\, ikt} \sum_n f(t-n) = \sum_n \int_0^1 dt\, e^{-2\pi\, ik(t-n)}\, f(t-n) \\ &= \int_{-\infty}^{\infty} dt\ e^{-2\pi\, ikt}\, f(t) = \hat{f}(k). \end{aligned} \tag{7.131}$$

But (a) implies that $\hat{f}(k) = u(e^{-2\pi\, ik})\hat{\phi}(k) = \delta_k^0$, hence

$$P(t) = \hat{f}(0) = \int_{-\infty}^{\infty} dt\ f(t), \tag{7.132}$$

proving (b). Assertion (c) now follows by choosing $f = \phi$. ■

Note: The term "partition of unity" refers to the fact that an arbitrary (but sufficiently reasonable) function can be written as a sum "localized" functions: $f(t) = \sum_n f_n(t)$, where $f_n(t) \equiv \phi_n(t) f(t)$.

Example 7.10: The Ideal Low-Pass Filter. Let

$$\hat{\phi}(\omega) = \begin{cases} 1 & |\omega| \leq .5 \\ 0 & |\omega| > .5. \end{cases} \tag{7.133}$$

In the time domain, ϕ is given by

$$\phi(t) = \frac{\sin(\pi t)}{\pi t}. \tag{7.134}$$

Since (7.127) holds, so does the orthonormality condition, which can also be easily verified directly. ϕ cannot be integrable, since $\hat{\phi}$ is discontinuous. Nevertheless, $\hat{\phi}(\omega)$ is bounded for all ω and continuous in a neighborhood of $\omega = 0$, and it satisfies $\hat{\phi}(k) = \delta_k^0$ for all $k \in \mathbf{Z}$. Hence ϕ generates a multiresolution analysis (see the comments below Proposition 7.1). The function defined by (7.116) is

$$F(t) = \int_{-.5}^{.5} d\omega\ e^{2\pi\, i\omega t}\ \hat{f}(\omega), \tag{7.135}$$

which is the band-limited function obtained by passing f through the low-pass filter ϕ. The projection of f to V_0 is (using $\phi_n^* f = F(n)$)

$$(P_0 f)(t) \equiv \sum_n \phi_n(t)\, \phi_n^* f = \sum_n \frac{\sin(\pi t - n\pi)}{\pi t - n\pi} F(n) = F(t), \tag{7.136}$$

by Shannon's sampling theorem of Section 5.1. (The simple relation $F = P_0 f$ does not hold in general; it is special to MRAs generated by ideal filters.)

Up to now, we have only considered sampling and reconstruction with ϕ_n, hence in V_0. Only the orthogonality and averaging properties of ϕ were taken into account, both of which relate to a single scale. A similar analysis can be made for V_m, and then it is seen that the role of a low-pass de-aliasing weight function is played by $|\hat{\phi}(2^m\omega)|^2$. To see how the different scales interact in the frequency domain, we now add the third (and most fundamental) property of the scaling function, namely the dilation equation $D\phi = h(T)\phi$. Taking the Fourier transform and using (7.119), this becomes

$$\hat{\phi}(2\omega) = m_0(\omega)\hat{\phi}(\omega), \tag{7.137}$$

where

$$m_0(\omega) \equiv \frac{1}{\sqrt{2}}\, h(z(\omega)) = m_0(\omega + 1). \tag{7.138}$$

Equation (7.98), which was a consequence of orthonormality and the dilation equations, stated that $|h(z)|^2 + |h(-z)|^2 = 2$. For $m_0(\omega)$, this reads

$$|m_0(\omega)|^2 + |m_0\,(\omega + .5)\,|^2 = 1, \tag{7.139}$$

since $z(\omega + .5) = -z(\omega)$. The averaging property implied that $h(1) = \sqrt{2}$ and $h(-1) = 0$, or equivalently

$$m_0(0) = 1, \quad m_0(.5) = 0. \tag{7.140}$$

(When the Fourier transform is defined with the complex exponentials $e^{\pm i\omega t}$, $m_0(\omega)$ is 2π-periodic instead of 1-periodic. Then (7.139) and (7.140) read $|m_0(\omega)|^2 + |m_0\,(\omega + \pi)\,|^2 = 1$ and $m_0(0) = 1, m_0(\pi) = 0$, respectively.)

The mother wavelet was defined in (7.70) by $D\psi = g(T)\phi$. This implies

$$\hat{\psi}(2\omega) = \frac{1}{\sqrt{2}}\, g(z(\omega))\hat{\phi}(\omega) \equiv m_1(\omega)\hat{\phi}(\omega). \tag{7.141}$$

If $g(T) = -T^{\lambda}h(-T)^*$, with λ an odd integer, then

$$m_1(\omega) = -\, e^{-2\pi\, i\lambda\omega}\, \overline{m_0(\omega + .5)}. \tag{7.142}$$

Thus (7.137), (7.139), and (7.141) imply

$$|\hat{\phi}(2\omega)|^2 + |\hat{\psi}(2\omega)|^2 = |\hat{\phi}(\omega)|^2. \tag{7.143}$$

Equation (7.143) can be viewed, roughly, as describing how the "spectral energy" of $f \in V_0$ is split up between V_1 and W_1. For example, if ϕ is an ideal low-pass filter, then ψ is an ideal band-pass filter, as shown in the next example, and the spectral energy is cut up *sharply* between V_1 and W_1. In general, the splitting is not "sharp" in the frequency domain (see Example 7.12). In any case, (7.143) suggests that the energy tends to be split *evenly* between V_1 and W_1, since $\int d\omega\, |\hat{\phi}(2\omega)|^2 = \int d\omega\, |\hat{\psi}(2\omega)|^2 = \frac{1}{2}$. (The actual amount going to V_1 and W_1 depends, of course, on f.)

Example 7.11: Shannon Wavelets. We have noted that the ideal low-pass filter of Example 7.10 defines a multiresolution analysis. To construct the corresponding wavelets, we must find $m_0(\omega)$ so that (7.137) holds. Since $\hat{\phi}(2\omega)$ is the characteristic function of $[-.25, .25]$, (7.137) implies that $m_0(\omega) = \hat{\phi}(2\omega)$ on $[-.5, .5]$. By periodicity,

$$m_0(\omega) = \sum_k \hat{\phi}(2\omega + 2k). \tag{7.144}$$

Since $\hat{\phi}$ is real-valued, (7.142) gives

$$m_1(\omega) = e^{-2\pi i\lambda\omega} \sum_k \hat{\phi}(2\omega + 2k + 1). \tag{7.145}$$

The support of $m_1(\omega)$ intersects with that of $\hat{\phi}(\omega)$ only in $[-.5, -.25]$ (for $k = 0$) and in $[.25, .5]$ (for $k = -1$). Thus (7.141) gives

$$\begin{aligned} \hat{\psi}(2\omega) &= -\, e^{-2\pi i\lambda\omega} \left(\hat{\phi}(2\omega + 1) + \hat{\phi}(2\omega - 1)\right) \hat{\phi}(\omega) \\ &= -\, e^{-2\pi i\lambda\omega} \left(\chi_{[-.5,-.25]}(\omega) + \chi_{[.25,.5]}(\omega)\right). \end{aligned} \tag{7.146}$$

Thus $\hat{\psi}(2\omega)$ is an *ideal band-pass filter* for the frequency bands $[-.5, -.25]$ and $[.25, .5]$. Similarly, $\hat{\psi}(\omega)$ is an ideal band-pass filter (up to the normalization factor $\sqrt{2}$) for the bands $[-1, -.5]$ and $[.5, 1]$. ψ is sometimes called the *Shannon wavelet.* In the time domain it is given (taking $\lambda = 1$) by

$$\psi(t) = \cos\left(1.5\pi(t - .5)\right) \frac{\sin\left(.5\pi(t - .5)\right)}{.5\pi(t - .5)}. \tag{7.147}$$

This wavelet is not very useful in practice, since it has very slow decay in the time domain due to the discontinuities in the frequency domain. In a sense, the Shannon wavelet and the Haar wavelet are at the opposite extremes of multiresolution analysis: The former gives maximal localization in time, the latter, in frequency. Each gives poor resolution in the opposite domain by a discrete version of the uncertainty principle. (They are analogous to the delta function and $f(t) \equiv 1$, which are at the extremes of continuous Fourier analysis.) All other examples we will encounter are intermediate between these two.

Example 7.12: Meyer Wavelets. The slow decay of the ideal low-pass filter in the time domain makes it far from "ideal" from the viewpoint of multiresolution analysis, since it gives rise to poor time localization. The Meyer scaling function solves this problem by smoothing out $\hat{\phi}(\omega)$. Choose a smooth function $\theta(u)$ with

$$\theta(u) = 0 \quad \text{for } u \le \frac{1}{3} \quad \text{and} \quad \theta(u) = \frac{\pi}{2} \quad \text{for } u \ge \tfrac{2}{3}, \tag{7.148}$$

with the additional property

$$\theta(u) + \theta(1-u) = \frac{\pi}{2} \quad \text{for all } u. \tag{7.149}$$

"Smooth" means, e.g., that θ is d times continuously differentiable, for some integer $d \geq 1$. Such functions are easily constructed, as shown in Exercise 7.1. Given $\theta(u)$, define $\phi(t)$ through its Fourier transform by

$$\hat{\phi}(\omega) = \cos\theta(|\omega|). \tag{7.150}$$

Then $\hat{\phi}(\omega)$ is also $d \geq 1$ times continuously differentiable, with $\hat{\phi}(\omega) = 1$ for $|\omega| \leq 1/3$ and $\hat{\phi}(\omega) = 0$ for $|\omega| \geq 2/3$. Hence the support of $\hat{\phi}$ is $[-2/3, 2/3]$, which exceeds a single period of m_0. It is easily seen that $\hat{\phi}$ satisfies (7.127). Only neighboring terms $|\hat{\phi}(\omega + k)|^2$ in the sum (i.e., with $\Delta k = 1$) have overlapping supports. For simplicity we concentrate on the two terms $|\hat{\phi}(\omega)|^2 + |\hat{\phi}(\omega - 1)|^2$. In the interval $[-1/3, 1/3]$, the supports do not overlap and the sum equals 1. For $1/3 \leq \omega \leq 2/3$, $\theta(|\omega - 1|) = \theta(1 - \omega) = \pi/2 - \theta(\omega)$ by (7.149). Hence

$$\hat{\phi}(\omega - 1) = \cos\left(\frac{\pi}{2} - \theta(\omega)\right) = \sin\theta(\omega) \tag{7.151}$$

and

$$|\hat{\phi}(\omega)|^2 + |\hat{\phi}(\omega - 1)|^2 = \cos^2\theta(\omega) + \sin^2\theta(\omega) = 1, \quad \tfrac{1}{3} \leq \omega \leq \tfrac{2}{3}. \tag{7.152}$$

This proves (7.127) for $-1/3 \leq \omega \leq 2/3$ and hence for all ω by periodicity. Clearly, $\hat{\phi}(k) = \delta_k^0$ holds as well. Since $\hat{\phi}$ is smooth, $\phi(t)$ has reasonable decay. The smoother that ϕ is (the higher the choice of d), the more rapid the decay is. Next, we verify that ϕ has the scaling property. Just as in Example 7.11, (7.137) is satisfied with

$$m_0(\omega) \equiv \sum_k \hat{\phi}(2\omega + 2k). \tag{7.153}$$

(This is a consequence of $\hat{\phi}(2\omega + 2k)\hat{\phi}(\omega) = \delta_k^0\, \hat{\phi}(2\omega)$, which is not true in general but holds in both examples.) Equation (7.145) also holds in this case, as does the top expression in (7.146) for the associated mother wavelet:

$$\hat{\psi}(2\omega) = -\, e^{-2\pi\, i\lambda\omega} \left(\hat{\phi}(2\omega - 1) + \hat{\phi}(2\omega + 1)\right) \hat{\phi}(\omega). \tag{7.154}$$

The arguments about overlapping supports are similar (exercise).

7.4 Frequency Inversion Symmetry and Complex Structure†

The MRA formalism described in Section 7.2 exhibits an as yet unexplained symmetry between low-frequency and high-frequency phenomena. Recall that we can start with the filter coefficients h_n, subject to certain constraints, or equivalently with $h(T) \in \mathcal{P}$. $h(T)$ determines $g(T)$ by

$$g(T) = -T^{\lambda} h(-T)^*, \quad \lambda \in \mathbf{Z} \text{ odd}, \tag{7.155}$$

and the scaling function and mother wavelet are determined (up to normalization) by

$$D\phi = h(T)\phi, \qquad D\psi = g(T)\phi. \tag{7.156}$$

The quadrature mirror filters $H : V_0 \to V_0$, $G : V_0 \to W_0$ and their adjoints are then given by

$$\begin{aligned} H^* u(T)\phi &= h(T)u(T^2)\phi \\ G^* u(T)\psi &= g(T)u(T^2)\phi \\ Hu(T)\phi &= S_{\Downarrow} h(T)^* u(T)\phi \\ Gu(T)\phi &= S_{\Downarrow} g(T)^* u(T)\psi. \end{aligned} \tag{7.157}$$

The symmetry of these relations with respect to $h(T) \leftrightarrow g(T)$, $\phi \leftrightarrow \psi$, and $H \leftrightarrow G$ is clearly evident, although it certainly has *not* been put into the formalism by design. Can this be an accident, with no fundamental basis and no significant applications? Note that this symmetry is one between the *low-frequency objects* $h(T), \phi, H$ and the *high-frequency objects* $g(T), \psi, G$. Its origin is somewhat mysterious, although it clearly has to do with the fact that V_0 is a space of (approximately) band-limited functions and the frequency spectrum in V_0 is divided evenly between V_1 and W_1, as suggested by (7.143). To my knowledge, this symmetry has not been named, so I will call it *frequency inversion symmetry.* It is certainly worthy of study, since without it multiresolution analysis might be incapable of defining wavelets.

The purpose of this section is to offer some insight into the mathematical basis for the above symmetry. For more details, see Kaiser (1992a). Our starting point is a generalization of the relation between $h(T)$ and $g(T)$. Define the operator $J : V_0 \to V_0$ by

$$Ju(T)\phi = -T^{\lambda} u(-T)^* \phi, \tag{7.158}$$

where λ is the odd integer used in the definition of $g(T)$. The next theorem shows that (a) J behaves much like multiplication by i, so it is a kind of "90° rotation" in V_0; (b) J "rotates" $D\phi \in V_1$ to $D\psi \in W_1$; and (c) it "rotates" the whole subspace V_1 onto W_1 and W_1 onto V_1. The frequency inversion symmetry

---

† This section is optional.

is therefore explained as an abstract rotation in V_0. A similar operator J_m can be defined on V_m for every $m \in \mathbf{Z}$ by $J_m = D^m J D^{-m}$.

Theorem 7.13.

$$\begin{aligned} &\text{(a)} \quad J^2 = -I_{V_0} \\ &\text{(b)} \quad JD\phi = D\psi, \quad JD\psi = -D\phi \\ &\text{(c)} \quad JV_1 = W_1, \quad JW_1 = V_1. \end{aligned} \tag{7.159}$$

Proof:

$$J^2 u(T)\phi = -JT^\lambda u(-T)^*\phi = T^\lambda \left[(-T)^\lambda\right]^* u(t)\phi = -u(T)\phi, \tag{7.160}$$

since λ is odd and $(T^\lambda)^* = T^{-\lambda}$. This proves (a). (b) follows from

$$JD\phi = Jh(T)\phi = -T^\lambda h(-T)^*\phi = g(T)\phi = D\psi \tag{7.161}$$

and $JD\psi = J^2 D\phi = -D\phi$, by (a). To prove (c), recall that a typical vector in V_1 has the form $Du(T)\phi = h(T)u(T^2)\phi$, while a typical vector in W_1 has the form $Du(T)\psi = g(T)u(T^2)\phi$. Hence

$$\begin{aligned} JDu(T)\phi &= Jh(T)u(T^2)\phi = -T^\lambda \left[h(-T)u(T^2)\right]^* \phi \\ &= g(T)u(T^2)^*\phi \in W_1. \end{aligned} \tag{7.162}$$

Since $\{Du(T)\phi : u(T) \in \mathcal{P}\}$ is dense in V_1 and $\{g(T)u(T^2)^*\phi : u(T) \in \mathcal{P}\}$ is dense in W_1, this proves that $JV_1 \equiv \{Jf : f \in V_1\} = W_1$ as claimed. It now follows from (a) that

$$JW_1 = J^2 V_1 \equiv \{J^2 f : f \in V_1\} = \{-f : f \in V_1\} = V_1. \quad \blacksquare \tag{7.163}$$

The relation between the filters is also implicit in (7.162):

$$JH^*u(T)\phi = Jh(T)u(T^2)\phi = g(T)u(T^2)^*\phi = G^*u(T)^*\psi. \tag{7.164}$$

Replacing $u(T)$ by $u(T)^*$ and applying J to both sides then gives

$$JG^*u(T)\psi = J^2 H^*u(T)^*\phi = -H^*u(T)^*\phi. \tag{7.165}$$

J is *real-linear*, i.e., $J[c\,u(T)\phi + v(T)\phi] = cJu(T)\phi + Jv(T)\phi$ for $c \in \mathbf{R}$. A real-linear mapping J on a real vector space V whose square is minus the identity is called a *complex structure*. For example, consider the operator $J : \mathbf{R}^2 \to \mathbf{R}^2$ defined by

$$J\begin{bmatrix} x \\ y \end{bmatrix} \equiv \begin{bmatrix} -y \\ x \end{bmatrix}, \quad \text{i.e.,} \quad J = \begin{bmatrix} 0 & -1 \\ 1 & 0 \end{bmatrix}, \tag{7.166}$$

which satisfies $J^2 = -I$. If we identify $\mathbf{R}^2$ with $\mathbf{C}$ by $\begin{bmatrix} x \\ y \end{bmatrix} \leftrightarrow x + iy$, then J becomes multiplication by i. Returning to our $J : V_0 \to V_0$, note that for *complex* c we have $[cu(T)]^* = \bar{c}\,u(T)^*$. Hence

$$Jc\,u(T)\phi = \bar{c}Ju(T)\phi, \quad \text{i.e.,} \quad Jc = \bar{c}\,J. \tag{7.167}$$

That means that J is *antilinear* with respect to complex scalar multiplication. This results in an interesting extension of the complex structure defined by J. Define $K : V_0 \to V_0$ by

$$K = iJ, \quad \text{i.e.,} \quad Ku(T)\phi \equiv iJu(T)\phi = -iT^\lambda u(-T)^*\phi, \tag{7.168}$$

where $i = \sqrt{-1}$. According to (7.167), we have

$$K^2 = iJiJ = -i^2J^2 = -I_{V_0} \tag{7.169}$$

and

$$JK = JiJ = -iJ^2 = i, \quad Ki = iJi = -i^2J = J. \tag{7.170}$$

Hence $\{i, J, K\}$ defines a *quaternionic structure* on V_0! This means that we can extend the "rotations" generated by J to ones very similar to general rotations in $\mathbf{R}^3$. I do not know at present what use can be made of these, but it seems to be an interesting fact.

The above implementation of frequency-inversion symmetry by the complex structure seems rather abstract, since no explanation was given as to *why* J should interchange low-frequency and high-frequency phenomena. To get a better understanding, we now view the action of J in $L^2(\mathbf{T})$, using the diagram of Figure 7.5. The operator corresponding to J is $\hat{\jmath} : L^2(\mathbf{T}) \to L^2(\mathbf{T})$, defined by

$$(\hat{\jmath}u)(z) \equiv -z^\lambda\, \overline{u(-z)}. \tag{7.171}$$

Since $z = e^{-2\pi i\omega}$ with period $0 \le |\omega| \le .5$, low frequencies are represented by $0 \le |\omega| < .25$ and high frequencies, by $.25 \le |\omega| \le .5$. (Recall Example 7.10.) Clearly, $\hat{\jmath}$ exchanges the low- and the high-frequency parts of $u(z)$. The corresponding operator $\jmath : \ell^2(\mathbf{Z}) \to \ell^2(\mathbf{Z})$ is given by

$$\jmath\{u_n\}_{n\in\mathbf{Z}} = \{(-1)^n\, \bar{u}_{\lambda-n}\}_{n\in\mathbf{Z}} \tag{7.172}$$

(see Figure 7.6).

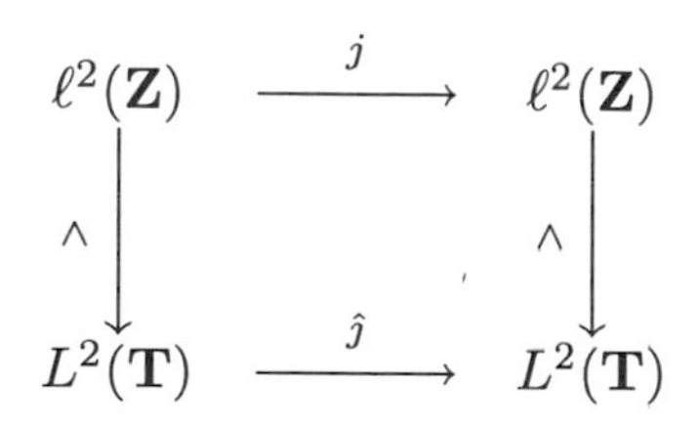

Figure 7.6. The complex structures in $\ell^2(\mathbf{Z})$ and $L^2(\mathbf{T})$.

Note that $u(-T)^* = \sum_n \bar{u}_n(-T^{-1})^n \equiv \bar{u}(-T^{-1})$, where $\bar{u}(T)$ is the polynomial defined by $\bar{u}(T) = \sum_n \bar{u}_n T^n$. In particular, if the u_n's are real, then $u(-T)^* = u(-T^{-1})$. Hence $(\hat{\jmath}u)(z) = -z^\lambda u(-\bar{z})$ and the mapping $\hat{\jmath}$ *amounts to a reflection about the y-axis in the z-plane, followed by multiplication by* $-z^\lambda$.

As an extreme example, consider sampling $f(t) \equiv 1$. ($f \notin L^2(\mathbf{R})$, hence this example does not fit into the L^2 theory, but it does make sense in terms of distributions!) Then $u_n = \phi_n^* f = 1$ for all n, and

$$j\{u_n\}_{n\in\mathbf{Z}} = \{(-1)^n\}_{n\in\mathbf{Z}}. \tag{7.173}$$

That is, j transforms the *least* oscillating sequence $\{1\}_{n\in\mathbf{Z}}$ into the sequence with the *most* oscillations.

The quaternionic structure can likewise be transfered to $\ell^2(\mathbf{Z})$ and $L^2(\mathbf{T})$ by defining $k = ij$ on $\ell^2(\mathbf{Z})$ and $\hat{k} = i\hat{\jmath}$ on $L^2(\mathbf{T})$.

7.5 Wavelet Packets†

One of the main reasons that wavelet analysis is so efficient is that the frequency domain is divided *logarithmically*, by octaves, instead of linearly by bands of equal width. This is accomplished by using scalings rather than modulations to sample different frequency bands. Thus wavelets sampling high frequencies have no more oscillations than ones sampling low frequencies – very unlike the situation in the windowed Fourier transform. However, it is sometimes desirable to further divide the octaves. That amounts to splitting the wavelet subspaces W_m, and it can actually be accomplished quite easily *within* the wavelet framework. The resulting formalism is a kind of hybrid of wavelet theory and the windowed Fourier transform, since it uses *repeating* wavelets to achieve higher frequency resolution, much as the WFT uses modulated windows. Wavelet packets were originally constructed by Coifman and Meyer; see Coifman, Meyer and Wickerhauser (1992).

To construct wavelet packets, let us assume that V_0 is real, i.e., only real coefficients are allowed in $u(T) \in \mathcal{P}$. (Then $V_m \equiv D^m V_0$ is also real.) This implies that the complex structure defined in Section 7.4 (which is now linear) is *skew-Hermitian*, i.e., $J^* = -J$ (exercise). Define the operators $F_0 : V_0 \to V_0$ and $F_1 : V_0 \to V_0$ by

$$F_0 = H, \quad F_1 = -HJ. \tag{7.174}$$

Essentially, F_1 performs like the high-pass filter G, extracting detail. In fact, by (7.40) and (7.41),

$$\begin{aligned} F_1 V_1 &= -HJV_1 = -HW_1 = \{0\} \\ F_1 W_1 &= -HJW_1 = -HV_1 = V_0 \\ F_1^* V_0 &= JH^* V_0 = JV_1 = W_1. \end{aligned} \tag{7.175}$$

† This section is optional.

The actions of F_1 and F_1^* on sequences are

$$\begin{aligned} F_1^*\{u_k\}_{k\in\mathbf{Z}} &= \left\{\sum_k g_{n+2k}\, u_k\right\}_{n\in\mathbf{Z}} \\ F_1\{u_n\}_{n\in\mathbf{Z}} &= \left\{\sum_n g_{n+2k}\, u_n\right\}_{k\in\mathbf{Z}}, \end{aligned} \tag{7.176}$$

(exercise) which are almost identical with the actions of G^* and G, as given by (7.90). (Keep in mind that we now assume everything to be real.) The main difference is that, unlike $G : V_0 \to W_0$, F_1 maps V_0 to itself. That makes it possible to compose any combination of F_0, F_1, F_0^*, and F_1^*, which leads directly to wavelet packets. The next proposition shows that F_0 and F_1 can be used exactly like H, G to implement the orthogonal decomposition of $V_0 = V_1 \oplus W_1$.

Proposition 7.14.

$$\begin{aligned} &\text{(a)} \quad F_a F_b^* = \delta_b^a I_{V_0} \quad (a, b = 0, 1) \\ &\text{(b)} \quad F_0^* F_0 + F_1^* F_1 = I_{V_0}. \end{aligned} \tag{7.177}$$

Proof: We already know from (7.91) that $F_0 F_0^* = HH^* = I_{V_0}$. Since $JJ^* = -J^2 = I_{V_0}$, it follows that $F_1 F_1^* = I_{V_0}$. By (7.41), (7.159)(c), and (7.40),

$$F_1 F_0^* V_0 = -HJH^* V_0 = -HJV_1 = -HW_1 = \{0\}. \tag{7.178}$$

Hence $F_1 F_0^* = 0$, which also implies $F_0 F_1^* = 0$. That proves (a). For the proof of (b) and the exact relation between F_1 and G, see Kaiser (1992a). ■

To consider arbitrary compositions of the new filters, it is convenient to introduce the following notation: Let $A = \{a_1, a_2, \ldots, a_m\}$ and $B = \{b_1, b_2, \ldots, b_m\}$ be *multi-indices*, i.e., ordered sets of indices with $a_k, b_k = 0, 1$. We also write the *length* of A as $|A| = m$. Define

$$\delta_B^A = \delta_{b_1}^{a_1} \delta_{b_2}^{a_2} \cdots \delta_{b_m}^{a_m}, \quad |A| = |B| = m. \tag{7.179}$$

Then we have the following generalization of Proposition 7.14.

Proposition 7.15.

$$\begin{aligned} &\text{(a)} \quad F_A F_B^* = \delta_B^A\, I_{V_0} \quad \textit{if} \quad |A| = |B| \ge 1. \\ &\text{(b)} \quad \sum_{|A|=m} F_A^* F_A = I_{V_0} \quad \textit{for any } m \ge 1, \end{aligned} \tag{7.180}$$

where the sum in (b) is over all 2^m multi-indices with $|A| = m$.

Proof: Follows immediately from (7.177) (exercise). ■

Proposition 7.15 states that the operators

$$\Pi_A \equiv F_A^* F_A, \tag{7.181}$$

with $|A| = m$ any fixed positive integer, are a complete set of orthogonal projections on V_0. That is,

$$\Pi_A^* = \Pi_A = \Pi_A^2, \quad \Pi_A \Pi_B = \delta_B^A \, \Pi_A, \quad \sum_{|A|=m} \Pi_A = I_{V_0}. \tag{7.182}$$

It follows that any $f \in V_0$ can be decomposed as

$$f = \sum_{|A|=m} \Pi_A f \equiv \sum_{|A|=m} f_A, \tag{7.183}$$

with $\langle f_A, f_B \rangle = 0$ whenever $B \neq A$. The subband filtering scheme in (7.92) and (7.93) can now be generalized. In the first recursion, an arbitrary $u \in V_0$ is decomposed by

$$u = \sum_{|A|=m} F_A^* u^A, \quad \text{where} \quad u^A \equiv F_A u. \tag{7.184}$$

Repeating this gives

$$\begin{aligned} u^{A_1 A_2 \cdots A_r} &= \sum_{|A_{r+1}|=m} F_{A_{r+1}}^* u^{A_1 A_2 \cdots A_{r+1}}, \\ \text{where} \quad u^{A_1 A_2 \cdots A_{r+1}} &\equiv F_{A_{r+1}} u^{A_1 A_2 \cdots A_r}. \end{aligned} \tag{7.185}$$

The essential difference between the wavelet decomposition (7.92)–(7.93) and the above wavelet packet decomposition is that in the former, only the "averages" get decomposed further, whereas in the latter, both averages and detail are subjected to further decomposition.

Exercises

7.1. In this exercise it is shown how to construct functions $\theta(u)$ such as those used in the definition of the Meyer MRA and wavelet basis. Let k be a positive integer, and let $P_k(x)$ be an odd polynomial of degree $2k+1$:

$$P_k(x) = a_0 x + a_1 x^3 + \cdots + a_k x^{2k+1}. \tag{7.186}$$

We impose the conditions $P_k(1) = 1$ and $P_k^{(m)}(1) = 0$ for $m = 1, 2, \ldots, k$. This gives $k+1$ linear equations in the $k+1$ coefficients, with a unique solution. For example,

$$P_1(x) = \tfrac{3}{2}\, x - \tfrac{1}{2}\, x^3, \quad P_2(x) = \tfrac{15}{8}\, x - \tfrac{5}{4}\, x^3 + \tfrac{3}{8}\, x^5. \tag{7.187}$$

Now define

$$\theta(u) = \frac{\pi}{4} \times \begin{cases} 0 & \text{if } u \leq \frac{1}{3} \\ \Big[1 + P_k(6u-3)\Big] & \text{if } \frac{1}{3} \leq u \leq \frac{2}{3} \\ 2 & \text{if } u \geq \frac{2}{3}. \end{cases} \tag{7.188}$$

(a) Prove that $\theta(u)$ is k times continuously differentiable.

(b) Verify that $\theta(u)$ satisfies (7.149).

(c) Find $P_3(x)$.

7.2. Verify (7.153).

7.3. Prove that if V_0 is real, then the complex structure of Section 7.4 is skew-adjoint, i.e., $J^* = -J$. (Use the orthonormality condition.)

7.4. Prove (7.176).

7.5. Prove (7.180).

Chapter 8

Daubechies' Orthonormal Wavelet Bases

Summary: In this chapter we construct finite filter sequences leading to a family of scaling functions ϕ^N and wavelets ψ^N, where $N = 1, 2, \ldots$. $N = 1$ gives the Haar system, and $N = 2, 3, \ldots$ give multiresolution analyses with orthonormal bases of continuous, compactly supported wavelets of increasing support width and increasing regularity. Each of these systems, first obtained by Daubechies, is optimal in a certain sense. We then examine some ways in which the dilation equations with the above filter sequences determine the scaling functions, both in the frequency domain and in the time domain. In the last section we propose a new algorithm for computing the scaling function in the time domain, based on the statistical concept of *cumulants*. We show that this method also provides an alternative scheme for the construction of filter sequences.

Prerequisites: Chapters 1 and 7.

8.1 Construction of FIR Filters

In Chapter 7, we saw that a multiresolution analysis (MRA) is completely determined by a scaling function satisfying the orthonormality and averaging conditions and the dilation equation $D\phi = h(T)\phi$, with $h(T)$ given. ϕ, in turn, is determined up to normalization by $h(T)$ or, equivalently, by the filter coefficients h_n. So it makes sense to begin with $\{h_n\}$ in order to construct the MRA and the associated wavelet basis. When all but a finite number of the h_n's vanish, the same is true of the g_n's. Then $h(T)$ and $g(T)$ are called *finite impulse response* (FIR) *filters.* The reason for this terminology is as follows: Since $h(T)$ and $g(T)$ operate by multiplication in the frequency domain, they are filters in the usual engineering sense, i.e., convolution operators. When applied to ϕ, which corresponds (under the sampling operator $S_0 : L^2(\mathbf{R}) \to \ell^2(\mathbf{Z})$) to the "impulse" sequence $\{\delta_n^0\}_{n\in\mathbf{Z}}$, these operators give the sequences $\{h_n\} = S_0 h(T)\phi$ and $\{g_n\} = S_0 g(T)\phi$, respectively. Hence $\{h_n\}$ and $\{g_n\}$ are called the *impulse responses* of $h(T)$ and $g(T)$.

The scaling function and wavelet corresponding to a FIR filter are compactly supported; hence they give very good localization in the time domain. The simplest example of a multiresolution analysis with FIR filters is that associated with the Haar system (Examples 7.2 and 7.7). Although (or *because*) it is extremely well localized in time, the Haar system is far from ideal. Its discontinuities give rise to very slow decay in frequency, consequently to poor frequency resolution. We have already seen examples of MRAs with good frequency resolution (the Shannon and Meyer systems), but they were at the other extreme of the uncertainty principle since they had very poor time localization. In particular, they have infinite impulse responses. Until recently, the Haar MRA was the *only* example of an orthonormal FIR system. In this chapter, following Daubechies (1988), we construct a whole *sequence* of low-pass FIR filters $h^N(T)$, $N = 1, 2, 3, \ldots$, and their associated high-pass FIR filters $g^N(T)$. The associated scaling functions and wavelets, obtained by solving $D\phi^N = h^N(T)\phi^N$ and $D\psi^N = g^N(T)\phi^N$, are supported on $[0, 2N-1]$. The first member of this sequence gives the Haar system ϕ^1, ψ^1. For $N \geq 2$, the Daubechies filters are generalizations of the Haar system possessing more and more regularity in the time domain with increasing N. For example, ϕ^2 is continuous and ϕ^3 is continuously differentiable. (However, ϕ^4 is *not* twice continuously differentiable. The number of continuous derivatives increases only fractionally with N, in the sense of Hölder exponents; see below.)

To avoid the frequent occurrence of $\sqrt{2}$, it is convenient to introduce $c_n = \sqrt{2}\, h_n$ and $d_n = \sqrt{2}\, g_n$, and let

$$\begin{aligned} P(T) &\equiv \tfrac{1}{2} \sum_n c_n T^n = \frac{1}{\sqrt{2}}\, h(T) \\ Q(T) &\equiv \tfrac{1}{2} \sum_n d_n T^n = \frac{1}{\sqrt{2}}\, g(T) = -T^{\lambda} P(-T)^* \end{aligned} \tag{8.1}$$

(see (7.58)). The Haar filters are now $P(T) = (I+T)/2$, $Q(T) = (I-T)/2$. For $N \geq 2$, the strategy is to find polynomials $P(T)$ satisfying the averaging property (7.99) and orthonormality condition (7.98), which now take the form

$$P(1) = 1, \qquad |P(z)|^2 + |P(-z)|^2 = 1, \qquad z = e^{-2\pi i\omega}. \tag{8.2}$$

By translating ϕ if necessary, we may assume that the first nonzero coefficient is c_0, so

$$P(z) = \tfrac{1}{2} \sum_{n=0}^{M} c_n z^n, \tag{8.3}$$

with $c_0, c_M \neq 0$. We assume that the c_n's are real, which implies that ϕ and ψ are real. By (8.2),

$$\sum_{n=0}^{M} c_n = 2, \qquad \sum_{n=0}^{M-2k} c_n c_{n+2k} = 2\delta_k^0, \quad k = 0, 1, \ldots, [M/2], \tag{8.4}$$

where $[M/2]$ denotes the largest integer $\leq M/2$. If M is even, then the second equality with $k = M/2$ gives $c_0 c_M = 0$, which contradicts the assumption that $c_0, c_M \neq 0$. Hence M *must be odd*, and we take $M = 2N-1$ with $N \geq 1$, so that $[M/2] = N-1$. Then (8.4) gives $N+1$ equations in the $2N$ unknowns

$c_0, c_1, \ldots, c_{2N-1}$. When $N = 1$ (so $N+1 = 2N$), the unique solution is $c_0 = c_1 = 1$, which gives the Haar system. When $N \geq 2$, we can add $N-1$ more constraints to (8.4). This freedom can be used to obtain wavelets with better regularity in the time domain, which then give better frequency resolution than the Haar system. (This is in accordance with the uncertainty principle: The support widths of ϕ and ψ will be seen to be $2N-1$, hence the larger we take N, the more decay $\hat{\phi}$ and $\hat{\psi}$ can have in the frequency domain. In turn, this means more smoothness for ϕ and ψ.)

To find an appropriate set of additional constrains, note that the averaging condition $P(1) = 1$, together with orthonormality, implies that $P(-1) = 0$. By (7.137) and (7.138), $m_0(\omega) = P(e^{-2\pi i\omega})$ and the dilation equation implies

$$\hat{\phi}(k) = P(e^{-i\pi k})\hat{\phi}(k/2) = 0, \qquad k \in \mathbf{Z} \text{ odd}. \tag{8.5}$$

When k is even but not zero, then $k = 2^\ell k'$ for some integers ℓ, k' with k' odd, and (8.5) (applied repeatedly) implies $\hat{\phi}(k) = \hat{\phi}(k') = 0$. Thus $\hat{\phi}(k) = \delta_k^0$ for all $k \in \mathbf{Z}$. (See also Proposition 7.9.) Recall that this property allowed us to interpret ϕ as a kind of "scattered" low-pass filter, since it meant that $\hat{\phi}$ tends to concentrate near $\omega = 0$, within the constraints allowed by the orthogonality condition $\sum_k |\hat{\phi}(\omega + k)|^2 = 1$. The additional constrains will now be chosen to improve this concentration by assuming that the zero of $P(z)$ at $z = -1$ is actually of order N. Then $P(z)$ can be factored as

$$P(z) = \left(\frac{1+z}{2}\right)^N W(z) = \left(\frac{1+z}{2}\right)^N \sum_{n=0}^{N-1} w_n z^n. \tag{8.6}$$

We have chosen the normalization so that the averaging condition reads $W(1) = 1$. An argument similar to the above now shows that the derivatives of $\hat{\phi}$ satisfy

$$\hat{\phi}^{(m)}(k) = 0, \qquad m = 0, 1, \ldots, N-1,\ 0 \neq k \in \mathbf{Z}. \tag{8.7}$$

That means that $\hat{\phi}$ is very *flat* at integers $k \neq 0$, further improving the separation of low frequencies from high frequencies. The larger N is, the better the expected frequency resolution. (Note that the ideal low-pass filter of Example 7.10 satisfies (8.7) with $N = \infty$.)

Equation (8.6) has several equivalent and important consequences, the most well known of which is that *the wavelet ψ associated with ϕ has N zero moments*, i.e.,

$$\int_{-\infty}^{\infty} dt \cdot t^m \psi(t) = 0, \qquad m = 0, 1, \ldots, N-1. \tag{8.8}$$

For $m = 0$ this is, of course, just the admissibility condition, which can be interpreted as stating that ψ has at least *some* oscillation, unlike a Gaussian (or ϕ, for that matter). The vanishing of the first N moments of ψ implies more and more oscillation with increasing N. See Daubechies (1992); see also

Beylkin, Coifman, and Rokhlin (1991) for applications to signal compression. To see what (8.6) means for the filter coefficients, note that

$$(z\partial_z)^m P(z) = \sum_n n^m c_n \, z^n .$$

Hence by (8.6),

$$(z\partial_z)^m P(-1) = \sum_{n=0}^{2N-1} (-1)^n n^m c_n = 0, \quad m = 0, 1 \ldots, N-1. \tag{8.9}$$

This is a simple set of equations generalizing the admissibility condition, which is the special case $m = 0$. It can be restated in terms of the high-pass filter coefficients $d_n = (-1)^n c_{2N-1-n}$ as

$$\sum_n n^m d_n = 0 \quad m = 0, 1 \ldots, N-1.$$

In this form, it is a discrete version of (8.8).

Our task now is to find polynomials of the form (8.6) satisfying the orthonormality condition. The set of all such polynomials can be found using some rather sophisticated theorems of algebra. We follow a shortcut suggested in Strichartz (1993), showing that the solution for $N = 1$, which is the Haar system, actually generates solutions for all $N \geq 2$. For $N = 1$, the unique solution is

$$\begin{aligned} P(z) &= (1+z)/2 = e^{-i\pi\omega}\cos(\pi\omega) \\ P(-z) &= (1-z)/2 = ie^{-i\pi\omega}\sin(\pi\omega). \end{aligned} \tag{8.10}$$

To find a solution for any $N \geq 2$, introduce the abbreviations

$$C(z) \equiv \cos(\pi\omega) = \frac{\bar{z}^{1/2} + z^{1/2}}{2}, \quad S(z) \equiv \sin(\pi\omega) = \frac{\bar{z}^{1/2} - z^{1/2}}{2i}, \tag{8.11}$$

and raise the identity $C^2 + S^2 = 1$ to the power $2N-1$:

$$1 = (C^2 + S^2)^{2N-1} = \sum_{k=0}^{2N-1} \binom{2N-1}{k} C^{4N-2-2k} S^{2k}, \tag{8.12}$$

where $\binom{M}{k} = M!/k!(M-k)!$ are the binomial coefficients. The sum on the right contains $2N$ terms. Let $\mathcal{P}_N(C)$ be the sum of the *first* N terms (using $S^2 = 1 - C^2$), a polynomial of degree $4N-2$:

$$\mathcal{P}_N(C) \equiv \sum_{k=0}^{N-1} \binom{2N-1}{k} C^{4N-2-2k}(1-C^2)^k \geq 0. \tag{8.13}$$

If we can find a polynomial $P(z)$ such that $|P(z)|^2 = \mathcal{P}_N(C)$, then $|P(-z)|^2$ is automatically the sum of the *last* N terms in (8.12) and the orthonormality condition in (8.2) is satisfied. (This follows from the fact that $C(-z) = -S(z)$ and $S(-z) = C(z)$.) Note that (8.13) can be factored as

$$\mathcal{P}_N(C) = C^{2N}\mathcal{W}_N(S) = \left|\frac{1+z}{2}\right|^{2N} \mathcal{W}_N(S), \tag{8.14}$$

where

$$\mathcal{W}_N(S) = \sum_{k=0}^{N-1} \binom{2N-1}{k} (1-S^2)^{N-1-k} S^{2k} \geq 0 \tag{8.15}$$

and $\mathcal{W}_N(0) = 1$. In order to obtain a filter $P(z)$ of the form (8.6), it only remains to find a "square root" $W(z)$ of $\mathcal{W}_N(S)$, i.e.,

$$W(z) = \sum_{n=0}^{N-1} w_n z^n \quad \text{such that} \quad |W(z)|^2 = \mathcal{W}_N(S), \tag{8.16}$$

with $W(1) = 1$ (which is consistent with $\mathcal{W}_N(0) = 1$) and real coefficients w_n (since c_n must be real). The existence of such $W(z)$'s is guaranteed by a Lemma due to F. Riesz, but the idea will become clear by working out the construction of $W(z)$ in the cases $N = 2$ and $N = 3$, where it will also emerge that $W(z)$ is not unique if $N > 1$.

Example 8.1: Daubechies filters with $N = 2$. For $N = 2$, (8.15) gives

$$\mathcal{W}_2(S) = 1 + 2S^2 = 2 - \frac{z + \bar{z}}{2}. \tag{8.17}$$

$W(z)$ must be a first-order polynomial $W(z) = a + bz$, with $a, b \in \mathbf{R}$. Thus $|W(z)|^2 = \mathcal{W}_2(S)$ implies

$$a^2 + b^2 = 2, \ 2ab = -1, \ \Rightarrow \ (a+b)^2 = 1, \ (a-b)^2 = 3. \tag{8.18}$$

Two independent choices of sign can be made, giving four possibilities, but only two of these satisfy $W(1) = 1$:

$$W_\pm(z) = \tfrac{1}{2}\left[(1 \pm \sqrt{3}) + (1 \mp \sqrt{3})z\right]. \tag{8.19}$$

The Daubechies filters with $N = 2$ are therefore

$$P_\pm(z) = \tfrac{1}{2}\left(\frac{1+z}{2}\right)^2 \left[(1 \pm \sqrt{3}) + (1 \mp \sqrt{3})z\right]. \tag{8.20}$$

Aside from the zero of order 2 at $z = -1$, $P_\pm(z)$ has a simple zero at $z_\pm = 2 \pm \sqrt{3}$. z_- is inside the unit circle, while z_+ is outside.

This "square-root process" $|W(z)|^2 \to W(z)$ is called *spectral factorization* in the electrical engineering literature, and the multiplicity in the choice of solutions is typical. The filter whose zeros (not counting those at $z = -1$) are all chosen to be *outside* the unit circle is called the *minimum-phase filter*, and the filter with all the zeros *inside* the unit circle is called the *maximum-phase filter.* These two filters are related by inversion in the complex z-plane:

$$P_-(z) = z^{2N-1}P_+(z^{-1}) = \left(\frac{1+z}{2}\right)^N z^{N-1}W_+(z^{-1}), \tag{8.21}$$

and the filter coefficients are related by

$$c_n^- = c_{2N-1-n}^+, \qquad w_n^- = w_{N-1-n}^+. \tag{8.22}$$

This means that the corresponding scaling functions are related by reflection about the midpoint $t = N - 1/2$ of their support:

$$\phi^-(t) = \phi^+(2N-1-t). \tag{8.23}$$

Because of this simple relation, there is no essential difference between the maximum and minimum phase filters. Both are therefore commonly called *extremal phase filters.* Thus we see that, up to inversion, there is just *one* Daubechies filter with $N = 2$.

Suppose that $\phi(t)$ is symmetric about $t = t_0$, i.e., $\phi(2t_0 - t) = \phi(t)$. In the frequency domain this means that

$$e^{-4\pi i\omega t_0}\hat{\phi}(-\omega) = \hat{\phi}(\omega). \tag{8.24}$$

Since $\hat{\phi}(-\omega) = \overline{\hat{\phi}(\omega)}$ because $\phi(t)$ is real, it follows that the phase of $\hat{\phi}(\omega)$ is

$$\hat{\phi}(\omega) = e^{-2\pi\, i\omega t_0}\, |\hat{\phi}(\omega)|. \tag{8.25}$$

For this reason, filters with a symmetric impulse response are called *linear phase filters.* In some applications (image coding, for example), linear phase filters are prefered. The Haar scaling function is a linear phase filter with $t_0 = 1/2$, as can be seen directly from $\phi(t) = \chi_{[0,1]}(t)$ or from $\hat{\phi}(\omega) = e^{-i\pi\omega}\ \mathrm{sinc}\ (\pi\omega)$. For a symmetric Daubechies scaling function with support $[0, 2N-1]$, we necessarily have $t_0 = N - 1/2$. By (8.22), the filter coefficients would have to satisfy $c_{2N-1-n} = c_n$. It is shown in Daubechies (1992), Theorem 8.1.4, that this cannot happen when $N > 1$. On the other hand, it is possible to construct symmetric, compactly supported *biorthogonal* wavelet bases, as done in Daubechies, Chapter 8. *The combination of orthogonality, compact support, and reality is incompatible with linearity of phase*, except in the trivial case $N = 1$.

Example 8.2: Daubechies filters with $N = 3$. Equation (8.15) gives

$$\mathcal{W}_3(S) = 1 + 3S^2 + 6S^4 = \tfrac{1}{8}\left[38 - 18(\bar{z} + z) + 3(\bar{z}^2 + z^2)\right]. \tag{8.26}$$

Setting $W(z) = a + bz + cz^2$ with $a, b, c \in \mathbf{R}$ and solving $|W(z)|^2 = \mathcal{W}_3(S)$ gives

$$P_\pm(z) = \left(\frac{1+z}{2}\right)^3 W_\pm(z), \qquad W_\pm(z) = a_\pm + bz + c_\pm z^2 \tag{8.27}$$

(exercise), where

$$\begin{aligned} a_\pm &= \tfrac{1}{4}\left(1+\sqrt{10}\pm\sqrt{5+2\sqrt{10}}\right) \\ b &= \tfrac{1}{2}\left(1-\sqrt{10}\right) \\ c_\pm &= \tfrac{1}{4}\left(1+\sqrt{10}\mp\sqrt{5+2\sqrt{10}}\right). \end{aligned} \tag{8.28}$$

Only these two solutions (out of eight solutions) survive the requirement that $W(z)$ be real with $W(1)=1$. Again, $P_+(z)$ has minimum phase and $P_-(z)$ has maximum phase, so there is effectively only one Daubechies filter with $N=3$. For $N\geq 4$, however, the spectral factorization also gives "in-between filters" that have neither minimum nor maximum phase. It is then possible to choose the zeros of $W(z)$ so that $\phi(t)$ is "least asymmetric" about $t=N-1/2$, although complete symmetry is impossible, as already mentioned. (See Daubechies (1992), Figure 6.4 and Section 8.1.1.)

The reader will have guessed by now that numerical methods must be used to obtain the Daubechies filters with large N. The values of the coefficients for the minimum-phase filters are listed in Daubechies' book on page 195.

Actually, the above procedure does not give the most general solution of (8.2) and (8.6). If $\mathcal{W}_N(S)$ is replaced by

$$\mathcal{W}(S)\equiv\mathcal{W}_N(S)+S^{2N}R\left(2-4S^2\right)=\mathcal{W}_N(S)+S^{2N}R\left(\bar{z}+z\right), \tag{8.29}$$

with R any *odd* polynomial such that the positivity condition $\mathcal{W}(S)\geq 0$ is retained for $|S|\leq 1$, then solving $\mathcal{W}(S)=|W(z)|^2$ for $W(z)$ by spectral factorization still leads to a solution $P(z)$, since

$$\begin{aligned} |P(z)|^2+|P(-z)|^2 &= C^{2N}\mathcal{W}_N(S)+C^{2N}S^{2N}R(\bar{z}+z) \\ &\quad + S^{2N}\mathcal{W}_N(C)+S^{2N}C^{2N}R(-\bar{z}-z), \end{aligned} \tag{8.30}$$

and the second and fourth terms on the right cancel. Indeed, this gives the *most general* solution. If the degree of R is M, then that of $\mathcal{W}(S)$ is $2N+2M$ and that of $W(z)$ is $N+M$. Hence the degree of $P(z)$ is $2N+M$, which is odd as necessary since M is odd by assumption. The support width of both ϕ and ψ is then also $2N+M$. On the other hand, we could have chosen a filter with $2\tilde{N}\equiv 2N+M+1$ coefficients and a zero of order $\tilde{N}$ at $z=-1$, in which case any of the resulting scaling functions and wavelets $\tilde{\phi}$ and $\tilde{\psi}$ have support width $\tilde{2N}-1=2N+M$, the same as ϕ and ψ. That is, *scaling functions and wavelets with $R\neq 0$ do not have the minimal support width for the given order of their zero at $z=-1$.* The additional freedom gained by requiring fewer than N factors of $(1+z)/2$ for a filter with $2N$ coefficients can be used to make ϕ and ψ more "regular," i.e., smoother, than $\tilde{\psi}$. (Again, this is the uncertainty principle at work: We can gain smoothness by widening the support width.) For a more detailed discussion of these ideas, we refer the reader to Sections 6.1, 7.3, and 7.4 in Daubechies' book.

A completely different and beautifully simple method of arriving at $|P(z)|^2$ has been developed recently by Strang (1992). Yet another new way is the main topic in Section 8.4.

8.2 Recursive Construction of ϕ in the Frequency Domain

Except for the trivial case $N = 1$, none of the Daubechies scaling functions and wavelets can be displayed explicitly. All are determined recursively from their filter coefficients, and in no case does this lead to a closed-form expression for ϕ or ψ. In fact, these functions display a *fractal geometry*, even though they are continuous for $N > 1$. (For example, the set R_α of all $t \in \mathbf{R}$ at which ϕ^2 has a given Hölder exponent α, with $.55 \leq \alpha \leq 1$, is a fractal set; see Daubechies, Section 7.1. This fact alone indicates that no formula can exist that gives ϕ or ψ in terms of elementary or special functions.) There are several recursive methods for constructing ϕ up to any desired precision. Once ϕ is known to a certain scale, ψ can be computed to the next finer scale by $\psi = D^{-1}g(T)\phi$, i.e.,

$$\psi(t) = \sum_n d_n \phi(2t - n). \tag{8.31}$$

In this section we describe a recursive scheme in the frequency domain. This uses the polynomial $P(e^{-2\pi i\omega}) \equiv m_0(\omega)$ to build up $\hat{\phi}(\omega)$ scale by scale. When the desired accuracy is obtained in the frequency domain, the inverse Fourier transform must still be applied to obtain the corresponding approximation to $\phi(t)$. By Plancherel's theorem, the global error in the time domain (measured by the norm) is the same as that in the frequency domain.

A repeated application of the dilation equation (7.137) gives

$$\hat{\phi}(\omega) = m_0(\omega/2)\hat{\phi}(\omega/2) = \prod_{m=1}^{M} m_0(\omega/2^m)\hat{\phi}(\omega/2^M). \tag{8.32}$$

As $M \to \infty$, $\hat{\phi}(\omega/2^M) \to \hat{\phi}(0) = 1$,† and we obtain $\hat{\phi}(\omega)$ formally as an infinite product:

$$\hat{\phi}(\omega) = \prod_{m=1}^{\infty} m_0(\omega/2^m) = \prod_{m=1}^{\infty} P\left(e^{-2\pi i\omega/2^m}\right). \tag{8.33}$$

This shows that a given set of filter coefficients can determine *at most one* scaling function (given the normalization $\hat{\phi}(0) = 1$). The sense in which the infinite product converges must now be examined to determine the nature of $\hat{\phi}$.

† Recall that $\hat{\phi}(\omega)$ is an entire-analytic function since ϕ is compactly supported.

It must also be shown that $\hat{\phi}(\omega)$, now *defined* by (8.33), is actually the Fourier transform of a function in the time domain, and the nature of this function must be investigated. Obviously, $m_0(0) = P(1) = 1$ is a necessary condition for *any* kind of convergence, since otherwise the factors do not approach 1. If $m_0(0) = 1$, then

$$m_0(\omega) = 1 + \tfrac{1}{2}\sum_n c_n(e^{-2\pi in\omega} - 1) = 1 - i\sum_n c_n e^{-\pi in\omega}\sin(\pi n\omega). \tag{8.34}$$

Hence

$$|m_0(\omega)| \le 1 + \sum_n |c_n||\sin(\pi n\omega)| \le 1 + A|\omega| \le e^{A|\omega|}, \tag{8.35}$$

where $A = \pi\sum_n |nc_n|$ (since $|\sin x| \le |x|$), and

$$\prod_{m=1}^{\infty} |m_0(\omega/2^m)| \le \exp\left(A\,|\omega|\sum_{n=1}^{\infty} 2^{-m}\right) = e^{A|\omega|}. \tag{8.36}$$

This shows that the infinite product converges absolutely and uniformly on compact subsets. In fact, it was shown by Deslauriers and Dubuc (1987) to converge to an entire function of exponential type that (by the Payley-Wiener theorem for distributions) is the Fourier transform of a distribution supported in $[0, 2N-1]$. However, this in itself is not very reassuring from a practical standpoint, since ϕ could still be very singular in the time domain. (Consider the case $P(z) \equiv 1$, which satisfies $P(1) = 1$ but not the orthogonality condition (8.2). It gives $\hat{\phi}(\omega) \equiv 1$, hence $\phi(t) = \delta(t)$.) Mallat (1989) proved that if $P(z)$ furthermore satisfies the orthogonality condition (8.2), then $\hat{\phi}$ belongs to $L^2(\mathbf{R})$ and has norm $\|\hat{\phi}\| \le 1$. Hence the averaging and orthonormality properties for $P(z)$ imply that $\phi \in L^2(\mathbf{R})$ (by Plancherel's theorem). For ψ to generate an orthonormal basis in $L^2(\mathbf{R})$ in the context of multiresolution analysis, we only need to know that ϕ satisfies the orthonormality condition $\phi_n^*\phi = \delta_n^0$ (Chapter 7), although some additional regularity would be welcome. However, this is *not* the case in general! The reason is that although the condition (8.2) on the filter $P(z)$ was *necessary* for orthonormality, it may not be *sufficient.* (This is somewhat reminiscent of the situation in differential equations, where a local solution need not extend to a global one; roughly speaking, $P(z)$ is a "local" version of $\hat{\phi}(\omega)$!) As a counterexample (borrowed from Daubechies' book), consider

$$P(z) = \left(\frac{1+z}{2}\right)(1 - z + z^2) = \frac{1+z^3}{2} = e^{-3\pi i\omega}\cos(3\pi\omega). \tag{8.37}$$

The averaging and orthonormality conditions (8.2) are both satisfied, and the infinite product is easy to compute in closed form. (This fact alone is enough to raise the suspicion that something is wrong!) Using

$$P(z) = \frac{1-z^6}{2(1-z^3)}, \tag{8.38}$$

we have

$$\prod_{m=1}^{M} m_0(\omega/2^m) = \prod_{m=1}^{M} \frac{1 - e^{-12\pi i\omega/2^m}}{2(1 - e^{-6\pi i\omega/2^m})} = \frac{1 - e^{-6\pi i\omega}}{2^M(1 - e^{-6\pi i\omega/2^M})}. \tag{8.39}$$

Hence

$$\hat{\phi}(\omega) = \lim_{M\to\infty} \frac{1 - e^{-6\pi i\omega}}{2^M(1 - e^{-6\pi i\omega/2^M})} = e^{-3\pi i\omega}\,\mathrm{sinc}\,(3\pi\omega) \tag{8.40}$$

and

$$\phi(t) = \begin{cases} 1/3 & 0 \le t \le 3 \\ 0 & \text{otherwise.} \end{cases} \tag{8.41}$$

This is just a stretched version of the Haar scaling function. Obviously, the translates $\phi_n(t)$ are *not* orthonormal. In fact, it turns out that

$$\sum_k |\hat{\phi}(\omega + k)|^2 = \tfrac{1}{3} + \tfrac{4}{9}\cos(2\pi\omega) + \tfrac{2}{9}\cos(4\pi\omega), \tag{8.42}$$

which vanishes for $\omega = 1/3$. This shows that $\{\phi_n\}$ is not even a "Riesz basis" for its linear span V_0, which means that, although the ϕ_n's are linearly independent, instabilities can occur: It is possible to have linear combinations $\sum_n u_n\phi_n$ with small norm for which $\sum_n |u_n|^2$ is arbitrarily large (exercise). (Furthermore, the coefficients u_n that produce a given $f \in V_0$ cannot be obtained simply by taking the inner products with ϕ_n, due to the lack of orthonormality; we must use a reciprocal basis instead.) Consequently, ϕ does not generate a reasonable multiresolution analysis. If the recipe for the MRA construction of the wavelet basis is applied, the result is not a basis but still a *tight frame*, with frame constant 3. This is just the redundancy equal to the stretch factor in (8.40).

The above example shows that additional conditions are required on $P(z)$ (aside from (8.2)) to ensure that the ϕ_n's are orthonormal and, consequently, that $\{\psi_{m,n}\}$ is an orthonormal basis. Although conditions have been found by Cohen (1990) and Lawton (1991) that are both necessary and sufficient, they are somewhat technical. A useful *sufficient* condition, also due to Cohen, is the following:

Theorem 8.3. *Let $P(z)$ satisfy the averaging and orthonormality conditions (8.2), and let ϕ be the corresponding scaling function defined by (8.32). Then the translates $\phi_n(t) \equiv \phi(t-n)$ are orthonormal if $m_0(\omega) \equiv P(e^{-2\pi i\omega})$ has no zeros in the interval $|\omega| \le 1/6$.*

See Daubechies' book (Corollary 6.3.2) for the proof. Note that in the above counterexample, $m_0(\pm 1/6) = 0$. This proves that the bound $1/6$ on $|\omega|$ is optimal, i.e., no lower value exists that guarantees orthonormality.

The above considerations are unnecessarily general for our present purpose, since they do not use the special form (8.6) of $P(z)$. We now examine the regularity of the Daubechies scaling functions ϕ^N, $N \geq 2$, constructed from the FIR filters $P(z)$ given by (8.6), with $W(z)$ obtained from (8.15) and (8.16) or (8.29) by spectral factorization.

How does the factor $A(z)^N \equiv ((1+z)/2)^N$ affect the regularity of ϕ? Before going into technicalities, note that the dilation equation in the time domain is now

$$\begin{aligned} D\phi &= \sqrt{2}\, P(T)\phi = \sqrt{2}\, A(T)^N W(T)\phi \\ &= \sqrt{2}\, W(T) A(T)^N \phi. \end{aligned} \tag{8.43}$$

Since each factor $A(T)$ performs an *averaging operation* on ϕ, i.e., $\phi(t) \to \frac{1}{2}[\phi(t) + \phi(t-1)]$, $A(T)^N\phi$ can be expected to be quite *smooth.* Furthermore, if the support of ϕ is $[a, b]$, then $A(T)^N\phi$ will be supported in $[a, b+N]$. The polynomial $W(T)$ (being of degree $N-1$) further spreads the support to $[a, b+2N-1]$. Since the support of $D\phi$ is $[2a, 2b]$, we have $2a = a$ and $2b = b + 2N - 1$, which shows that $a = 0$ and $b = 2N-1$; hence the support of ϕ is $[0, 2N-1]$ as claimed earlier. This is independent of the special form (8.43) of $P(T)$, depending only on the fact that $P(T)$ is a polynomial of degree $2N-1$. The point of interest now is the smoothing operation performed by $A(T)^N$. The larger N is, the smoother we can expect $D\phi$ (and hence also ϕ) to be, *provided* that $W(T)$ does not undo the smoothing effect of $A(T)^N$ through "differencing" by alternating terms. To get a quantitative picture, we go to the frequency domain, where T acts simply as multiplication by $z = e^{-2\pi i\omega}$. We must examine the effect of $A(z)^N$ on the infinite product in (8.33):

$$\hat{\phi}(\omega) = \left[\prod_{m=1}^{\infty} \left(\frac{1 + e^{-2\pi i\omega/2^m}}{2}\right)\right]^N \prod_{m=1}^{\infty} W(e^{-2\pi i\omega/2^m}). \tag{8.44}$$

The first factor has already been computed in (8.38) through (8.40), with z replaced by z^3. Letting $\omega \to \omega/3$ in (8.40), we have

$$\prod_{m=1}^{\infty} \left(\frac{1 + e^{-2\pi i\omega/2^m}}{2}\right) = e^{-i\pi\omega}\,\mathrm{sinc}\,(\pi\omega) = \frac{e^{-i\pi\omega}\sin(\pi\omega)}{\pi\omega}. \tag{8.45}$$

Let

$$F(\omega) \equiv \prod_{m=1}^{\infty} W(e^{-2\pi i\omega/2^m}). \tag{8.46}$$

Since $W(1) = 1$, the estimates (8.34) through (8.36) show that the infinite product (8.46) converges absolutely (and uniformly on compact subsets). Since $W(z)$ is a polynomial of degree $N-1$, the theorem of Deslauriers and Dubuc (1987) invoked earlier for $P(z)$ shows that $F(\omega)$ is an entire function of exponential type that is the Fourier transform of a distribution $\check{F}(t)$ with support in $[0, N-1]$. We have

$$\hat{\phi}(\omega) = \frac{e^{-i\pi N\omega}\sin^N(\pi\omega)}{\pi^N\omega^N}\, F(\omega). \tag{8.47}$$

Now everything boils down to the behavior of $F(\omega)$. Since both factors in (8.47) are analytic, only the behavior of $F(\omega)$ as $|\omega| \to \infty$ concerns us, since this will determine the regularity of the inverse Fourier transform of $\hat{\phi}$. Suppose that $|F(\omega)|$ grows like $|\omega|^\beta$ as $|\omega| \to \infty$, for some $\beta \in \mathbf{R}$, which we denote by $|F(\omega)| \sim |\omega|^\beta$. Then $|\hat{\phi}(\omega)| \sim |\omega|^{\beta-N}$. If ϕ is to be continuous in the time domain, it suffices for $\hat{\phi}$ to be absolutely integrable, and this will be the case if $\beta - N < -1$, i.e., $\beta = N - 1 - \varepsilon$ for some $\varepsilon > 0$. Similarly, if $\beta = N - 1 - k - \varepsilon$ for some integer $k \geq 1$, then $\omega^k \hat{\phi}(\omega)$ is absolutely integrable. Since the k-th derivative of $\phi(t)$ is related to $\hat{\phi}(\omega)$ by

$$\phi^{(k)}(t) = \int_{-\infty}^{\infty} d\omega \; (2\pi i\omega)^k \, e^{2\pi\, i\omega t} \, \hat{\phi}(\omega), \tag{8.48}$$

it follows that $\phi^{(k)}$ is continuous, denoted as $f \in C^k$. What if $\beta = N - 1 - \alpha - \varepsilon$ for some positive *noninteger* α? Let $\alpha = k + \gamma$ with k being a nonnegative integer and $0 < \gamma < 1$. Then $\omega^k \hat{\phi}$ is absolutely integrable, and hence $\phi^{(k)}(t)$ is continuous. Although $\phi^{(k+1)}(t)$ is not continuous, it is clear that ϕ is more regular than the mere statement $\phi \in C^k$ would allow. This suggests a notion something like a "fractional derivative," which can be made precise as follows: A function $f(t)$ is said to be *Hölder continuous* (also called *Lipschitz continuous*) with exponent γ if

$$|f(t+h) - f(t)| \leq C|h|^\gamma \quad \text{for all } t, h \in \mathbf{R}. \tag{8.49}$$

Obviously f is then also uniformly continuous. We denote the (vector) space of all such functions by C^γ. This definition applies only for $0 < \gamma < 1$. The choice $\gamma = 1$ (which is not allowed in (8.49)) would make f almost differentiable, but not quite. (Take $f(t) = |t|$, for example.) To see the relation between $f \in C^\gamma$ (defined in the frequency domain) and differentiability (defined in the time domain), note that

$$\frac{f(t+h) - f(t)}{|h|^\gamma} = \int_{-\infty}^{\infty} d\omega \; e^{2\pi\, i\omega t} \, \frac{e^{2\pi\, i\omega h} - 1}{|h|^\gamma} \, \hat{f}(\omega). \tag{8.50}$$

It is not difficult to show that if $\hat{f}(\omega)$ is absolutely integrable and if $|\hat{f}(\omega)| \sim |\omega|^{-1-\gamma-\varepsilon}$, then the integral on the right-hand side is absolutely convergent and bounded by a constant C independent of h; hence $f \in C^\gamma$. That gives a close relation between the rate of decay of f in frequency and its smoothness in time. The definition (8.49) can be extended naturally to arbitrary $\alpha \geq 0$ as follows. If $\alpha = k + \gamma$ with $k \in \mathbf{N}$ and $0 \leq \gamma < 1$, then f is said to be Hölder continuous with exponent α if $f \in C^k$ and $f^{(k)} \in C^\gamma$. (When $\gamma = 0$, this reduces to $f \in C^k$, since C^0 is the space of continuous functions.) We then write $f \in C^\alpha$.

The relevance of all this to the regularity of ϕ is as follows: Equation (8.47) shows that if the growth of $F(\omega)$ can be controlled, then $\hat{\phi}(\omega)$ will decay accordingly. This can then be used to give information about the regularity of ϕ

in the time domain, in the form of Hölder continuity. In order to prove that ϕ belongs to C^α, it is sufficient to show that $|F(\omega)| \sim |\omega|^{N-1-\alpha-\varepsilon}$ for some $\varepsilon > 0$. Thus, one looks for conditions on $W(z)$ that guarantee such growth estimates for $|F(\omega)|$. The following result is taken from Daubechies' book (Section 7.1.1).

Theorem 8.4. *Suppose that*

$$\max_{|z|=1} |W(z)| < 2^{N-1-\alpha} \tag{8.51}$$

for some $\alpha > 0$. Then $\phi \in C^\alpha$.

For example, let $N = 1$ and $W(z) \equiv 1$ (Haar case). Then $\max|W(z)| = 1$, and α would have to be negative in order for (8.51) to hold. Thus Theorem 8.4 does not even imply continuity for ϕ^1 (which is, in fact, discontinuous). More interesting is the case of ϕ^2, where $N = 2$ and

$$W(z) = a + bz, \quad a = \frac{1 \pm \sqrt{3}}{2}, \quad b = \frac{1 \mp \sqrt{3}}{2}. \tag{8.52}$$

With $z = \cos(2\pi\omega) + i\sin(2\pi\omega)$, we obtain

$$|W(z)|^2 = 2 - \cos(2\pi\omega), \quad \text{hence} \quad \max|W(z)| = \sqrt{3}. \tag{8.53}$$

To find α from (8.51), first solve $\sqrt{3} = 2^{2-1-\mu} = \exp[(1-\mu)\ln 2]$, which gives

$$\mu = 1 - \frac{\ln 3}{2\ln 2} \approx .2075. \tag{8.54}$$

Thus (8.51) holds with any $\alpha < \mu$, showing that ϕ^2 is (at least) continuous. This does not prove that ϕ^2 is not differentiable, since (8.51) is only a *sufficient condition* for $\phi \in C^\alpha$. Actually, much better estimates can be obtained by more sophisticated means, but these are beyond the scope of this book. (See Daubechies' book, Section 7.2.) It is even possible to find the *optimal* value of α, which turns out to be $\approx .550$ for ϕ^2. Furthermore, it can be shown that ϕ^2 possesses a whole hierarchy of *local* Hölder exponents. A local exponent $\gamma(t)$ for f at $t \in \mathbf{R}$ satisfies (8.49) with t fixed; clearly the best local exponent is greater than or equal to any global exponent, and it can vary from point to point, since f may be smoother at some points than others. It turns out that the best local Hölder exponent $\alpha(t)$ for ϕ^2 ranges between .55 and 1. Especially interesting are the points $t = k/2^m$ with $k, m \in \mathbf{Z}$ and $m \geq 0$, called *diadic rationals.* At such points, the Hölder exponent flips from its maximum of 1 (as t is approached from the left) to its minimum of .55 (as t is approached from the right). In fact, ϕ^2 can be shown to be left-differentiable but not right-differentiable at all diadic rationals! This behavior can be observed by looking at the graph of ϕ^2 at $t = 1$ and $t = 2$, displayed at the end of the chapter.

By a variety of techniques involving refinements of Theorem 8.4 and detailed analyses of the polynomials $\mathcal{W}_N$ in (8.15), it is possible to show that all the Daubechies minimal-support scaling functions ϕ^N with $N \geq 2$ are continuous and their (global) Hölder exponents increase with N, so they become more and more smooth. In particular, it turns out that ϕ^N is continuously differentiable for $N \geq 3$. For large N, it can be shown that $\phi^N \in C^\alpha$ with

$$\alpha = \left(1 - \frac{\ln 3}{2\ln 2}\right) N = \mu N. \tag{8.55}$$

(Again, this does not mean that α is optimal!) The factor μ was already encountered in (8.54) in connection with ϕ^2. Thus, for large N, we gain about $\mu \approx 0.2$ "derivatives" every time N is increased by 1.

8.3 Recursion in Time: Diadic Interpolation

The infinite product obtained in the last section gives $\hat{\phi}(\omega)$, and the inverse Fourier transform must then be applied to obtain $\phi(t)$. In particular, that construction is highly *nonlocal*, since each recursion changes $\phi(t)$ at all values of t. If we are mainly intersted in finding ϕ near a particular value t_0, for example, we must nevertheless compute ϕ to the same precision at all t. In this section we show how ϕ can be constructed locally in time. This technique, like that in the frequency domain, is recursive. However, rather than giving a better and better approximation at every recursion, it gives the *exact values* of $\phi(t)$. Where is the catch? At every recursion, we have $\phi(t)$ only at a *finite set* of points: first at the integers, then at the half-integers, then at the quarter-integers, etc.

To obtain ϕ at the integers, note that for $t = n$ the dilation equation reads

$$\phi(n) = \sum_{k=0}^{2N-1} c_k \phi(2n-k) = \sum_{k=0}^{2N-1} c_{2n-k}\phi(k), \tag{8.56}$$

with the convention that $c_m = 0$ unless $0 \leq m \leq 2N-1$. Both sides involve the values of ϕ only at the integers; hence (8.56) is a linear equation for the unknowns $\phi(n)$. If ϕ is supported in $[0, 2N-1]$ and is continuous, then only $\phi(1), \phi(2), \ldots, \phi(2N-2)$ are (possibly) nonzero. Thus we have $M \equiv 2N-2$ unknowns, and (8.56) becomes an $M \times M$ matrix equation, stating that the column vector $u \equiv [\phi(1) \cdots \phi(M)]^*$ is an eigenvector of the matrix $A_{n,k} \equiv c_{2n-k}$, $n, k = 1, \ldots, M$. The normalization of u is determined by the partition of unity (Proposition 7.9). This gives the exact values of ϕ at the integers. From here on, the computation is completely straightforward. At $t = n + .5$, the dilation equation gives

$$\phi(n+.5) = \sum_{k=0}^{2N-1} c_k \phi(n+1-k), \tag{8.57}$$

which determines the exact values of ϕ at the half-integers, and so on. Any solution of $Au = u$ thus determines ϕ at all the diadic rationals $t = k/2^m$ ($k, m \in \mathbf{Z}$, $m \geq 0$), which form a dense set. If ϕ is continuous, that determines ϕ everywhere. Conversely, any solution of the dilation equation determines a solution of $Au = u$. Since we already know that the dilation equation has a unique normalized solution, it follows that u is unique as well, i.e., the eigenvalue 1 of A is nondegenerate.

Example 8.5: The Scaling Function with $N = 2$. Taking the minimum-phase filter of Example 8.1, we have

$$P(z) = \left(\frac{1+z}{2}\right)^2 (a + bz) \equiv \tfrac{1}{2} \sum_{n=0}^{3} c_n z^n, \tag{8.58}$$

where $a = (1+\sqrt{3})/2$ and $b = (1-\sqrt{3})/2$. This gives

$$c_0 = \frac{1+\sqrt{3}}{4}, \quad c_1 = \frac{3+\sqrt{3}}{4}, \quad c_2 = \frac{3-\sqrt{3}}{4}, \quad c_3 = \frac{1-\sqrt{3}}{4}. \tag{8.59}$$

$Au = u$ becomes

$$\begin{bmatrix} c_1 & c_0 \\ c_3 & c_2 \end{bmatrix} \begin{bmatrix} \phi(1) \\ \phi(2) \end{bmatrix} = \begin{bmatrix} \phi(1) \\ \phi(2) \end{bmatrix}. \tag{8.60}$$

Together with the normalization $\sum_k \phi(k) = 1$, this gives the unique solution

$$\phi(1) = \frac{1+\sqrt{3}}{2}, \qquad \phi(2) = \frac{1-\sqrt{3}}{2}. \tag{8.61}$$

Hence

$$\begin{aligned} \phi(.5) &= \sum_{k=0}^{3} c_k \phi(1-k) = c_0 \phi(1) = \frac{2+\sqrt{3}}{4} \\ \phi(1.5) &= \sum_{k=0}^{3} c_k \phi(3-k) = c_1 \phi(2) + c_2 \phi(1) = 0 \\ \phi(2.5) &= \sum_{k=0}^{3} c_k \phi(5-k) = c_3 \phi(2) = \frac{2-\sqrt{3}}{4}. \end{aligned} \tag{8.62}$$

The scaling functions ϕ^N with minimal support ($R = 0$) and minimal phase are plotted at the end of the chapter. Figures 8.1 through 8.4 show ϕ^N for $N = 2$, 3, 4, and 5, along with their wavelets ψ^N. Note that for $N = 2$ and 3, ϕ and ψ are quite rough at the diadic rationals, for example, the integers. As N increases and the support widens, these functions become smoother.

8.4 The Method of Cumulants

Since $\int dt\, \phi(t) = 1$, we may think of $\phi(t)$ as something like a probability distribution. This idea, although consistent with the interpretation of ϕ as an "averaging function," is not strictly valid because $\phi(t)$ can assume negative values. The probability picture will therefore serve merely as *motivation* for the approach suggested here. We have referred to $\phi_n^* f$ as "samples" of f. But since ϕ_n is supported on the interval $n \le t \le n+2N-1$, the above "probability" picture suggests that $\phi_n^* f$ should not be the sample of f at $t=n$ but at $t=\tau+n$, where

$$\tau \equiv \int_{-\infty}^{\infty} dt \cdot t\,\phi(t) \tag{8.63}$$

is the *mean time* with respect to ϕ. That is, the "probability" picture suggests

$$\text{Condition } E': \qquad \phi_n^* f = f(n+\tau), \quad n \in \mathbf{Z}. \tag{8.64}$$

Obviously, (8.64) cannot hold for all $f \in L^2(\mathbf{R})$, since f does not have well-defined values (Section 1.3). For $f = \phi$, it implies

$$\text{Condition } E: \qquad \phi(\tau+n) = \phi_n^* \phi = \delta_n^0, \quad n \in \mathbf{Z}. \tag{8.65}$$

Conversely, (8.65) implies (8.64) for every $f \in V_0$. For if $f = u(T)\phi$, then

$$\begin{aligned} f(\tau+n) &= \sum_k u_k \phi_k(\tau+n) \\ &= \sum_k u_k \phi(\tau+n-k) = u_n = \phi_n^* f. \end{aligned} \tag{8.66}$$

Definition 8.6. *A scaling function is* exact *if it satisfies Condition E.*

Note that the Haar scaling function $\chi_{[0,1)}(t)$ is exact, with $\tau = .5$. There is no obvious reason why (8.65) should hold for any $N > 1$, and we will see that in general it is only an approximation, although an unexpectedly good one. But for $N=2$ and $N=3$, there is a surprise: Condition E appears to hold *exactly!* At the time of this writing, I have only a partial proof of this (see below). Nevertheless, (8.65) is useful for the following reasons:

(a) Whether exact or not, Condition E gives a new *algorithm* for constructing and plotting ϕ. This algorithm is a refinement of diadic interpolation, and is somewhat simpler than the latter because no eigen-equation needs to be solved at the beginning.

(b) To the extent that it is exact, Condition E gives a simple and direct interpretation to the filter coefficients c_n of ϕ: They are *samples* of ϕ at $t = (\tau+n)/2$.

(c) Research into the validity of Condition E leads to a new method for constructing filter coefficients for orthonormal wavelets, and possibly other multiresolution analyses, based on the statistical concept of *cumulants*. This method is independent of the validity of Condition E.

This section gives a brief account of the cumulant method. The new algorithm for constructing ϕ from the filter coefficients begins by interpreting Condition E as defining the *initial values* of ϕ in a recursion scheme:

$$\phi^{(0)}(\tau + n) \equiv \delta_n^0. \tag{8.67}$$

This is analogous to finding $\phi(n)$ in the diadic interpolation scheme, except that no eigen-equation needs to be solved. If ϕ is exact, we are ahead of diadic interpolation. If not, (8.67) is still a very good initial guess leading to rapid convergence. From (8.67) we proceed exactly as in diadic interpolation, except that we cannot assume to have the exact values at any stage. The first iteration is

$$\phi^{(1)}\left(\frac{\tau + n}{2}\right) \equiv \sum_k c_k \phi^{(0)}(\tau + n - k) = c_n\,, \tag{8.68}$$

and each iteration is obtained from the previous one by

$$\phi^{(m+1)}\left(\frac{\tau + n}{2^{m+1}}\right) = \sum_k c_k \phi^{(m)}\left(\frac{\tau + n}{2^m} - k\right). \tag{8.69}$$

If Condition E holds, then $\phi^{(m)} = \phi$ for all $m \geq 0$ and (8.68) shows that the filter coefficients c_n are *values* of ϕ at $t = (\tau + n)/2$. (This gives the c_n's a nice intuitive interpretation!)

In order to use this scheme we need to find τ in terms of the filter coefficients. We claim that

$$\tau = \tfrac{1}{2} \sum_n n c_n\,, \tag{8.70}$$

which is just a *discrete* analogue of (8.63) using $\{c_n/2\}$ as a discrete probability distribution on $\mathbf{Z}$. (Recall that $\sum_n c_n = 2$.) Equation (8.70) will be seen to be a special case of a relation that leads, as it turns out, to the heart of the matter. Since ϕ is integrable with compact support, we can define

$$\Phi(s) = \int_{-\infty}^{\infty} dt\; e^{st}\, \phi(t) = \hat{\phi}\left(\frac{is}{2\pi}\right). \tag{8.71}$$

(Remember that $\hat{\phi}(\omega)$ is an entire function.) Then (8.33) becomes

$$\Phi(s) = \prod_{m=1}^{\infty} P\left(e^{s/2^m}\right), \tag{8.72}$$

and the dilation equation reads

$$\Phi(2s) = P(e^s)\Phi(s). \tag{8.73}$$

Since $\Phi(s)$ is an entire function with $\Phi(0) = 1$, there is a neighborhood D_Φ of $s = 0$ where a branch of $\log \Phi(s)$ can be defined. A similar argument shows that

$\log P(e^s)$ is also defined in some neighborhood D_P of $s=0$; hence $\log P\left(e^{s/2^m}\right)$ is defined in D_P for all $m \geq 0$. Equation (8.72) implies that

$$\log \Phi(s) = \sum_{m=1}^{\infty} \log P\left(e^{s/2^m}\right) \tag{8.74}$$

in $D_\Phi \cap D_P$. Define the *moments* of ϕ about $t=0$ by

$$M_k \equiv \int_{-\infty}^{\infty} dt\, t^k \phi(t), \qquad k = 0, 1, 2, \ldots. \tag{8.75}$$

In particular, $M_0 = 1$ and $M_1 = \tau$. $\Phi(s)$ is a *generating function* for these moments, since

$$\partial_s^k \Phi(s)\Big|_{s=0} = M_k\,. \tag{8.76}$$

The dilation equation (8.73) gives a recursion relation for the moments that allows their computation from the filter coefficients. But a more direct relation exists in terms of logarithms because of the multiplicative nature of (8.73). Define the *continuous and discrete cumulants* for $k \geq 0$ by

$$\begin{aligned} K_k &\equiv \partial_s^k \log \Phi(s)\Big|_{s=0} \\ \kappa_k &\equiv \partial_s^k \log P(e^s)\Big|_{s=0}, \end{aligned} \tag{8.77}$$

respectively. Note that

$$\begin{aligned} K_0 &= \log \Phi(0) = 0 \\ K_1 &= \frac{\Phi'(0)}{\Phi(0)} = \tau \\ K_2 &= M_2 - \tau^2 = \int_{-\infty}^{\infty} dt\, (t-\tau)^2 \phi(t). \end{aligned} \tag{8.78}$$

If ϕ were a true probability distribution (i.e., $\phi(t) \geq 0$), then K_2 would be its second central moment and $\sqrt{K_2}$ its standard deviation. Because ϕ can actually assume negative values, our use of cumulants stretches the usual concept. See Gnedenko and Kolmogorov (1954) for definitions and properties of cumulants, there called "semi-invariants." We now derive a relation that will allow us to find K_k in terms of the filter coefficients. Recall that

$$P(e^s) = \tfrac{1}{2} \sum_n c_n e^{ns}, \tag{8.79}$$

and define the *discrete moments* (of the c_n's) by analogy with (8.76):

$$\mu_k \equiv \partial_s^k P(e^s)\Big|_{s=0} = \tfrac{1}{2} \sum_n c_n n^k\,. \tag{8.80}$$

The discrete cumulants and discrete moments satisfy relations similar to (8.78), the first three of which are

$$\begin{aligned} \kappa_0 &= \log P(1) = 0 \\ \kappa_1 &= \frac{P'(1)}{P(1)} = \mu_1 \\ \kappa_2 &= \mu_2 - \mu_1^2 \,. \end{aligned} \tag{8.81}$$

This allows us to compute the κ's from the c's, and all that remains is to relate these to the K's.

Theorem 8.7. *The continuous cumulants are given in terms of the discrete cumulants by*

$$K_k = \frac{\kappa_k}{2^k - 1}, \qquad k \geq 0. \tag{8.82}$$

Proof: By (8.74),

$$\begin{aligned} K_k &= \sum_{m=1}^{\infty} \partial_s^k \log P(e^{s/2^m}) \Big|_{s=0} \\ &= \sum_{m=1}^{\infty} 2^{-mk} \partial_u^k \log P(e^u) \Big|_{u=0} \\ &= \sum_{m=1}^{\infty} 2^{-mk} \kappa_k = \frac{\kappa_k}{2^k - 1} \,. \quad \blacksquare \end{aligned} \tag{8.83}$$

For $k = 1$, (8.82) reduces to (8.70), and we can proceed with the modified diadic interpolation scheme. The results for the minimum-phase, minimum-support filters with $N = 2$, 3, 4, and 5 are shown in Figures 8.1 through 8.4. Figures 8.1 and 8.2 suggest that ϕ^2 and ϕ^3 are exact, while Figures 8.3 and 8.4 show very small deviations from exactness for ϕ^4 and ϕ^5. Similar computations for the minimum-phase, minimum-support filters with $N = 9$ and 10, and other Daubechies filters with nonminimal phase or nonminimal support, show that although obviously not exact, Condition E still gives a good set of initial values leading to rapid convergence.

The new derivation of the filter coefficients from cumulants is based on the following result:

Theorem 8.8. *Let* $P(z) = .5 \sum_n c_n z^n$ *be an entire function with* c_n *real. Suppose* $P(z)$ *satisfies*

$$|P(z)|^2 + |P(-z)|^2 = 1 \quad \text{for} \quad |z| = 1 \tag{8.84}$$

and has a zero of order N *at* $z = -1$*:*

$$P(z) = \left(\frac{1+z}{2} \right)^N W(z), \tag{8.85}$$

with $W(z)$ entire. Then

$$\kappa_{2k} \equiv \partial^{2k} \log P(e^s)\Big|_{s=0} = 0, \quad 0 \leq k < N. \tag{8.86}$$

Proof: Since the c_n's are real, $\overline{P(z)} = P(\bar{z})$. Equation (8.84) implies that

$$P(e^s)P(e^{-s}) + P(-e^s)P(-e^{-s}) = 1 \quad \text{for all } s \in \mathbf{C}. \tag{8.87}$$

(For $s = -2\pi i\omega$ with ω real, (8.87) reduces to (8.84); since the left-hand side of (8.87) is an entire function of s and equals 1 on the imaginary axis, it has to equal 1 identically for all s.) But

$$P(-e^s) = \left(\frac{1-e^s}{2}\right)^N W(-e^s). \tag{8.88}$$

Since

$$\frac{1-e^s}{2} = s\,U(s) \tag{8.89}$$

where $U(s)$ is entire, (8.88) gives

$$P(-e^s) = s^N\, U(s)^N W(-e^s) \equiv s^N V(s), \tag{8.90}$$

with $V(s)$ entire. Therefore

$$\begin{aligned} P(e^s)P(e^{-s}) &= 1 - P(-e^s)P(-e^{-s}) \\ &= 1 - s^{2N}\,(-1)^N V(s)V(-s) \\ &\equiv 1 + s^{2N} F(s) \end{aligned} \tag{8.91}$$

for some entire function $F(s)$. Now for $|w| < 1$,

$$\log(1+w) = w - \tfrac{1}{2}w^3 + \tfrac{1}{3}w^3 - \cdots \equiv wH(w), \tag{8.92}$$

where $H(w)$ is analytic in $|w| < 1$. Therefore there exists a neighborhood D of $s = 0$ where

$$\log\left[P(e^s)P(e^{-s})\right] = s^{2N} F(s)\, H\left(s^{2N} F(s)\right) \equiv s^{2N} G(s), \tag{8.93}$$

with $G(s)$ analytic in D. Recall that we can also define a branch of $\log P(e^s)$ in a neighborhood D_P of $s = 0$. We may assume that D_P is circular, so that $-s \in D_P$ whenever $s \in D_P$. Then

$$\log P(e^s) + \log P(e^{-s}) = s^{2N} G(s) \tag{8.94}$$

in $D \cap D_P$. Now apply ∂^k and set $s = 0$. The right-hand side vanishes because it still has factors of s left after the differentiation. Hence by the definition of κ_k,

$$\kappa_k + (-1)^k \kappa_k = 0, \qquad k < 2N. \tag{8.95}$$

For odd k, (8.95) is an identity; for $k = 0, 2, 4, \ldots, 2N-2$, it reduces to (8.86). ∎

Therefore, a necessary condition for orthonormality is that the first N even, discrete cumulants κ_{2k} vanish. Then Theorem 8.7 implies that the corresponding continuous cumulants also vanish. In particular, $K_2 = 0$ and our "probability distribution" $\phi(t)$ must have a vanishing "variance"!

Theorem 8.8 can be used as a basis for the derivation of orthonormal multiresolution filters. For example, if ϕ is to have a minimal compact support for a given value of N, then $W(z)$ in (8.85) must be a polynomial of degree $N-1$, which means it has N coefficients. But (8.86) gives N equations for these coefficients, so we can find the c_n's by solving (8.86). For $N = 2$ and $N = 3$, this gives the Daubechies filter coefficients.

For an orthonormal, minimal-support filter, the first N even cumulants thus determine the filter coefficients (by vanishing). The *odd* cumulants, on the other hand, play an important role in relation to exactness. It can be shown (Kaiser [1994d]) that if a filter satisfies (8.85), then

$$\begin{aligned} N \geq 1 &\Rightarrow \sum_n \phi(\tau+n) = 1 \\ N \geq 2 &\Rightarrow \sum_n n\phi(\tau+n) = 0 \\ N \geq 3 &\Rightarrow \sum_n n^2\phi(\tau+n) = 0 \\ N \geq 4 &\Rightarrow \sum_n n^3\phi(\tau+n) = K_3 = \frac{\kappa_3}{7}, \end{aligned} \tag{8.96}$$

with similar equations for $N \geq 5, N \geq 6$, etc. For any $N \geq 1$, we may regard the $2N-1$ values of $\phi(\tau+n)$ in the support of ϕ as unknowns. Then (8.96) gives N equations for these unknowns, leaving $N-1$ degrees of freedom. The case $N = 1$ is trivial, and implies that the Haar scaling function is exact. For $N = 2$, we find $\tau = .634$.[†] The first two equations in (8.96) are *consistent* with exactness for ϕ^2 but not quite enough to prove it, since there are three unknowns $\phi(\tau)$, $\phi(\tau+1)$, and $\phi(\tau+2)$. For $N = 3$ and minimum phase, we have $\tau = .817$. The first three equations are again consistent with exactness, but now there are five unknowns $\phi(\tau), \ldots, \phi(\tau+4)$, so we are again short of a proof. The fourth equation proves that ϕ^N *cannot be exact for* $N \geq 4$, *unless* $\kappa_3 = 0$. But $\kappa_3 = -.109$ for ϕ^4 (minimum-phase), so this filter is not exact. The fact that it is *close* to being exact, as evidenced by Figure 8.3, is consistent with the small value of $\kappa_3/7 = -.0156$ in (8.96). Higher-order odd cumulants further spoil the exactness of higher-order filters. One might say that they are *obstructions* to exactness.

[†] This is for the minimum-phase filter; for the maximum-phase filter we have $\phi(t) \rightarrow \phi(2N-1-t)$, so $\tau \rightarrow 2N-1-\tau = 2.366$.

Conjecture 8.9. *ϕ^2 and ϕ^3 are exact.*

Equations (8.96) can be derived from the Poisson summation formula in combination with (8.7); they are related to the beautiful Strang–Fix approximation theory (Strang and Fix [1973]; Strang [1989]), which has been generalized recently by de Boor, DeVore, and Ron (1994).

8.5 Plots of Daubechies Wavelets

Figures 8.1 through 8.4 show the results of computing the minimal-support, minimal-phase Daubechies scaling functions and wavelets by the cumulant algorithm, using four iterations. The initial values of ϕ are marked with $\circ$, and the first three iterations are marked with $*$, $\times$, and $+$. The fourth iteration for ϕ is a solid curve, and the wavelet ψ as computed from the third iteration of ϕ is a dashed curve. The vertical dotted line is $t = \tau$.

Exercises

8.1. Verify that $P(z)$, defined by (8.6), satisfies the orthonormality condition (8.2) if $W(z)$ is a solution of (8.16), with $\mathcal{W}_N(S)$ given by (8.15).

8.2. Prove (8.23).

8.3. Let ϕ be the scaling function (8.41). For any $M \in \mathbf{N}$, let $u_{3n} = 1, u_{3n+1} = -1$, $n = 0, 1, \ldots, M-1$, with all other $u_k = 0$. Compute and compare $\sum_k |u_k|^2$ and $\| \sum_k u_k \phi_k \|^2$. What happens as $M \to \infty$? Explain why ϕ is "bad."

8.4. Find the wavelet corresponding to the scaling function (8.41).

8.5. Compute the next recursion for ϕ from (8.62), i.e., find $\phi(.25), \phi(.75), \ldots, \phi(2.75)$.

8.6. Use a computer to find and plot $\phi^2(t)$ by the method of Section 8.2: Compute $\hat{\phi}$ to some accuracy, then take its inverse Fourier transform. (Experiment with the accuracy.)

8.7. Use a computer to find and plot $\phi^2(t)$ by diadic interpolation, to the scale $\Delta t = 2^{-4}$.

8.8. Plot the minimum-phase, minimum-support scaling function $\phi^{10}(t)$ by the method of cumulants, using three iterations. (The coefficients $h_n = c_n/\sqrt{2}$ are listed in Daubechies (1992), p. 195.)

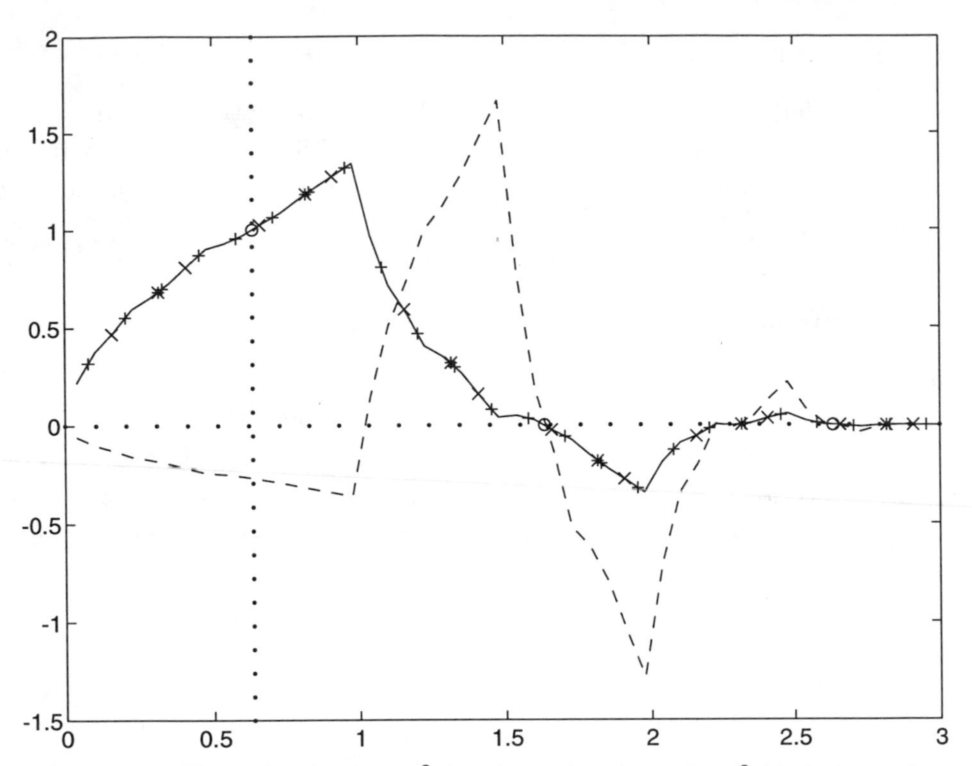

Figure 8.1. The scaling function ϕ^2 (*solid curve*) and wavelet ψ^2 (*dashed curve*).

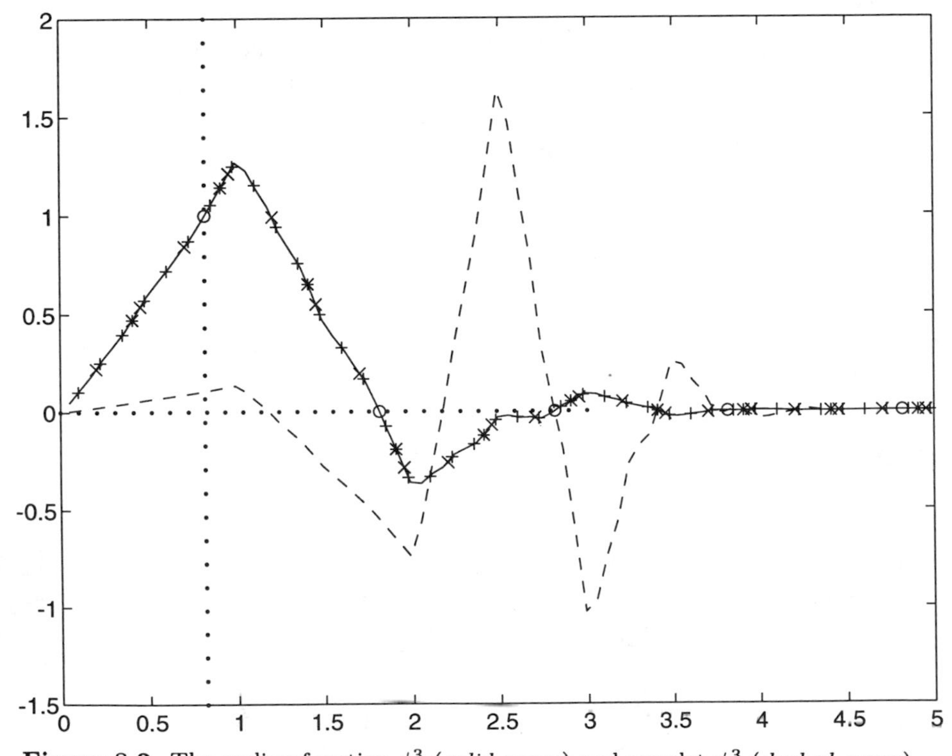

Figure 8.2. The scaling function ϕ^3 (*solid curve*) and wavelet ψ^3 (*dashed curve*).

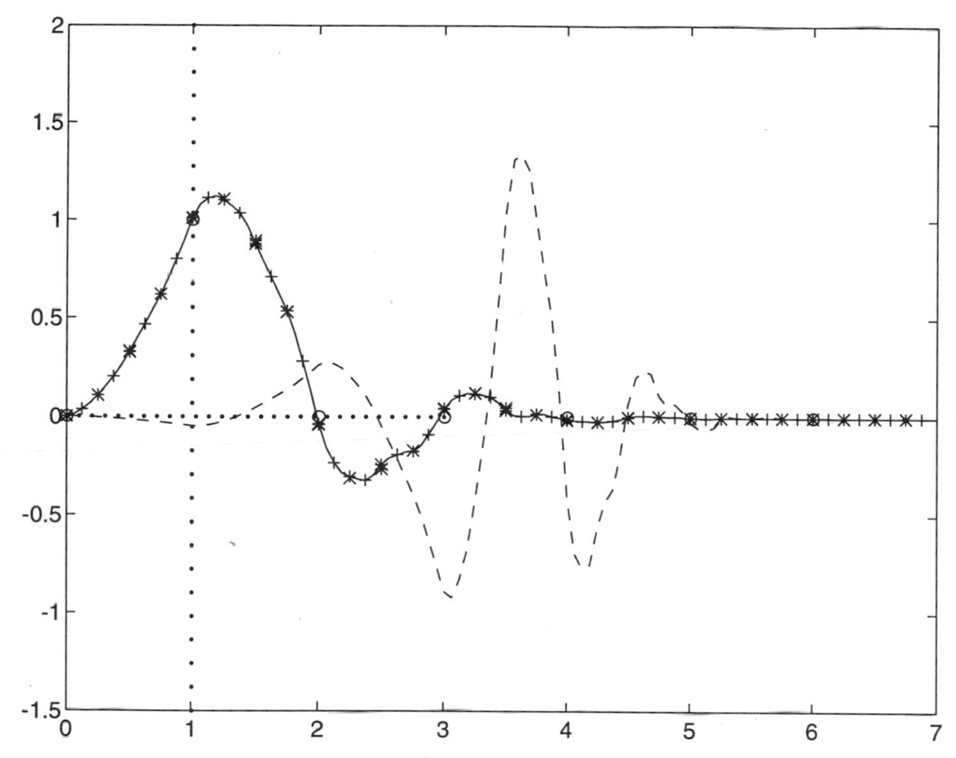

Figure 8.3. The scaling function ϕ^4 (*solid curve*) and wavelet ψ^4 (*dashed curve*).

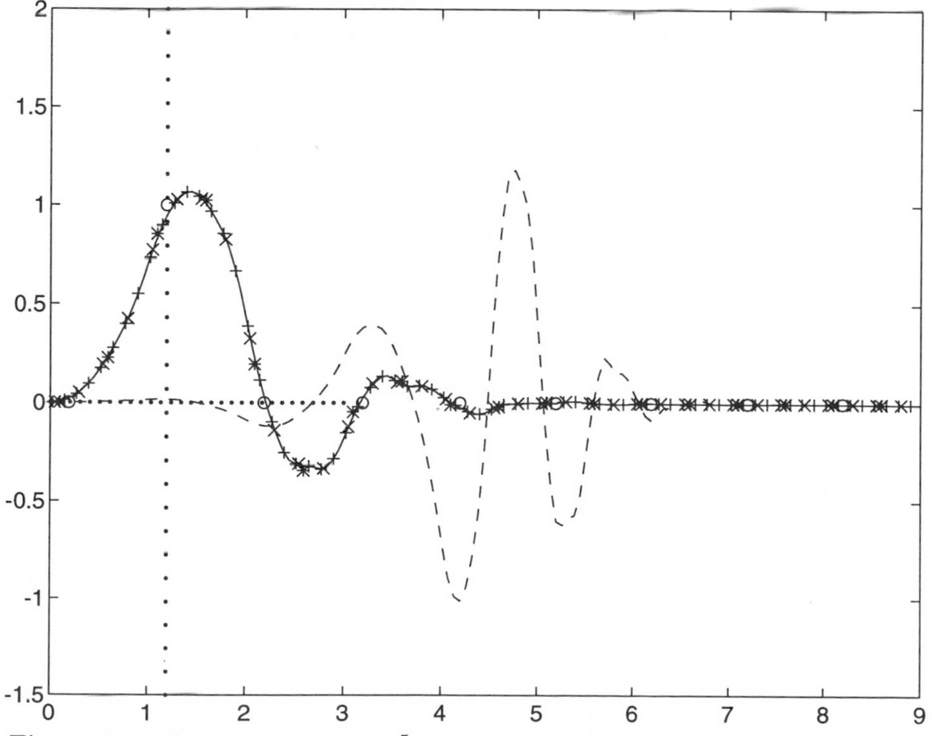

Figure 8.4. The scaling function ϕ^5 (*solid curve*) and wavelet ψ^5 (*dashed curve*).

PART II

Physical Wavelets

Chapter 9

Introduction to Wavelet Electromagnetics

Summary: In this chapter, we apply wavelet ideas to electromagnetic waves, i.e., solutions of Maxwell's equations. This is possible and natural because Maxwell's equations in free space are invariant under a large group of symmetries (the *conformal group* of space-time) that includes translations and dilations, the basic operations of wavelet theory. The structure of these equations also makes it possible to extend their solutions analytically to a certain domain $\mathcal{T} \subset \mathbf{C}^4$ in complex space-time, which renders their wavelet representation particularly simple. In fact, the extension of an electromagnetic wave to complex space-time *is* already its wavelet transforms, so no further transformation is needed beyond analytic continuation! The process of continuation uniquely determines a family of *electromagnetic wavelets* $\mathbf{\Psi}_z$, labeled by points z in the complex space-time $\mathcal{T}$. These points have the form $z = x + iy$, where $x \in \mathbf{R}^4$ is a point in real space-time and y belongs to the open cone V' consisting of all causal (past and future) directions in $\mathbf{R}^4$. $\mathbf{\Psi}_z$ itself is a matrix-valued solution of Maxwell's equations, representing a *triplet* of electromagnetic waves. This triplet is *focused* at the real space-time point $x = \operatorname{Re} z \in \mathbf{R}^4$, and its *scale, helicity*, and *center velocity* are vested in the imaginary space-time vector $y = \operatorname{Im} z \in V'$. An arbitrary electromagnetic wave can be represented as a superposition of wavelets $\mathbf{\Psi}_z$ parameterized by certain subsets of $\mathcal{T}$, for example, the set E of all points with real space and imaginary time coordinates called the *Euclidean region*. Such wavelet representations make it possible to design electromagnetic waves according to local specifications in space and scale at any given initial time.

Prerequisites: Chapters 1 and 3. Some knowledge of electromagnetics, complex analysis, and partial differential equations is helpful in this chapter.

9.1 Signals and Waves

Information, like energy, has the remarkable (and very fortunate!) property that it can repeatedly change forms without losing its essence. A "signal" usually refers to encoded information, such as a time-variable amplitude or frequency modulation imprinted on an electromagnetic wave acting as a "carrier." The information is taken for a "ride" on the wave, at the speed of light, to be decoded later by a receiver. Many of the most important signals do, in fact, have to ride an electromagnetic wave at some stage in their journey. But this ride may not be entirely free, since electromagnetic waves are subject to the physical laws of propagation and scattering. The effects of these laws on communication are often ignored in discussions of signal analysis and processing. But the physical laws are, in some sense, a code in themselves – in fact, an unchangeable, *absolute* code. Therefore it would seem that an understanding of the interaction between the two codes (the physical laws and the encoded information) can be helpful. In the transmission phase, it may suggest a preferred choice of both carrier and code. In the reception phase, it may suggest better decoding methods that take full advantage of the nature of the carrier and use it to understand the precise impact of the "medium" on the "message." This is especially important when the medium *is* the message, as it can be, for example, in optics and radar. In these cases, the initial signal is often trivial (for example, a spot light) and the information sought by the receiver is the distribution and nature of reflectors, absorbers, and so on.

In order to arrange a marriage between signal analysis and physics, it is first necessary to formulate both in the same language or code. For signals, this means any choice of scheme for analysis and synthesis. For electromagnetic waves, the "code" consists of the laws of propagation and scattering, so we have little choice. It therefore makes sense to begin with the physics and try to reformulate electromagnetics in a way that is most suitable for signal analysis. The two themes for signal analysis explored in this book have been time-frequency and time-scale analysis. In this chapter we will see that the propagation of electromagnetic waves can be formulated very naturally in terms of a generalized time-scale analysis. In fact, the laws of electromagnetics uniquely determine a family of wavelets that are specifically *dedicated* to electromagnetics. These wavelets already incorporate the physical laws of propagation, and they can be used to construct arbitrary electromagnetic waves by superposition. Since the physical laws are now built into the wavelets, we can concentrate on the "message" in the synthesis stage, i.e., when choosing the coefficient function for the superposition. The analysis stage consists of finding a coefficient function for a given wave; again this isolates the message by displaying the informational contents of the wave.

Time-scale analysis is natural to electromagnetics for the following fundamental reason: Maxwell's equations in free space are invariant under a group of transformations, called the *conformal group* $\mathcal{C}$, which includes space-time translations and dilations. That is, if any solution is translated by an arbitrary amount in space or time, then it remains a solution. Similarly, a solution can be scaled uniformly in space-time and still remain a solution. Since these are precisely the kind of operations used to construct wavelet analysis, it is reasonable to expect a similar analysis to exist for electromagnetics. The other ingredient in the construction of ordinary wavelet analysis is the Hilbert space $L^2(\mathbf{R})$, chosen somewhat arbitrarily. Translations and dilations were represented by unitary operators on $L^2(\mathbf{R})$, and the wavelets formed a (continuous or discrete) frame in $L^2(\mathbf{R})$. In the case of electromagnetics, we need a Hilbert space $\mathcal{H}$ consisting of solutions of Maxwell's equations instead of unconstrained functions (such as those in $L^2(\mathbf{R})$).

But we must not blindly apply the "hammer" of wavelet analysis to this new "nail" $\mathcal{H}$,[†] since this nail actually (and not surprisingly) turns out to have much more structure than $L^2(\mathbf{R})$. It is, in fact, more like a *screw* than a nail! The extra structure is related to the other symmetries contained in the conformal group $\mathcal{C}$. Besides translations and dilations, $\mathcal{C}$ contains rotations, Lorentz transformations, and special conformal transformations. In fact, these transformations taken together essentially *characterize* free electromagnetic waves: It is possible to begin with $\mathcal{C}$ and construct $\mathcal{H}$ as a *representation space* for $\mathcal{C}$, where conformal transformations act as unitary operators (Bargmann and Wigner [1948], Gross [1964]). When constructing a wavelet-style analysis for electromagnetics, it is important to keep in mind that *most of the standard continuous analysis-synthesis schemes are completely determined by groups.* In Fourier analysis, the group consists of translations (Katznelson [1976]). In windowed Fourier analysis, it is the *Weyl-Heisenberg group* $\mathcal{W}$ consisting of translations, modulations and phase factors (Folland [1989]). In ordinary wavelet analysis, it is the *affine group* $\mathcal{A}$ consisting of translations and dilations (Aslaksen and Klauder [1968, 1969]). Since both $\mathcal{W}$ and $\mathcal{A}$ contain translations, the associated analyses are related to Fourier analysis but contain more detail. The extra detail appears in the parameters labeling the frame vectors: The (unnormalizable) "frame vectors" $e_\omega(t) = e^{2\pi i\omega t}$ in Fourier analysis are parameterized only by frequency; the frame vectors $g_{\omega,t}$ of windowed Fourier analysis are parameterized by frequency and time, and the frame vectors $\psi_{s,t}$ of ordinary wavelet analysis are parameterized by scale and time. Since the conformal group $\mathcal{C}$ contains $\mathcal{A}$, we expect that the electromagnetic wavelet analysis will be related to, but more detailed than, the standard wavelet analysis. In fact, its frame vectors will be seen to be *electromagnetic pulses* $\mathbf{\Psi}_z$ parameterized by points $z = x + iy$ in a *complex space-time domain* $\mathcal{T} \subset \mathbf{C}^4$, thus having *eight* real parameters. The real space-time point $x = (\mathbf{x}, t) \in \mathbf{R}^4$ represents the *focal point* of $\mathbf{\Psi}_z$. This means that at times prior to t, $\mathbf{\Psi}_z$ converges toward the space point $\mathbf{x}$, and at later times it diverges away from $\mathbf{x}$. At time t, $\mathbf{\Psi}_z$ is *localized* around $\mathbf{x} \in \mathbf{R}^3$. Since it is a free electromagnetic wave, it does not remain localized. But it does have a well-defined *center* that is described by the other half of z, the imaginary space-time vector $y = (\mathbf{y}, s)$. y must be *causal*, meaning that $|\mathbf{y}| < c|s|$, where

[†] I am thinking of the wonderful saying: *If the only tool you have is a hammer, then everything looks like a nail.* Until recently, of course, the only tool available to most signal analysts was the Fourier transform or its windowed version, so everything looked like either time or frequency. Now some wavelet enthusiasts claim that everything is either time or a scale.

c is the speed of light, and it has the following interpretation: $\mathbf{v} = \mathbf{y}/s$ is the *velocity* of the center of $\mathbf{\Psi}_z$, and $|s|$ is the *time scale (duration)* of $\mathbf{\Psi}_z$. Equivalently, $c|s|$ is the *spatial extent* of $\mathbf{\Psi}_z$ at its time of localization t. The sign of the imaginary time s is interpreted as the *helicity* of $\mathbf{\Psi}_z$, which is related to its polarization.

In Section 9.2 we solve Maxwell's equations by Fourier analysis. Solutions are represented as superpositions of *plane waves* parameterized by wave number and frequency. We define a Hilbert space $\mathcal{H}$ of such solutions by integration over the light cone in the Fourier domain. In Section 9.3 we introduce the *analytic-signal transform* (AST). This is simultaneously a generalization to n dimensions of two objects: the ordinary wavelet transform, and Gabor's idea of "analytic signals." The AST extends any function or vector field from $\mathbf{R}^n$ to $\mathbf{C}^n$, and we call the extended function the *analytic signal* of the original function. In general, such analytic signals are not analytic functions since there may not exist any analytic extensions. We prove that the analytic signal of any solution in $\mathcal{H}$ *is* analytic in a certain domain $\mathcal{T} \subset \mathbf{C}^4$, the *causal tube.* This analyticity is due to the light-cone structure of electromagnetic waves in the Fourier domain, just as the analyticity of Gabor's one-dimensional analytic signals is due to their having only positive-frequency components. In Section 9.4 we use the analytic signals of solutions to generate the electromagnetic wavelets $\mathbf{\Psi}_z$. We derive a resolution of unity in $\mathcal{H}$ in terms of these wavelets, which gives a means of writing an arbitrary electromagnetic wave in $\mathcal{H}$ as a superposition of wavelets. In Section 9.5 we compute the wavelets explicitly, which also gives their reproducing kernel. In Section 9.6 this explicit form of $\mathbf{\Psi}_z$ is used to give the wavelet parameters $z = x + iy$ a complete physical and geometric interpretation, as explained above. In Section 9.7 we show how general electromagnetic waves can be constructed from wavelets, with initial data specified locally and by scale.

The wavelet analysis of electromagnetics is based entirely on the homogeneous Maxwell equations. *Acoustic* waves, on the other hand, obey the scalar wave equation, which has a very similar structure to Maxwell's equations in that its symmetry group is the same conformal group $\mathcal{C}$. Therefore a similar wavelet analysis exists for acoustics as well as for electromagnetics. In the language of groups, the two analyses are just different *representations* of $\mathcal{C}$. Acoustic wavelet representations and the corresponding wavelets Ψ_z^α are studied in Chapter 11.

9.2 Fourier Electromagnetics

An electromagnetic wave in free space (without sources or boundaries) is described by a pair of vector fields depending on the space-time variables $x = (\mathbf{x}, x_0)$, namely, the electric field $\mathbf{E}(x)$ and the magnetic field $\mathbf{B}(x)$. (Here $\mathbf{x}$ is the position and $x_0 = t$ is the time.) These are subject to Maxwell's equations,

$$\begin{aligned} \nabla \times \mathbf{E} + \partial_0 \mathbf{B} = \mathbf{0}, \qquad \nabla \cdot \mathbf{E} = 0, \\ \nabla \times \mathbf{B} - \partial_0 \mathbf{E} = \mathbf{0}, \qquad \nabla \cdot \mathbf{B} = 0, \end{aligned} \tag{9.1}$$

where ∇ is the gradient with respect to the space variables and ∂_0 is the time derivative. We have set the speed of light $c = 1$ for convenience. The parameter c may be reinserted at any stage by dimensional analysis. In keeping with the style of this book, we are not concerned here with the exact nature of the "functions" $\mathbf{E}(x)$ and $\mathbf{B}(x)$ (they may be regarded as tempered distributions; see Kaiser [1994c]). Note that the equations are symmetric under the linear mapping defined by $J : \mathbf{E} \mapsto \mathbf{B}$, $\mathbf{B} \mapsto -\mathbf{E}$, and that J^2 is minus the identity map. This means that J is a "90°-rotation" in the space of solutions, analogous to multiplication by i in the complex plane. Such a mapping on a real vector space is called a *complex structure.* The combinations $\mathbf{B} \pm i\,\mathbf{E}$ diagonalize J, since $J(\mathbf{B} \pm i\,\mathbf{E}) = \pm i\,(\mathbf{B} \pm i\,\mathbf{E})$. They each map Maxwell's equations to a form in which the concepts of helicity and polarization become very simple. It will suffice to consider only $\mathbf{F} \equiv \mathbf{B} + i\,\mathbf{E}$, since the other combination is equivalent. Equations (9.1) then become

$$\partial_0 \mathbf{F} = i\nabla \times \mathbf{F}, \qquad \nabla \cdot \mathbf{F} = 0. \tag{9.2}$$

Note that the first of these equations is an evolution equation (initial-value problem), while the second is a constraint on the initial values. Taking the divergence of the first equation shows that the constraint is conserved by time evolution. Note also that it is the factor i in (9.2) (i.e., the complex structure) that couples the time evolution of the electric field $\mathbf{E}$ to that of the magnetic field $\mathbf{B}$. Equation (9.2) implies

$$-\partial_0^2 \mathbf{F} = \nabla \times (\nabla \times \mathbf{F}) = \nabla(\nabla \cdot \mathbf{F}) - \nabla^2 \mathbf{F} = -\nabla^2 \mathbf{F} \tag{9.3}$$

in Cartesian coordinates, hence the components of $\mathbf{F}$ become decoupled and each satisfies the wave equation

$$\Box\, \mathbf{F} \equiv (-\partial_0^2 + \nabla^2)\mathbf{F} = \mathbf{0}. \tag{9.4}$$

To solve (9.2), write $\mathbf{F}$ as a Fourier transform[†]

$$\mathbf{F}(x) = (2\pi)^{-4} \int_{\mathbf{R}^4} d^4 p\, e^{ip\cdot x}\, \hat{\mathbf{F}}(p), \tag{9.5}$$

[†] In this chapter we change our convention for Fourier transforms. The four-vector p now denotes wave number and frequency in *radians* per unit length or unit time. Also, we are identifying $\mathbf{R}^4$ with its dual $\mathbf{R}_4$ by means of the Lorent-invariant inner product, so the pairing between x and p is expressed as an inner product. See the discussion in Section 1.4.

where $p = (\mathbf{p}, p_0) \in \mathbf{R}^4$ with $\mathbf{p} \in \mathbf{R}^3$ as the spatial wave vector and p_0 as the frequency. We use the Lorentz-invariant inner product

$$p \cdot x \equiv p_0 x_0 - \mathbf{p} \cdot \mathbf{x}.$$

(Note also that we have reverted to the convention for the Fourier transform with the 2π factors in front of integrals instead of in the exponents. This is the convention used in the physics literature.) The wave equation (9.4) implies that $p^2 \hat{\mathbf{F}}(p) = 0$, where

$$p^2 \equiv p \cdot p = p_0^2 - |\mathbf{p}|^2.$$

Thus $\hat{\mathbf{F}}$ must be supported on the *light cone* $C = \{p : p^2 = 0, p \neq 0\}^\dagger$ shown in Figure 9.1, i.e., on

$$\begin{aligned} C &= \{(\mathbf{p}, p_0) \in \mathbf{R}^4 : p_0^2 = |\mathbf{p}|^2, \mathbf{p} \neq \mathbf{0}\} = C_+ \cup C_- \\ C_+ &= \{(\mathbf{p}, p_0) : p_0 = |\mathbf{p}| > 0\} = \text{positive-frequency light cone} \\ C_- &= \{(\mathbf{p}, p_0) : p_0 = -|\mathbf{p}| < 0\} = \text{negative-frequency light cone.} \end{aligned} \tag{9.6}$$

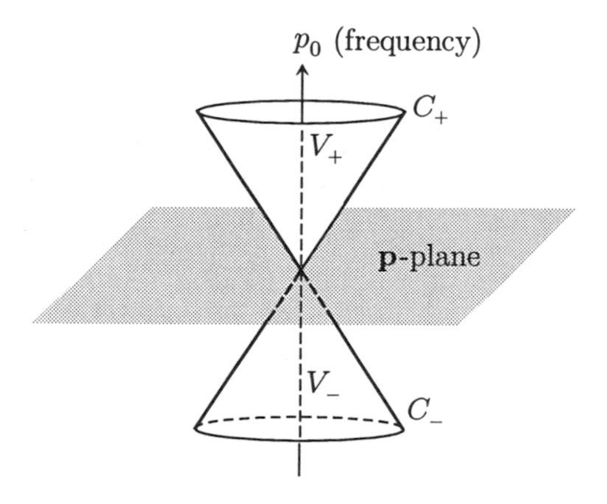

Figure 9.1. The positive- and negative-frequency cones $V_\pm$ and their boundaries, and the positive- and negative-frequency *light* cones $C_\pm = \partial V_\pm$.

Hence $\hat{\mathbf{F}}$ has the form

$$\hat{\mathbf{F}}(p) = 2\pi\delta(p^2)\, \mathbf{f}(p),$$

where $\mathbf{f} : C \to \mathbf{C}^3$ is a function on C or, equivalently, the *pair* of functions on $\mathbf{R}^3$ given by

$$\mathbf{f}_\pm(\mathbf{p}) = \mathbf{f}(\mathbf{p}, \pm\omega), \quad \text{where} \quad \omega \equiv |\mathbf{p}|$$

† We assume that the fields decay at infinity, so $\hat{\mathbf{F}}(p)$ contains no term proportional to $\delta(p)$, i.e., $\mathbf{F}(x)$ has no "DC component." To eliminate this possibility, we exclude $p = 0$ from the light cone.

is the absolute value of the frequency. But

$$\delta(p^2) = \delta((p_0 - \omega)(p_0 + \omega)) = \frac{\delta(p_0 - \omega) + \delta(p_0 + \omega)}{2\omega}. \tag{9.7}$$

Hence (9.5) becomes

$$\begin{aligned} \mathbf{F}(x) &= (2\pi)^{-3} \int_{\mathbf{R}^3} \frac{d^3\mathbf{p}}{2\omega}\, e^{-i\mathbf{p}\cdot\mathbf{x}} \left[e^{i\omega t}\, \mathbf{f}_+(\mathbf{p}) + e^{-i\omega t}\, \mathbf{f}_-(\mathbf{p})\right] \\ &= \int_C d\tilde{p}\; e^{ip\cdot x}\, \mathbf{f}(p), \end{aligned} \tag{9.8}$$

where

$$d\tilde{p} \equiv (2\pi)^{-3} d^3\mathbf{p}/2\,\omega$$

is a Lorentz-invariant measure on C. In order for (9.8) to give a solution of (9.2), $\mathbf{f}(p)$ must further satisfy the algebraic conditions

$$i\, p_0\, \mathbf{f}(p) = \mathbf{p} \times \mathbf{f}(p), \qquad \mathbf{p} \cdot \mathbf{f}(p) = 0 \tag{9.9}$$

for all $p \in C$, and the first of these equations suffices since it implies the second. Let

$$\mathbf{n}(p) = \frac{\mathbf{p}}{p_0},$$

so that $p \in C$ if and only if $|\mathbf{n}(p)| = 1$. Define the operator $\mathbf{\Gamma} \equiv \mathbf{\Gamma}(p)$ on arbitrary functions $\mathbf{g} : C \to \mathbf{C}^3$ by

$$(\mathbf{\Gamma}\mathbf{g})(p) \equiv -i\, \mathbf{n}(p) \times \mathbf{g}(p), \qquad p \in C. \tag{9.10}$$

$\mathbf{\Gamma}(p)$ is represented by the Hermitian matrix

$$\mathbf{\Gamma}(p) = \frac{i}{p_0} \begin{bmatrix} 0 & p_3 & -p_2 \\ -p_3 & 0 & p_1 \\ p_2 & -p_1 & 0 \end{bmatrix}, \tag{9.11}$$

with matrix elements $\Gamma_{mn}(p) = i p_0^{-1} \sum_{k=1}^{3} \varepsilon_{mnk}\, p_k$, where ε_{mnk} is the totally antisymmetric tensor with $\varepsilon_{123} = 1$. In terms of $\mathbf{\Gamma}(p)$, (9.9) becomes

$$\mathbf{\Gamma}\mathbf{f}(p) = \mathbf{f}(p). \tag{9.12}$$

Now for any $\mathbf{g} : C \to \mathbf{C}^3$,

$$\mathbf{\Gamma}^2\mathbf{g} = -\mathbf{n} \times (\mathbf{n} \times \mathbf{g}) = \mathbf{g} - (\mathbf{n} \cdot \mathbf{g})\mathbf{n}, \tag{9.13}$$

so $\mathbf{\Gamma}(p)^2$ is the orthogonal projection to the subspace of $\mathbf{C}^3$ orthogonal to $\mathbf{n}(p)$, and it follows that

$$\mathbf{\Gamma}^3\mathbf{g} = \mathbf{\Gamma}\mathbf{g}. \tag{9.14}$$

The eigenvalues of $\mathbf{\Gamma}(p)$, for each $p \in C$, are therefore $1, 0$ and -1, and (9.12) states that $\mathbf{f}(p)$ is an eigenvector with eigenvalue 1. Since $\overline{\mathbf{\Gamma}(p)} = -\mathbf{\Gamma}(p)$, (9.12) implies that $\mathbf{\Gamma}\bar{\mathbf{f}} = -\bar{\mathbf{f}}$. A similar operator was defined and studied in much more detail by Moses (1971) in connection with fluid mechanics as well as electrodynamics.

Consider a single component of (9.8), i.e., the plane-wave solution

$$\mathbf{F}_p(x) \equiv e^{ip\cdot x}\,\mathbf{f}(p) = \mathbf{B}_p(x) + i\,\mathbf{E}_p(x), \tag{9.15}$$

with arbitrary but fixed $p \in C$ and $\mathbf{f}(p) \neq \mathbf{0}$. The electric and magnetic fields are obtained by taking the real and imaginary parts. Now $\mathbf{\Gamma}(p)\,\mathbf{f}(p) = \mathbf{f}(p)$ and $\mathbf{\Gamma}(p)\,\overline{\mathbf{f}(p)} = -\overline{\mathbf{f}(p)}$ imply

$$\mathbf{\Gamma}(p)\mathbf{F}_p(x) = \mathbf{F}_p(x), \qquad \mathbf{\Gamma}(p)\,\overline{\mathbf{F}_p(x)} = -\overline{\mathbf{F}_p(x)}. \tag{9.16}$$

Since $\mathbf{\Gamma}(p)^* = \mathbf{\Gamma}(p)$, these eigenvectors of $\mathbf{\Gamma}(p)$ with eigenvalues 1 and -1 must be orthogonal:

$$\overline{\mathbf{F}_p(x)}^*\mathbf{F}_p(x) = \mathbf{F}_p(x)\cdot\mathbf{F}_p(x) = 0,$$

where the asterisk denotes the Hermitian transpose. Taking real and imaginary parts, we get

$$|\mathbf{B}_p(x)|^2 = |\mathbf{E}_p(x)|^2, \qquad \mathbf{B}_p(x)\cdot\mathbf{E}_p(x) = 0. \tag{9.17}$$

The first equation shows that neither $\mathbf{B}_p(x)$ nor $\mathbf{E}_p(x)$ can vanish at any x (since $\mathbf{f}(p) \neq \mathbf{0}$). Furthermore, (9.16) implies that

$$\mathbf{p} \times \mathbf{E}_p(x) = p_0\mathbf{B}_p(x).$$

Thus, for any x, $\{\mathbf{p}, \mathbf{E}_p(x), \mathbf{B}_p(x)\}$ is a *right-handed orthogonal basis* if $p_0 > 0$ (i.e., $p \in C_+$) and a *left-handed orthogonal basis* if $p_0 < 0$ ($p \in C_-$). Taking the real and imaginary parts of (9.15) and using $\mathbf{f}(p) = \mathbf{B}_p(0) + i\,\mathbf{E}_p(0)$, we have

$$\begin{aligned} \mathbf{B}_p(x) &= \cos(p\cdot x)\mathbf{B}_p(0) - \sin(p\cdot x)\mathbf{E}_p(0), \\ \mathbf{E}_p(x) &= \cos(p\cdot x)\mathbf{E}_p(0) + \sin(p\cdot x)\mathbf{B}_p(0). \end{aligned} \tag{9.18}$$

An observer at any fixed location $\mathbf{x} \in \mathbf{R}^3$ sees these fields rotating as a function of time in the plane orthogonal to $\mathbf{p}$. If $p \in C_+$, the rotation is that of a right-handed corkscrew, or helix, moving in the direction of $\mathbf{p}$, whereas if $p \in C_-$, it is that of a left-handed corkscrew. Hence $\mathbf{F}_p(x)$ is said to have *positive helicity* if $p \in C_+$ and *negative helicity* if $p \in C_-$.

A general solution of the form (9.8) has positive helicity if $\mathbf{f}(p)$ is supported in C_+ and negative helicity if $\mathbf{f}(p)$ is supported in C_-. Other states of polarization, such as linear or elliptic, are obtained by mixing positive and negative helicities. The significance of the complex combination $\mathbf{F}(x) = \mathbf{B}(x) + i\,\mathbf{E}(x)$ therefore seems to be that, *in Fourier space, the sign of the frequency p_0 gives the helicity of the solution!* (Usually in signal analysis, the sign of the frequency is not given any physical interpretation, and negative frequencies are regarded as a necessary mathematical artifact due to the choice of complex exponentials over sines and cosines.) In other words, the combination $\mathbf{B} + i\,\mathbf{E}$ "polarizes" the helicity, with positive and negative helicity states being represented in C_+ and C_-, respectively. Had we used the opposite combination $\mathbf{B} - i\,\mathbf{E}$, C_+ and C_- would have parameterized the plane-wave solutions with opposite helicities. Nothing new seems to be gained by considering this alternative.†

In order to eliminate the constraint, we now proceed as follows: Let

$$\mathbf{\Pi}(p) \equiv \tfrac{1}{2}\left[\mathbf{\Gamma}(p) + \mathbf{\Gamma}^2(p)\right]. \tag{9.19}$$

Explicitly,

$$\mathbf{\Pi}(p) = \frac{1}{2p_0^2}\begin{bmatrix} p_0^2 - p_1^2 & -p_1p_2 + ip_0p_3 & -p_1p_3 - ip_0p_2 \\ -p_1p_2 - ip_0p_3 & p_0^2 - p_2^2 & -p_2p_3 + ip_0p_1 \\ -p_1p_3 + ip_0p_2 & -p_2p_3 - ip_0p_1 & p_0^2 - p_3^2 \end{bmatrix}. \tag{9.20}$$

The established properties $\mathbf{\Gamma}^* = \mathbf{\Gamma} = \mathbf{\Gamma}^3$ imply that $\mathbf{\Pi}^* = \mathbf{\Pi} = \mathbf{\Pi}^2$ and $\mathbf{\Gamma\Pi} = \mathbf{\Pi}$, which proves that $\mathbf{\Pi}(p)$ is the orthogonal projection to eigenvectors of $\mathbf{\Gamma}(p)$ with eigenvalue 1. Thus, to satisfy the constraint (9.12), we need only replace the constrained function $\mathbf{f}(p)$ in (9.8) by $\mathbf{\Pi}(p)\mathbf{f}(p)$, where $\mathbf{f}(p)$ is now unconstrained.

Proposition 9.1. *Solutions of Maxwell's have the following Fourier representation:*

$$\mathbf{F}(x) = \int_C d\tilde{p}\; e^{ip\cdot x}\,\mathbf{\Pi}(p)\,\mathbf{f}(p)\,, \tag{9.21}$$

with $\mathbf{f} : C \to \mathbf{C}^3$ *unconstrained.*

Consequently, the mapping $\mathbf{f} \mapsto \mathbf{F}$ is not one-to-one since $\mathbf{\Pi}$ is a projection operator. In fact, $\mathbf{f}$ is closely related to the *potentials* for $\mathbf{F}$, which consist of a real 3-vector potential $\mathbf{A}(x)$ and a real scalar potential $A_0(x)$ such that

$$\mathbf{B} = \nabla \times \mathbf{A}, \qquad \mathbf{E} = -\partial_0\mathbf{A} - \nabla A_0. \tag{9.22}$$

The combination $(\mathbf{A}(x), A_0(x))$ is called a "4-vector potential" for the field. We can assume without loss of generality that the potential satisfies the Lorentz condition (Jackson [1975])

$$\nabla \cdot \mathbf{A} + \partial_0 A_0 = 0.$$

† Maxwell's equations are invariant under the continuous group of *duality rotations*, of which the complex structure J mapping $\mathbf{E}$ to $\mathbf{B}$ and $\mathbf{B}$ to $-\mathbf{E}$ is a special case. In the complexified solution space, the combinations $\mathbf{B} \pm i\,\mathbf{E}$ form invariant subspaces with respect to the duality rotations. That gives the choice of $\mathbf{B}+i\mathbf{E}$ an interpretation in terms of group representation theory.

Since $\mathbf{A}$ and A_0 also satisfy the wave equation (9.4), they have Fourier representations similar to (9.8):

$$\mathbf{A}(x) = \int_C d\tilde{p}\ e^{ip\cdot x}\,\mathbf{a}(p), \quad A_0(x) = \int_C d\tilde{p}\ e^{ip\cdot x}\,a_0(p). \tag{9.23}$$

The Lorentz condition means that $p_0a_0(p) = \mathbf{p}\cdot\mathbf{a}(p)$, or $a_0(p) = \mathbf{n}\cdot\mathbf{a}(p)$, so a_0 is determined by $\mathbf{a}$. Equations (9.22) will be satisfied provided that the Fourier representatives $\mathbf{e}(p)$, $\mathbf{b}(p)$ of $\mathbf{E},\mathbf{B}$ satisfy

$$\mathbf{b} = -i\,\mathbf{p}\times\mathbf{a} = p_0\,\mathbf{\Gamma}\,\mathbf{a}, \qquad \mathbf{e} = -ip_0\mathbf{a} + i\,\mathbf{p}a_0 = -ip_0\,\mathbf{\Gamma}^2\,\mathbf{a}. \tag{9.24}$$

Hence $\mathbf{F} = \mathbf{B} + i\,\mathbf{E}$ is represented in Fourier space by

$$\mathbf{b}(p) + i\,\mathbf{e}(p) = p_0\left[\mathbf{\Gamma}(p) + \mathbf{\Gamma}(p)^2\right]\,\mathbf{a}(p) = 2p_0\mathbf{\Pi}(p)\,\mathbf{a}(p). \tag{9.25}$$

This shows that we can interpret the unconstrained function $\mathbf{f}(p)$ in (9.21) as being directly related to the 3-vector potential by

$$\mathbf{f}(p) = 2p_0\,\mathbf{a}(p), \tag{9.26}$$

modulo terms annihilated by $\mathbf{\Pi}(p)$, which correspond to eigenvalues -1 and 0 of $\mathbf{\Gamma}(p)$. Viewed in this light, the nonuniqueness of $\mathbf{f}$ in (9.21) is an expression of *gauge freedom* in the $\mathbf{B} + i\,\mathbf{E}$ representation, as seen from Fourier space. In the space-time domain, ($\mathbf{B}$, $\mathbf{E}$) are the components of a 2-form F in $\mathbf{R}^4$ and ($\mathbf{A}$, A_0) are the components of a 1-form A. Then equations (9.1) become $dF = 0$ and $\delta F = 0$ (where δ is the divergence with respect to the Lorentz inner product), Equations (9.22) become unified as $F = dA$, the Lorentz condition reads $\delta A = 0$, and the gauge freedom corresponds to the invariance of F under $A \to A + d\chi$, where $\chi(x)$ is a scalar solution of the wave equation.

Maxwell's equations are invariant under a large group of space-time transformations. Such transformations produce new solutions from known ones by acting on the underlying space-time variables (possibly with a multiplier to rotate or scale the vector fields). Some trivial examples are space and time translations. Obviously, a translated version of a solution is again a solution, since the equations have constant coefficients. Similarly, a rotated version of a solution is a solution. A less obvious example is Lorentz transformations, which are interpreted as transforming to a uniformly *moving* reference frame in space-time. (In fact, it was in the study of the Lorentz invariance of Maxwell's equations that the Special Theory of Relativity originated; see Einstein et al. (1923).) The scale transformations $x \to ax$, $a \neq 0$, also map solutions to solutions, since Maxwell's equations are homogeneous in the space-time variables. Finally, the equations are invariant under "special conformal transformations" (Bateman [1910], Cunningham [1910]), which can be interpreted as transforming to a uniformly *accelerating* reference frame (Page [1936]; Hill [1945, 1947, 1951]).

Altogether, these transformations form a 15-dimensional Lie group called the *conformal group*, which is locally isomorphic to $SU(2,2)$ and will here be denoted by $\mathcal{C}$. Whereas wavelets in one dimension are related to one another by translations and scalings, electromagnetic wavelets will be seen to be related by conformal transformations, which include translations and scalings. (A study of the action of $SU(2,2)$ on solutions of Maxwell's equations has been made by Rühl (1972).)

To construct the machinery of wavelet analysis, we need a Hilbert space. That is, we must define an inner product on solutions. It is important to choose the inner product to be invariant under the largest possible group of symmetries since this allows the largest set of solutions in $\mathcal{H}$ to be generated by unitary transformations from any one known solution. (In quantum mechanics, invariance of the inner product is also an expression of the fundamental invariance of the laws of nature with respect to the symmetries in question.) Let $\mathbf{f}(p)$ satisfy (9.12), and let $\mathbf{a}(p)$ be a vector potential for $\mathbf{f}$ satisfying the Lorentz condition so that the scalar potential is determined by $a_0(p) = \mathbf{n}(p) \cdot \mathbf{a}(p)$. By (9.25),

$$\begin{aligned} |\mathbf{f}(p)|^2 &= 4p_0^2\, |\mathbf{\Pi}(p)\, \mathbf{a}(p)|^2 = 4\omega^2\, \overline{\mathbf{a}(p)} \cdot \mathbf{\Pi}(p)\, \mathbf{a}(p) \\ &= 2\omega^2\, \overline{\mathbf{a}(p)} \cdot \mathbf{\Gamma}(p)\, \mathbf{a}(p) + 2\omega^2\, \overline{\mathbf{a}(p)} \cdot \mathbf{\Gamma}(p)^2\, \mathbf{a}(p). \end{aligned} \tag{9.27}$$

The first term is

$$-2i\omega^2\, \overline{\mathbf{a}(p)} \cdot (\mathbf{n} \times \mathbf{a}(p)) = 2i\omega^2\, \mathbf{n} \cdot (\overline{\mathbf{a}(p)} \times \mathbf{a}(p)), \tag{9.28}$$

which cancels its counterpart with $p \to -p$ on account of the reality condition $\overline{\mathbf{a}(-p)} = \mathbf{a}(p)$. Thus

$$\begin{aligned} \int_C \frac{d\tilde{p}}{\omega^2}\, |\mathbf{f}(p)|^2 &= 2 \int_C d\tilde{p}\;\; \overline{\mathbf{a}(p)} \cdot \big[\mathbf{a}(p) - \mathbf{n}(\mathbf{n} \cdot \mathbf{a}(p))\big] \\ &= 2 \int_C d\tilde{p}\;\; \Big[|\mathbf{a}(p)|^2 - |a_0(p)|^2\Big]. \end{aligned} \tag{9.29}$$

The integrand in the last expression is the negative of the Lorentz-square of the 4-potential $(\mathbf{a}(p), a_0(p))$. Consequently, the integral can be shown to be invariant under Lorentz transformations. (Note that $|\mathbf{a}|^2 - |a_0|^2 \geq 0$, vanishing only when $\mathbf{a}(p)$ is a multiple of $\mathbf{p}$, in which case $\mathbf{f} = \mathbf{0}$. This corresponds to "longitudinal polarization.") Hence (9.29) defines a norm on solutions that is invariant under Lorentz transformations as well as space-time translations. In fact, the norm (9.29) is uniquely determined, up to a constant factor, by the requirement that it be so invariant. Moreover, Gross (1964) has shown it to be invariant under the full conformal group $\mathcal{C}$. Again we eliminate the constraint by replacing $\mathbf{f}(p)$ with $\mathbf{\Pi}(p)\, \mathbf{f}(p)$. Thus, let $\mathcal{H}$ be the set of all solutions $\mathbf{F}(x)$

defined by (9.21) with $\mathbf{f} : C \to \mathbf{C}^3$ square-integrable in the sense that

$$\begin{aligned} \|\mathbf{F}\|^2 &= \int_C \frac{d\tilde{p}}{\omega^2} \, |\mathbf{\Pi}(p)\, \mathbf{f}(p)|^2 \\ &= (2\pi)^{-3} \int_C \frac{d^3\mathbf{p}}{2\omega^3} \, |\mathbf{\Pi}(p)\, \mathbf{f}(p)|^2 < \infty. \end{aligned} \tag{9.30}$$

$\mathcal{H}$ is a Hilbert space under the inner product obtained by polarizing the norm (9.30) and using $(\mathbf{\Pi f})^* \mathbf{\Pi g} = \mathbf{f}^* \mathbf{\Pi}^* \mathbf{\Pi g} = \mathbf{f}^* \mathbf{\Pi g}$.

Definition 9.2. *$\mathcal{H}$ is the Hilbert space of solutions (9.21) with inner product*

$$\langle\, \mathbf{F}, \mathbf{G} \,\rangle = \int_C \frac{d\tilde{p}}{\omega^2} \, \mathbf{f}(p)^* \mathbf{\Pi}(p)\, \mathbf{g}(p). \tag{9.31}$$

$\mathcal{H}$ will be our main arena for developing the analysis and synthesis of solutions in terms of wavelets. Note that when (9.12) holds and $\mathbf{f}_\pm(\mathbf{p}) \equiv \mathbf{f}(\mathbf{p}, \pm\omega)$ are continuous, then

$$\mathbf{f}(0) = \mathbf{f}_\pm(\mathbf{0}) = \mathbf{0} \tag{9.32}$$

must hold in order for (9.30) to be satisfied. This is a kind of *admissibility condition* that applies to *all* solutions in $\mathcal{H}$: There must not be a "DC component." In fact, the measure $d^3\mathbf{p} \,/ |\mathbf{p}|^3$ is a three-dimensional version of the measure $d\xi / |\xi|$ that played a role in the admissibity condition of one-dimensional wavelets.

To show the invariance of (9.30) under conformal transformations, Gross derived an equivalent norm expressed directly in terms of the values of the fields in space at any particular time $x_0 = t$:

$$\|\mathbf{F}\|^2_{\text{Gross}} \equiv \frac{1}{\pi^2} \int_{\mathbf{R}^6} \frac{d^3\mathbf{x}\, d^3\mathbf{y}}{|\mathbf{x} - \mathbf{y}|^2} \, \mathbf{F}(\mathbf{x}, t)^* \mathbf{F}(\mathbf{y}, t). \tag{9.33}$$

The right-hand side is independent of t due to the invariance of Maxwell's equations under time translations (which is, in turn, related to the conservation of energy). A disadvantage of the expression (9.33) is that it is *nonlocal* since it uses the values of the field simultaneously at the space points $\mathbf{x}$ and $\mathbf{y}$. In fact, it is known that *no local expression for the inner product can exist in terms of the field values in (real) space-time* $\mathbf{R}^4$ (Bargmann and Wigner [1948]). In Section 4, we derive an alternate expression for the inner product directly in terms of the values of the electromagnetic fields, extended analytically to *complex* space-time. This expression is "local" in the space-scale domain (rather than in space alone). But first we must introduce the tool that implements the extension to complex space-time.

9.3 The Analytic-Signal Transform

We now introduce a multidimensional generalization of the wavelet transform that will be used below to analyze electromagnetic fields. To motivate the construction, imagine that we have a (possibly complex) vector field $\mathbf{F}:\mathbf{R}^4\to\mathbf{C}^m$ on space-time. For example, $\mathbf{F}$ could be an electromagnetic wave (in which case $m=3$ and $\mathbf{F}$ is given by (9.21)), but this need not be assumed. Suppose we want to measure† this field with a small instrument, one that may be regarded as a point in space at any time. Suppose further that the impulse response of the instrument is a (possibly) complex-valued function $h:\mathbf{R}\to\mathbf{C}$. If the support width of h is small, then we can assume that the velocity of the instrument is approximately constant while the measurement takes place. Then the trajectory on the instrument in space-time may be parameterized as

$$x(\tau)=x+\tau y,\quad \text{i.e.,}\quad \begin{cases}\mathbf{x}(\tau)=\mathbf{x}+\tau\mathbf{y}\\ x_0(\tau)=x_0+\tau y_0,\end{cases} \tag{9.34}$$

where $x=(\mathbf{x},x_0)\in\mathbf{R}^4$ is the "initial" point and $y=(\mathbf{y},y_0)$ is "velocity." The actual three-dimensional velocity in space is

$$\mathbf{v}=\frac{\mathbf{y}}{y_0}, \tag{9.35}$$

so we require

$$|\mathbf{v}|<c,\quad \text{i.e.,}\quad |\mathbf{y}|<c\,|y_0|. \tag{9.36}$$

We *model* the result of measuring the field $\mathbf{F}$ with the instrument h, moving along the above trajectory, as

$$\mathbf{F}_h(x,y)\equiv\int_{-\infty}^{\infty}d\tau\;\bar{h}(\tau)\,\mathbf{F}(x+\tau y). \tag{9.37}$$

This is a multidimensional generalization of the wavelet transform (Kaiser [1990a, b]; Kaiser and Streater [1992]). To see this, suppose for a moment that we replace space-time by time alone, and the field is scalar-valued, i.e., $F:\mathbf{R}\to\mathbf{C}$. Then $x,y\in\mathbf{R}$ and (9.36) reduces to $y\neq 0$. By changing variables in (9.37), the transform becomes

$$F_h(x,y)=\int_{-\infty}^{\infty}du\;\frac{1}{|y|}\,\bar{h}\left(\frac{u-x}{y}\right)F(u). \tag{9.38}$$

---

† I do not claim that this is a realistic description of electromagnetic field measurement. It is only meant to motivate the generalized wavelet transforms developed in this section by relating them to the standard wavelet transforms.

This is recognized as essentially identical to the usual wavelet transform, with the wavelets now represented by

$$h_{x,y}(u) \sim \frac{1}{|y|} h\left(\frac{u-x}{y}\right). \tag{9.39}$$

(The symbol $\sim$ instead of equality will be explained below.) In the one-dimensional case treated in Chapter 3, we assumed, without any particular reason, that F belonged to $L^2(\mathbf{R})$. Inspired by the electromagnetic norm (9.30), let us now suppose instead that F belongs to the Hilbert space

$$\mathcal{H}_\alpha = \{F : \|F\|_\alpha^2 < \infty\}, \text{ where } \|F\|_\alpha^2 \equiv \frac{1}{2\pi}\int_{-\infty}^{\infty} \frac{dp}{|p|^\alpha} |\hat{F}(p)|^2 \tag{9.40}$$

for some $\alpha \ge 0$. The inner product in $\mathcal{H}_\alpha$ is obtained, as usual, by polarizing the expression for $\|F\|_\alpha^2$. Battle (1992) calls $\mathcal{H}_{2\alpha}$ a "massless Sobolev space of degree α." We now derive a reconstruction formula for this case to prepare ourselves for the more complicated case of electromagnetics. The main points of interest in this "practice run" are the modifications forced on the definition of the wavelets and the admissibility condition by the presence of the weight factor $|p|^{-\alpha}$ in the frequency domain. Since

$$\int_{-\infty}^{\infty} du\; e^{-ipu} \frac{1}{|y|} h\left(\frac{u-x}{y}\right) = e^{-ipx}\hat{h}(py), \tag{9.41}$$

Parseval's formula applied to (9.38) gives

$$\begin{aligned} F_h(x,y) &= (2\pi)^{-1}\int_{-\infty}^{\infty} dp\; e^{ipx}\,\overline{\hat{h}}(py)\hat{F}(p) \\ &= (2\pi)^{-1}\int_{-\infty}^{\infty} \frac{dp}{|p|^\alpha}\, |p|^\alpha e^{ipx}\,\overline{\hat{h}}(py)\hat{F}(p) \\ &\equiv \langle\, h_{x,y}\,, F\,\rangle_\alpha \equiv h_{x,y}^*\, F\,, \end{aligned} \tag{9.42}$$

where

$$\hat{h}_{x,y}(p) \equiv |p|^\alpha e^{-ipx}\,\hat{h}(py). \tag{9.43}$$

This shows that the wavelets are given in the time domain by

$$h_{x,y}(u) = \frac{1}{2\pi}\int_{-\infty}^{\infty} dp\; |p|^\alpha e^{ip(u-x)}\,\hat{h}(py). \tag{9.44}$$

If we take $\alpha = 0$, then $\mathcal{H}_\alpha = L^2(\mathbf{R})$ and (9.44) reduces to (9.39). For $\alpha > 0$, (9.39) is not valid. (That explains the symbol $\sim$ in (9.39).) Now let us see how to reconstruct F from its transform F_h. Applying Plancherel's formula to (9.42) gives

$$\int_{-\infty}^{\infty} dx\; |F_h(x,y)|^2 = (2\pi)^{-1}\int_{-\infty}^{\infty} dp\; |\hat{h}(py)|^2 |\hat{F}(p)|^2. \tag{9.45}$$

Integrating both sides over y with the weight function $|y|^{\alpha-1}$ gives

$$\begin{aligned}\int_{-\infty}^{\infty} dx \int_{-\infty}^{\infty} dy\ |y|^{\alpha-1}\,|F_h(x,y)|^2 \\ &= (2\pi)^{-1}\int_{-\infty}^{\infty} dp\ |\hat{F}(p)|^2 \int_{-\infty}^{\infty} dy\ |y|^{\alpha-1}\,|\hat{h}(py)|^2 \\ &= C_\alpha \|F\|_\alpha^2\,,\end{aligned} \tag{9.46}$$

where we have changed y to $\xi = py$, and let

$$C_\alpha \equiv \int_{-\infty}^{\infty} d\xi\ |\xi|^{\alpha-1}|\hat{h}(\xi)|^2, \tag{9.47}$$

which now is assumed to be finite. Therefore we define an inner product for transforms by

$$\begin{aligned}\langle F_h\,, G_h\,\rangle &\equiv C_\alpha^{-1}\int_{-\infty}^{\infty} dx \int_{-\infty}^{\infty} dy\ |y|^{\alpha-1}\,\overline{F_h(x,y)}\,G_h(x,y) \\ &= C_\alpha^{-1}\int_{-\infty}^{\infty} dx \int_{-\infty}^{\infty} dy\ |y|^{\alpha-1}\,F^* h_{x,y}\,h_{x,y}^*\,G,\end{aligned} \tag{9.48}$$

so that (9.46) becomes a Plancherel-type relation $\|F_h\|^2 = \|F\|_\alpha^2$, and by polarization,

$$\langle F_h\,, G_h\,\rangle = F^* G. \tag{9.49}$$

This means that (9.48) gives a resolution of unity:

$$C_\alpha^{-1}\int_{-\infty}^{\infty} dx \int_{-\infty}^{\infty} dy\ |y|^{\alpha-1}\,h_{x,y}\,h_{x,y}^* = I \quad \text{weakly in } \mathcal{H}_\alpha\,. \tag{9.50}$$

As usual, this means that vectors in $\mathcal{H}_\alpha$ can be expanded in terms of the wavelets $h_{x,y}$, and all the theorems in Chapter 4 apply concerning the consistency condition for transforms, least-squares approximations, and so on. Note that the admissibility condition is now $C_\alpha < \infty$, which reduces to the usual admissibility condition only when $\alpha = 0$.

Having explained the relation of (9.37) to the usual one-dimensional wavelet transform, we now return to the four-dimensional case. Inserting the Fourier representation of $\mathbf{F}$, we obtain

$$\begin{aligned}\mathbf{F}_h(x,y) &= (2\pi)^{-4}\int_{-\infty}^{\infty} d\tau\ \bar{h}(\tau)\int_{\mathbf{R}^4} d^4p\, e^{ip\cdot(x+\tau y)}\hat{\mathbf{F}}(p) \\ &= (2\pi)^{-4}\int_{\mathbf{R}^4} d^4p\, e^{ip\cdot x}\,\hat{\mathbf{F}}(p)\int_{-\infty}^{\infty} d\tau\ e^{i\tau p\cdot y}\,\bar{h}(\tau) \\ &= (2\pi)^{-4}\int_{\mathbf{R}^4} d^4p\, e^{ip\cdot x}\,\overline{\hat{h}}(p\cdot y)\,\hat{\mathbf{F}}(p).\end{aligned} \tag{9.51}$$

That is, $\mathbf{F}_h(x,y)$ is a windowed inverse Fourier transform of $\hat{\mathbf{F}}(p)$, where the window in the Fourier domain is $\hat{h}(p\cdot y)$. Note that all Fourier components

$\hat{\mathbf{F}}(p)$ in a given three-dimensional hyperplane $P_a \equiv \{p : p \cdot y = a\}$ are given equal weights by this window.[†] To construct wavelets from h, as was done in the one-dimensional case by (9.44), we need an inner product on the vector fields $\mathbf{F}$. This will be done in the next section, where the $\mathbf{F}$'s will be assumed to be electromagnetic waves. But now we specialize our transform by choosing

$$\bar{h}(\tau) = \frac{1}{\pi i} \frac{1}{\tau - i} . \tag{9.52}$$

With this choice, we denote $\mathbf{F}_h(x, y)$ by $\tilde{\mathbf{F}}(x + iy)$ and call it the analytic signal of $\mathbf{F}$. We state the definition in the general case, where $\mathbf{R}^4$ is replaced by $\mathbf{R}^n$.

Definition 9.3. *The* analytic signal *of* $\mathbf{F} : \mathbf{R}^n \to \mathbf{C}^m$ *is the function* $\tilde{\mathbf{F}} : \mathbf{C}^n \to \mathbf{C}^m$ *defined formally by*

$$\tilde{\mathbf{F}}(x + iy) \equiv \frac{1}{\pi i} \int_{-\infty}^{\infty} \frac{d\tau}{\tau - i} \, \mathbf{F}(x + \tau y). \tag{9.53}$$

The mapping $\mathbf{F} \mapsto \tilde{\mathbf{F}}$ *will be called the* analytic-signal transform *(AST).*

In general $\tilde{\mathbf{F}}$ is not analytic, so it may actually depend on both $z \equiv x + iy$ and $\bar{z} = x - iy$. The reason for the above notation will become clear below. By contour integration, we find

$$\hat{h}(\xi) = -\frac{1}{\pi i} \int_{-\infty}^{\infty} \frac{d\tau}{\tau + i} \, e^{-i\xi\tau} = 2\theta(\xi) e^{-\xi}, \tag{9.54}$$

where θ is the step function

$$\theta(\xi) \equiv \begin{cases} 1 & \xi > 0 \\ .5 & \xi = 0 \\ 0 & \xi < 0. \end{cases} \tag{9.55}$$

Thus we have shown the following:

Proposition 9.4. *The analytic-signal transform is given in the Fourier domain by*

$$\tilde{\mathbf{F}}(x + iy) = (2\pi)^{-4} \int_{\mathbf{R}^4} d^4p \, 2\theta(p \cdot y) \, e^{ip \cdot (x+iy)} \, \hat{\mathbf{F}}(p). \tag{9.56}$$

The factor $\theta(p \cdot y)$ plays two important roles in (9.56):

[†] Equation (9.37) defines a one-dimensional *windowed Radon transform.* A similar transform can be defined modeled on field measurements where the instrument is assumed to occupy $k \leq 3$ space dimensions. (For a point instrument, $k = 0$; for a wire antenna, $k = 1$; for a dish antenna, $k = 2$.) This gives a $k + 1$-dimensional windowed Radon transform in $\mathbf{R}^4$. See Kaiser (1990a, b), and Kaiser and Streater (1992).)

(a) Suppose that $\hat{\mathbf{F}}(p)$ is absolutely integrable, so that $\mathbf{F}(x)$ is continuous. If the factor $\theta(p \cdot y)$ were absent in (9.56), then $e^{-p \cdot y}$ can grow without bounds, and the above integral can diverge unless $\hat{\mathbf{F}}(p)$ happens to decay sufficiently rapidly to balance this growth. This option is not available for general electromagnetic waves in the Hilbert space $\mathcal{H}$ of Section 9.2, since there $\hat{\mathbf{F}}(p)$ is supported in the light cone $C = C_+ \cup C_-$; so whenever p belongs to the support of $\hat{\mathbf{F}}$, then $-p$ may possibly also belong to the support of $\hat{\mathbf{F}}$, and there is no reason for $\hat{\mathbf{F}}$ to have the exponential decay required to make the integral converge.

(b) The factor $\theta(p \cdot y)$ tends to spoil the analyticity of $\tilde{\mathbf{F}}$. If this factor were absent, and the integral somehow converged, then the resulting function depends only on $z = x + iy$ rather than on both z and $\bar{z}$, and $\tilde{\mathbf{F}}(z)$ is analytic (formally, at least).

We will now show that under certain conditions, which hold in particular when $\mathbf{F}$ belongs to the Hilbert space $\mathcal{H}$, $\tilde{\mathbf{F}}(z)$ is indeed analytic, provided z is restricted to a certain domain $\mathcal{T} \subset \mathbf{C}^4$ in complex space-time. Again, we revert temporarily from four dimensions (space-time) to one dimension (*time*), in order to clarify the ideas. Suppose that $F : \mathbf{R} \to \mathbf{C}$ is such that $\hat{F}(p)$ is absolutely integrable. Then (9.56) is replaced by

$$\tilde{F}(x + iy) = (2\pi)^{-1} \int_{-\infty}^{\infty} dp \;\, 2\theta(py)\, e^{ip(x+iy)}\, \hat{F}(p). \tag{9.57}$$

If $y > 0$, then $\theta(py) = \theta(p)$ and

$$\tilde{F}(x + iy) = \pi^{-1} \int_{0}^{\infty} dp\, e^{ip(x+iy)}\, \hat{F}(p), \qquad y > 0. \tag{9.58}$$

The integral converges absolutely and, as $z = x + iy$ varies over the upper-half z-plane, it defines an analytic function. Therefore $\tilde{F}(z)$ *is analytic in the upper-half plane* $\mathbf{C}_+$, *and for* $z \in \mathbf{C}_+$, $\tilde{F}$ *contains only positive-frequency components.* Similarly, when $y < 0$, then $\theta(py) = \theta(-p)$ and

$$\tilde{F}(x + iy) = \pi^{-1} \int_{-\infty}^{0} dp\, e^{ip(x+iy)}\, \hat{F}(p), \qquad y < 0. \tag{9.59}$$

$\tilde{F}$ *is analytic in the lower-half plane* $\mathbf{C}_-$, *and for* $z \in \mathbf{C}_-$, $\tilde{F}(z)$ *contains only negative-frequency components.* Thus, $\tilde{F}$ is analytic in $\mathbf{C}_+ \cup \mathbf{C}_-$. It is *not* analytic on the real axis, in general. The relation between $\tilde{F}(z)$ and the original function $F(x)$ can be summarized as follows:

$$\tfrac{1}{2} \lim_{y \to 0+} \left[\tilde{F}(x + iy) + \tilde{F}(x - iy)\right] = (2\pi)^{-1} \int_{-\infty}^{\infty} dp\; e^{ipx}\, \hat{F}(p) = F(x). \tag{9.60}$$

Thus $F(x)$ is a "two-sided boundary value" of $\tilde{F}(z)$, as z approaches the real axis simultaneously from above and below. On the other hand, the discontinuity across $\mathbf{R}$ is given by

$$\begin{aligned}\tfrac{1}{2} \lim_{y \to 0^+} \left[\tilde{F}(x+iy) - \tilde{F}(x-iy) \right] &= (2\pi)^{-1} \int_{-\infty}^{\infty} dp \ \text{sign}\,(p) e^{ipx} \, \hat{F}(p) \\ &= iHF(x),\end{aligned} \tag{9.61}$$

where $H : L^2(\mathbf{R}) \to L^2(\mathbf{R})$ is the *Hilbert transform.* When F is real, then $\overline{\tilde{F}(z)} = \tilde{F}(\bar{z})$ for all z and

$$\begin{aligned} F(x) &= \lim_{y \to 0^+} \operatorname{Re} \tilde{F}(x+iy) \\ HF(x) &= \lim_{y \to 0^+} \operatorname{Im} \tilde{F}(x+iy). \end{aligned} \tag{9.62}$$

For example,

$$F(x) = \cos x \ \Rightarrow \ \hat{F}(p) = \pi\delta(p-1) + \pi\delta(p+1), \tag{9.63}$$

so

$$\tilde{F}(z) = \begin{cases} e^{iz}, & z \in \mathbf{C}_+ \\ \cos z, & z \in \mathbf{R} \\ e^{-iz}, & z \in \mathbf{C}_- \end{cases}, \qquad HF(x) = \sin x. \tag{9.64}$$

The function $\tilde{F}$ in (9.58) was introduced by Gabor (1946), who called it the *analytic signal* of F. It has proved to be very useful in optics, radar, and many other fields. (For optics, see Born and Wolf (1975), Klauder and Sudarshan (1968). For radar, see Rihaczek (1968); also Cook and Bernfeld (1967), where analytic signals are called "complex waves.") When F is real, then the Gabor analytic signal (defined in $\mathbf{C}_+$ only) suffices to determine F by (9.62), because the negative-frequency part of F is simply the complex conjugate of the positive-frequency part. When F is complex, the two parts are independent and both (9.58) and (9.59) are needed in order to recover F.

Our definition (9.53) is a generalization of Gabor's idea to any number of dimensions and to functions that are not necessarily real-valued. That explains the name "analytic-signal transform" given to the mapping $\mathbf{F} \mapsto \tilde{\mathbf{F}}$ in (9.53). To extend the relations (9.60) and (9.61) to the general case, we return to the original definition (9.53) with $z = x + i\varepsilon y$, $y \neq 0$, and consider its limit as $\varepsilon \to 0^+$:

$$
\begin{aligned}
\lim_{\varepsilon\to 0+} \tilde{\mathbf{F}}(x+i\varepsilon y) &= \frac{1}{\pi i} \lim_{\varepsilon\to 0+} \int_{-\infty}^{\infty} \frac{d\tau}{\tau - i} \mathbf{F}(x+\varepsilon\tau y) \\
&= \frac{1}{\pi i} \lim_{\varepsilon\to 0+} \int_{-\infty}^{\infty} \frac{du}{u-i\varepsilon} \mathbf{F}(x+uy) \\
&= \frac{1}{\pi i} \left[i\pi \mathbf{F}(x) + P\int_{-\infty}^{\infty} \frac{du}{u} \mathbf{F}(x+uy) \right] \qquad (9.65) \\
&= \mathbf{F}(x) + \frac{1}{\pi i} P \int_{-\infty}^{\infty} \frac{du}{u} \mathbf{F}(x+uy) \\
&= \mathbf{F}(x) + \frac{i}{\pi} P \int_{-\infty}^{\infty} \frac{du}{u} \mathbf{F}(x-uy),
\end{aligned}
$$

where $P\int\cdots$ denotes the Cauchy principal value. The *multivariate Hilbert transform in the direction of* $y \neq 0$ is defined by

$$
H_y \mathbf{F}(x) \equiv \frac{1}{\pi} P \int_{-\infty}^{\infty} \frac{du}{u} \mathbf{F}(x-uy). \qquad (9.66)
$$

It actually depends only on the direction of y, since

$$
H_{ay}\,\mathbf{F} = \mathrm{sign}\,(a)\, H_y \mathbf{F}, \qquad a \neq 0, \qquad (9.67)
$$

so the usual definition (Stein [1970]) assumes that y is a unit vector with respect to an arbitrary norm in $\mathbf{R}^n$ (which Stein takes to be the Euclidean norm). Since in our case $\mathbf{R}^n = \mathbf{R}^4$ is the Minkowskian space-time with the indefinite Lorentz scalar product, we prefer the unnormalized version. Equation (9.65) gives the generalizations of (9.60) and (9.61):

$$
\begin{aligned}
\mathbf{F}(x) &= \frac{1}{2} \lim_{\varepsilon\to 0+} \left[\tilde{\mathbf{F}}(x+i\varepsilon y) + \tilde{\mathbf{F}}(x-i\varepsilon y) \right] \\
H_y \mathbf{F}(x) &= \frac{1}{2i} \lim_{\varepsilon\to 0+} \left[\tilde{\mathbf{F}}(x+i\varepsilon y) - \tilde{\mathbf{F}}(x-i\varepsilon y) \right].
\end{aligned} \qquad (9.68)
$$

What about analyticity in the general case? Recall that for free electromagnetic waves, $\hat{\mathbf{F}}(p) = \delta(p^2)\mathbf{f}(p)$, so $\hat{\mathbf{F}}$ is supported on the light cone $C = \{p : p^2 = 0\}$. More generally, suppose $\hat{\mathbf{F}}$ is supported inside the *solid* double cone

$$
\begin{aligned}
V &= \{p = (\mathbf{p}, p_0) : p^2 = p_0^2 - c^2|\mathbf{p}|^2 \geq 0\} = V_+ \cup V_-\,, \\
V_+ &= \{p = (\mathbf{p}, p_0) : p_0 \geq c\,|\mathbf{p}|\}, \qquad (9.69) \\
V_- &= \{p = (\mathbf{p}, p_0) : p_0 \leq -c\,|\mathbf{p}|\},
\end{aligned}
$$

where we have temporarily reinserted the speed of light c. (This will clarify the difference between $V_\pm$ and their dual cones $V'_\pm$, to be introduced below.)

V_+ is the *positive-frequency cone* and V_- is the *negative-frequency cone.* The positive- and negative-frequency *light* cones are the *boundaries* of $V_\pm$: $C_\pm = \partial V_\pm \subset V_\pm$. (See Figure 9.1.) $V_\pm$ are both *convex*, a property not shared by $C_\pm$. (This means that arbitrary linear combinations of vectors in $V_\pm$ with positive coefficients also belong to $V_\pm$.) The four-dimensional positive- and negative-frequency *cones* $V_\pm$ take the place of the positive- and negative-frequency *axes* in the one-dimensional case, i.e.,

$$V_+ \leftrightarrow \{p \in \mathbf{R} : p > 0\} \equiv \mathbf{R}_+ , \quad V_- \leftrightarrow \{p \in \mathbf{R} : p < 0\} \equiv \mathbf{R}_- . \tag{9.70}$$

The *dual cones* of V_+ and V_- are defined by

$$\begin{aligned} V'_+ &\equiv \{y \in \mathbf{R}^4 : p \cdot y > 0 \text{ for all } p \in V_+\} \\ V'_- &\equiv \{y \in \mathbf{R}^4 : p \cdot y > 0 \text{ for all } p \in V_-\}. \end{aligned} \tag{9.71}$$

To find V'_+ explicitly, let $y \in V'_+$. Then we have

$$p \cdot y = \omega y_0 - \mathbf{p} \cdot \mathbf{y} > 0 \quad \text{for all } p \in V_+ . \tag{9.72}$$

If $\mathbf{y} = \mathbf{0}$, then (9.72) implies $y_0 > 0$. If $\mathbf{y} \neq \mathbf{0}$, taking $\mathbf{p} = \omega \mathbf{y}/(c\,|\mathbf{y}|)$ in (9.72) gives

$$c\,y_0 - |\mathbf{y}| > 0. \tag{9.73}$$

This condition, in turn, is seen to be sufficient for y to belong to V'_+. If it holds, then by the Schwarz inequality in $\mathbf{R}^3$,

$$p \in V_+ \;\Rightarrow\; p \cdot y = \omega y_0 - \mathbf{p} \cdot \mathbf{y} \geq \omega y_0 - |\mathbf{p}|\,|\mathbf{y}| = |\mathbf{p}|(c\,y_0 - |\mathbf{y}|) > 0. \tag{9.74}$$

Thus we have shown that

$$V'_+ = \{y = (\mathbf{y}, y_0) : c\,y_0 > |\mathbf{y}|\}. \tag{9.75}$$

Similarly,

$$V'_- = \{y = (\mathbf{y}, y_0) : c\,y_0 < -|\mathbf{y}|\}. \tag{9.76}$$

V'_+ and V'_- are called the *future cone* and the *past cone* in space-time, respectively. The reason for the name is that an event at the space-time point x can causally influence an event at x' if and only if it is possible to travel from x to x' with a speed less than c, which is the case if and only if $x' - x \in V'_+$. Similarly, x can *be* influenced by x'' if and only if $x'' - x \in V'_-$. For this reason we call the double cone

$$V' \equiv V'_+ \cup V'_- = \{y : y^2 \equiv c^2 y_0^2 - |\mathbf{y}|^2 > 0\}, \tag{9.77}$$

the *causal cone.* This name will be seen to be justified in the case of electromagnetics in particular; we will arrive at a family of electromagnetic wavelets labeled by $z = x + iy$ with $y \in V'$, and it will turn out that $\mathbf{v} \equiv \mathbf{y}/y_0$ is the velocity of their *center.* But $|\mathbf{v}| < c$ means exactly the same thing as $y = (\mathbf{y}, y_0) \in V'$!

Note that C_+ represents the *extreme points* in V_+ (since $C_+ = \partial V_+$); hence in order to construct V'_+, we need only $p \in C_+$ instead of *all* positive-frequency wave vectors $p \in V_+$:

$$V'_+ = \{y : p \cdot y > 0 \quad \text{for all } p \in C_+\}. \tag{9.78}$$

Similarly, V'_- is "generated" by C_- .

We come at last to analyticity. In view of the above, define the *future tube* as the complex space-time region

$$\mathcal{T}_+ \equiv \{z = x + iy \in \mathbf{C}^4 : y \in V'_+\} \tag{9.79}$$

and the *past tube* as

$$\mathcal{T}_- \equiv \{z = x + iy \in \mathbf{C}^4 : y \in V'_-\}. \tag{9.80}$$

The future tube and the past tube are the multidimensional generalizations of the upper-half and the lower-half complex time planes $\mathbf{C}_+$ *and* $\mathbf{C}_-$ *.* See Stein and Weiss (1971) for a mathematical treatment of this general concept. These domains play an important role in rigorous mathematical treatments of quantum field theory; see Streater and Wightman (1964), Glimm and Jaffe (1981). They are also related to the theory of twistors; see Ward and Wells (1990).

The union

$$\mathcal{T} = \mathcal{T}_+ \cup \mathcal{T}_- \tag{9.81}$$

will be called the *causal tube.* Note that the two halves of $\mathcal{T}$ are disjoint, like $\mathbf{C}_+$ and $\mathbf{C}_-$.

Proposition 9.5. *Let* $\mathbf{F}(x)$ *be any electromagnetic wave in* $\mathcal{H}$*. Then its analytic signal* $\tilde{\mathbf{F}}(z)$ *is analytic in the causal tube* $\mathcal{T}$*.*

Proof: Inserting $\hat{\mathbf{F}}(p) = \delta(p^2)\,\mathbf{\Pi}(p)\mathbf{f}(p)$ into (9.56) gives

$$\tilde{\mathbf{F}}(x + iy) = \int_C d\tilde{p}\; 2\theta(p \cdot y)\, e^{ip\cdot(x+iy)}\mathbf{\Pi}(p)\mathbf{f}(p), \tag{9.82}$$

exactly as in Section 9.2. If $z \in \mathcal{T}_+$, i.e., $y \in V'_+$, then $p \cdot y > 0$ for all $p \in C_+$ and $p \cdot y < 0$ for all $p \in C_-$. Therefore (9.82) reduces to

$$\tilde{\mathbf{F}}(x + iy) = 2\int_{C_+} d\tilde{p}\;\; e^{ip\cdot(x+iy)}\mathbf{\Pi}(p)\mathbf{f}(p) \equiv \tilde{\mathbf{F}}_+(x + iy), \tag{9.83}$$

which is analytic. Similarly, if $z \in \mathcal{T}_-$, i.e., $y \in V'_-$, then $p \cdot y < 0$ for $p \in V_+$ and $p \cdot y > 0$ for $p \in V_-$, and thus

$$\tilde{\mathbf{F}}(x + iy) = 2\int_{C_-} d\tilde{p}\;\; e^{ip\cdot(x+iy)}\mathbf{\Pi}(p)\mathbf{f}(p) \equiv \tilde{\mathbf{F}}_-(x + iy), \tag{9.84}$$

which is also analytic. ■

Note that, *for* $z \in \mathcal{T}_+$, $\tilde{\mathbf{F}}(z)$ *contains only positive-frequency components, and for* $z \in \mathcal{T}_-$, $\tilde{\mathbf{F}}(z)$ *contains only negative-frequency components.* This is a generalization of the one-dimensional case, where the analytic signals in the upper-half time plane contain only positive frequencies and those in the lower-half time plane contain only negative frequencies. The analytic-signal transform therefore "polarizes" $\mathbf{F}(x)$: its positive-frequency part goes to $\mathcal{T}_+$, and its negative-frequency part goes to $\mathcal{T}_-$. We have seen that $\tilde{\mathbf{F}}_+$ and $\tilde{\mathbf{F}}_-$ have direct physical interpretations as the positive-helicity and negative-helicity components of the wave.

9.4 Electromagnetic Wavelets

Fix any $z = x + iy \in \mathcal{T}$, and define the linear mapping

$$\mathcal{E}_z : \mathcal{H} \to \mathbf{C}^3 \quad \text{by} \quad \mathcal{E}_z \mathbf{F} \equiv \tilde{\mathbf{F}}(z) = \int_C d\tilde{p}\ 2\theta(p\cdot y)\, e^{ip\cdot z}\, \mathbf{\Pi}(p)\, \mathbf{f}(p). \tag{9.85}$$

$\mathcal{E}_z$ is an *evaluation map*, giving the *value* of the analytic signal $\tilde{\mathbf{F}}$ at the fixed point z. We will show that $\mathcal{E}_z$ is a bounded operator from the Hilbert space $\mathcal{H}$ to the Hilbert space $\mathbf{C}^3$ (equipped with the standard inner product). Therefore, $\mathcal{E}_z$ has a bounded adjoint.

Definition 9.6. *The electromagnetic wavelets* $\mathbf{\Psi}_z$*,* $z \in \mathcal{T}$*, are defined as the adjoints of the evaluation maps* $\mathcal{E}_z$*:*

$$\mathbf{\Psi}_z = \mathcal{E}_z^* : \mathbf{C}^3 \to \mathcal{H}. \tag{9.86}$$

If we deal with scalar functions F rather than vector fields $\mathbf{F}$, then $\tilde{F}$ is just a complex-valued function and the wavelets are linear maps $\Psi_z : \mathbf{C} \to \mathcal{H}$. As explained in Section 1.3, such maps are essentially vectors in $\mathcal{H}$; namely, we can identify Ψ_z with $\Psi_z 1 \in \mathcal{H}$. (This will indeed be done for the *acoustic* wavelets in Chapter 11, since there we are dealing with scalar solutions of the wave equation.)

To find the $\mathbf{\Psi}_z$'s explicitly, choose any orthonormal basis $\mathbf{u}_1, \mathbf{u}_2, \mathbf{u}_3$ in $\mathbf{C}^3$ and let

$$\mathbf{\Psi}_{z,k} \equiv \mathbf{\Psi}_z \mathbf{u}_k \in \mathcal{H}, \qquad k = 1, 2, 3. \tag{9.87}$$

This gives three solutions of Maxwell's equations, all of which will be wavelets "at" z. $\mathbf{\Psi}_z$ is a *matrix-valued* solution of Maxwell's equations, obtained by putting the three (column) vector solutions $\mathbf{\Psi}_{z,k}$ together. The k-th component of $\tilde{\mathbf{F}}(z)$ with respect to the basis $\{\mathbf{u}_k\}$ is

$$\begin{aligned}\tilde{F}_k(z) &\equiv \mathbf{u}_k^* \, \tilde{\mathbf{F}}(z) = \mathbf{u}_k^* \, \mathcal{E}_z \mathbf{F} \\ &= \mathbf{u}_k^* \, \mathbf{\Psi}_z^* \mathbf{F} = (\mathbf{\Psi}_z \mathbf{u}_k)^* \mathbf{F} \\ &= \mathbf{\Psi}_{z,k}^* \, \mathbf{F}.\end{aligned} \tag{9.88}$$

By (9.85),

$$\mathbf{u}_k^* \tilde{\mathbf{F}}(z) = \int_C \frac{d\tilde{p}}{\omega^2} \, 2\omega^2 \, \theta(p \cdot y) \, e^{ip \cdot z} \, \mathbf{u}_k^* \, \mathbf{\Pi}(p) \, \mathbf{f}(p) \,, \tag{9.89}$$

which shows that $\mathbf{\Psi}_{z,k}$ is given in the Fourier domain by

$$\psi_{z,k}(p) = 2\omega^2 \, \theta(p \cdot y) \, e^{-ip \cdot \bar{z}} \, \mathbf{\Pi}(p) \, \mathbf{u}_k \, . \tag{9.90}$$

Note that each $\psi_{z,k}(p)$ satisfies the constraint since $\mathbf{\Gamma}(p) \, \mathbf{\Pi}(p) = \mathbf{\Pi}(p)$. The matrix-valued wavelet $\mathbf{\Psi}_z$ in the Fourier domain is therefore

$$\psi_z(p) = 2\omega^2 \, \theta(p \cdot y) \, e^{-ip \cdot \bar{z}} \, \mathbf{\Pi}(p). \tag{9.91}$$

In the space-time domain we have (using $\mathbf{\Pi}(p) \, \psi_z(p) = \psi_z(p)$)

$$\begin{aligned}\mathbf{\Psi}_z(x') &\equiv \int_C d\tilde{p} \;\; e^{ip \cdot x'} \; \psi_z(p) \\ &= \int_C d\tilde{p} \;\; 2\omega^2 \, \theta(p \cdot y) \, e^{ip \cdot (x' - \bar{z})} \, \mathbf{\Pi}(p).\end{aligned} \tag{9.92}$$

Now that we have the wavelets, we want to make them into a "basis" that can be used to decompose and compose arbitrary solutions. This will be accomplished by constructing a resolution of unity in terms of these wavelets. To this end, we derive an expression for the inner product in $\mathcal{H}$ directly in terms of the values $\tilde{\mathbf{F}}(z)$ of the analytic signals. To begin with, it will suffice to consider the values of $\tilde{\mathbf{F}}$ only at points with an imaginary time coordinate $z_0 = is$ and real space coordinates $\mathbf{z} = \mathbf{x}$. Therefore define

$$E \equiv \{z = (\mathbf{x}, is) : \mathbf{x} \in \mathbf{R}^3, \;\; 0 \neq s \in \mathbf{R}\} \subset \mathcal{T}. \tag{9.93}$$

In physics, E is called the *Euclidean region* because the negative of the indefinite Lorentz metric restricts to the positive-definite Euclidean metric on E:

$$-z^2 = -(is)^2 + |\mathbf{x}|^2 = s^2 + |\mathbf{x}|^2. \tag{9.94}$$

The Euclidean region plays an important role in constructive quantum field theory; see Glimm and Jaffe (1981). Later $\mathbf{x}$ will be interpreted as the center of the wavelets $\mathbf{\Psi}_{z,k}$, and s as their helicity and scale combined. By (9.82),

$$\begin{aligned}\tilde{\mathbf{F}}(\mathbf{x}, is) &= \int_C d\tilde{p}\; 2\theta(p_0 s)\, e^{-p_0 s - i\mathbf{p}\cdot\mathbf{x}}\, \mathbf{\Pi}(p)\, \mathbf{f}(p) \\ &= 2\int_{\mathbf{R}^3} d\tilde{p}\; e^{-i\mathbf{p}\cdot\mathbf{x}} \Big[\theta(\omega s)\, e^{-\omega s}\, \mathbf{\Pi}(\mathbf{p},\omega)\, \mathbf{f}(\mathbf{p},\omega) \\ &\qquad + \theta(-\omega s)\, e^{\omega s}\, \mathbf{\Pi}(\mathbf{p},-\omega)\, \mathbf{f}(\mathbf{p},-\omega)\Big] \\ &= \mathcal{F}_{\mathbf{p}}^{-1}\Big[\omega^{-1}\theta(s)\, e^{-\omega s}\, \mathbf{\Pi}(\mathbf{p},\omega)\, \mathbf{f}(\mathbf{p},\omega) \\ &\qquad + \omega^{-1}\theta(-s)\, e^{\omega s}\, \mathbf{\Pi}(\mathbf{p},-\omega)\, \mathbf{f}(\mathbf{p},-\omega)\Big](\mathbf{x}),\end{aligned} \tag{9.95}$$

where $\mathcal{F}_{\mathbf{p}}^{-1}$ denotes the inverse Fourier transform with respect to $\mathbf{p}$. Hence by Plancherel's formula,

$$\begin{aligned}\int_{\mathbf{R}^3} d^3\mathbf{x}\, |\tilde{\mathbf{F}}(\mathbf{x}, is)|^2 &= \int_{\mathbf{R}^3} \frac{d^3\mathbf{p}}{(2\pi)^3\omega^2}\Big[\theta(s)\, e^{-2\omega s}\, |\mathbf{\Pi}(\mathbf{p},\omega)\, \mathbf{f}(\mathbf{p},\omega)|^2 \\ &\qquad + \theta(-s)\, e^{2\omega s}\, |\mathbf{\Pi}(\mathbf{p},-\omega)\, \mathbf{f}(\mathbf{p},-\omega)|^2\Big],\end{aligned} \tag{9.96}$$

where we used $\theta(u)^2 = \theta(u)$ and $\theta(u)\,\theta(-u) = 0$ for $u \neq 0$. Thus

$$\begin{aligned}\int_E d^3\mathbf{x}\, ds\, |\tilde{\mathbf{F}}(\mathbf{x}, is)|^2 &= \int_{\mathbf{R}^3} \frac{d^3\mathbf{p}}{2(2\pi)^3\omega^3}\Big[\, |\mathbf{\Pi}(\mathbf{p},\omega)\, \mathbf{f}(\mathbf{p},\omega)|^2 \\ &\qquad + |\mathbf{\Pi}(\mathbf{p},-\omega)\, \mathbf{f}(\mathbf{p},-\omega)|^2\Big] \\ &= \int_C \frac{d\tilde{p}}{\omega^2}\, |\mathbf{\Pi}(p)\, \mathbf{f}(p)|^2 \\ &= \int_C \frac{d\tilde{p}}{\omega^2}\, \mathbf{f}(p)^*\, \mathbf{\Pi}(p)\, \mathbf{f}(p) = \|\mathbf{F}\|^2,\end{aligned} \tag{9.97}$$

since $\mathbf{\Pi}^*\mathbf{\Pi} = \mathbf{\Pi}^2 = \mathbf{\Pi}$. Let $\tilde{\mathcal{H}}$ be the set of all analytic signals $\tilde{\mathbf{F}}$ with $\mathbf{F} \in \mathcal{H}$, i.e.,

$$\tilde{\mathcal{H}} \equiv \{\tilde{\mathbf{F}} : \mathbf{F} \in \mathcal{H}\}. \tag{9.98}$$

Definition 9.7. *The inner product in* $\tilde{\mathcal{H}}$ *is defined by*

$$\langle\, \tilde{\mathbf{F}}, \tilde{\mathbf{G}}\, \rangle = \int_E d\mu(z)\, \tilde{\mathbf{F}}(z)^*\, \tilde{\mathbf{G}}(z) \equiv \tilde{\mathbf{F}}^*\tilde{\mathbf{G}}, \tag{9.99}$$

where

$$d\mu(z) \equiv d^3\mathbf{x}\; ds, \qquad z = (\mathbf{x}, is) \in E. \tag{9.100}$$

Theorem 9.8. $\tilde{\mathcal{H}}$ *is a Hilbert space under the inner product (9.99), and the map* $\mathbf{F} \mapsto \tilde{\mathbf{F}}$ *is unitary from* $\mathcal{H}$ *onto* $\tilde{\mathcal{H}}$, *so*

$$\langle\, \tilde{\mathbf{F}}, \tilde{\mathbf{G}}\, \rangle = \langle\, \mathbf{F}, \mathbf{G}\, \rangle. \tag{9.101}$$

Proof: Most of the work has already been done. Equation (9.97) states that the norms in $\mathcal{H}$ and $\tilde{\mathcal{H}}$ are equal, i.e., that we have the Plancherel-like identity $\|\tilde{\mathbf{F}}\|^2 = \|\mathbf{F}\|^2$. By polarization, this implies the Parseval-like identity (9.101), so the map is an isometry. Its range is, by definition, all of $\tilde{\mathcal{H}}$, which proves that the map $\mathbf{F} \mapsto \tilde{\mathbf{F}}$ is indeed unitary. ■

Using star notation, the Hermitian transpose $\tilde{\mathbf{F}}(z)^* : \mathbf{C}^3 \to \mathbf{C}$ of the column vector $\tilde{\mathbf{F}}(z) \in \mathbf{C}^3$ can be expressed as the composition

$$\tilde{\mathbf{F}}(z)^* = (\boldsymbol{\Psi}_z^* \mathbf{F})^* = \mathbf{F}^* \boldsymbol{\Psi}_z . \tag{9.102}$$

In diagram form,

$$\mathbf{C}^3 \xrightarrow{\;\boldsymbol{\Psi}_z\;} \mathcal{H} \xrightarrow{\;\mathbf{F}^*\;} \mathbf{C}. \tag{9.103}$$

Hence the integrand in (9.99) is

$$\tilde{\mathbf{F}}(z)^* \, \tilde{\mathbf{G}}(z) = (\boldsymbol{\Psi}_z^* \mathbf{F})^* \, \boldsymbol{\Psi}_z^* \mathbf{G} = \mathbf{F}^* \, \boldsymbol{\Psi}_z \, \boldsymbol{\Psi}_z^* \mathbf{G}, \tag{9.104}$$

where $\boldsymbol{\Psi}_z \, \boldsymbol{\Psi}_z^* : \mathcal{H} \to \mathcal{H}$ is the composition

$$\mathcal{H} \xrightarrow{\;\boldsymbol{\Psi}_z^*\;} \mathbf{C}^3 \xrightarrow{\;\boldsymbol{\Psi}_z\;} \mathcal{H}. \tag{9.105}$$

Therefore (9.101) reads

$$\int_E d\mu(z) \, \mathbf{F}^* \, \boldsymbol{\Psi}_z \, \boldsymbol{\Psi}_z^* \mathbf{G} = \mathbf{F}^* \, \mathbf{G}, \qquad \mathbf{F}, \mathbf{G} \in \mathcal{H}. \tag{9.106}$$

Theorem 9.9.
(a) The wavelets $\boldsymbol{\Psi}_z$ with $z \in E$ give the following resolution of the identity I in $\mathcal{H}$:

$$\int_E d\mu(z) \, \boldsymbol{\Psi}_z \, \boldsymbol{\Psi}_z^* = I, \tag{9.107}$$

where the equality holds weakly in $\mathcal{H}$, i.e., (9.106) is satisfied.
(b) Every solution $\mathbf{F} \in \mathcal{H}$ can be written as a superposition of the wavelets $\boldsymbol{\Psi}_z$ with $z = (\mathbf{x}, is) \in E$, according to

$$\mathbf{F} = \int_E d\mu(z) \, \boldsymbol{\Psi}_z \, \boldsymbol{\Psi}_z^* \mathbf{F} = \int_E d\mu(z) \, \boldsymbol{\Psi}_z \, \tilde{\mathbf{F}}(z), \tag{9.108}$$

i.e.,

$$\mathbf{F}(x') = \int_E d\mu(z) \, \boldsymbol{\Psi}_z(x') \, \tilde{\mathbf{F}}(z) \quad a.e. \tag{9.109}$$

(9.108) holds weakly *in $\mathcal{H}$, i.e., the inner products of both sides with any member of $\mathcal{H}$ are equal. However, for the analytic signals, we have*

$$\tilde{\mathbf{F}}(z') = \boldsymbol{\Psi}_{z'}^* \, \mathbf{F} = \int_E d\mu(z) \, \boldsymbol{\Psi}_{z'}^* \, \boldsymbol{\Psi}_z \, \tilde{\mathbf{F}}(z) \tag{9.110}$$

pointwise, *for all $z' \in \mathcal{T}$.*

Proof: Only the pointwise convergence in (9.110) remains to be shown. This follows from the boundedness of $\boldsymbol{\Psi}_{z'}$, which will be proved in the next section. ■

The pointwise equality fails, in general, for the boundary values $\mathbf{F}(x)$ because the evaluation maps (or, equivalently, their adjoints $\mathbf{\Psi}_z$) become unbounded as $y \to 0$. This will be seen in the next section.

The opposite composition $\mathbf{\Psi}_{z'}^* \mathbf{\Psi}_z : \mathbf{C}^3 \to \mathbf{C}^3$, i.e.,

$$\mathbf{C}^3 \xrightarrow{\mathbf{\Psi}_z} \mathcal{H} \xrightarrow{\mathbf{\Psi}_z^*} \mathbf{C}^3, \tag{9.111}$$

is a matrix-valued function on $\mathcal{T} \times \mathcal{T}$:

$$\begin{aligned} \mathbf{K}(z' \,|\, z) \equiv \mathbf{\Psi}_{z'}^* \mathbf{\Psi}_z &= \int_C \frac{d\tilde{p}}{\omega^2}\, 4\omega^4\, \theta(p \cdot y')\, \theta(p \cdot y)\, e^{ip\cdot(z' - \bar{z})}\, \mathbf{\Pi}(p)^2 \\ &= 4 \int_C d\tilde{p}\ \ \omega^2\, \theta(p \cdot y')\, \theta(p \cdot y)\, e^{ip\cdot(z' - \bar{z})}\, \mathbf{\Pi}(p). \end{aligned} \tag{9.112}$$

Equation (9.110) shows that $\mathbf{K}(z' \,|\, z)$ is a *reproducing kernel* for the Hilbert space $\tilde{\mathcal{H}}$. (See Chapter 4 for a brief account of reproducing kernels. A very nice introduction can be found in Hille (1972).) The boundary value of $\mathbf{K}(z' \,|\, z)$ as $y' \to 0$ is, according to (9.68) and (9.92), given by

$$\begin{aligned} \mathbf{K}(x' \,|\, z) &\equiv \tfrac{1}{2} \lim_{\varepsilon \to 0+} [\mathbf{K}(x' + i\varepsilon y' \,|\, z) + \mathbf{K}(x' - i\varepsilon y' \,|\, z)] \\ &= 2 \int_C d\tilde{p}\ \ \omega^2\, [\theta(p \cdot y') + \theta(-p \cdot y')]\, \theta(p \cdot y)\, e^{ip\cdot(z' - \bar{z})}\, \mathbf{\Pi}(p) \\ &= 2 \int_C d\tilde{p}\ \ \omega^2\, \theta(p \cdot y)\, e^{ip\cdot(z' - \bar{z})}\, \mathbf{\Pi}(p) \\ &= \mathbf{\Psi}_z(x'). \end{aligned} \tag{9.113}$$

In the next section we compute $\mathbf{\Psi}_z(x')$ and therefore, by analytic continuation, also $\mathbf{K}(z' \,|\, z)$.

Let us examine the nature of the matrix solution $\mathbf{\Psi}_z$. Recall that $\mathbf{\Psi}_z$ may be regarded as three ordinary (column vector) wavelet solutions $\mathbf{\Psi}_{z,k}$ combined. Since $\mathbf{\Pi}(p)$ is the orthogonal projection to the eigenspace of $\mathbf{\Gamma}(p)$ with the nondegenerate eigenvalue 1, all the columns (as well as the rows) of $\mathbf{\Pi}(p)$ are all multiples of one another. But the factors are p-dependent, and the algebraic linear dependence in Fourier space translates to a *differential equation* in space-time. For the *rows*, this differential equation is just Maxwell's vector equation (9.1) relating the different components of each wavelet $\mathbf{\Psi}_{z,k}$. (Recall that the scalar Maxwell equation is then implied by the wave equation.) Since $\mathbf{\Pi}(p)$ is Hermitian, the same argument goes for the columns. Explicitly, (9.91) together with $\mathbf{\Gamma\Pi} = \mathbf{\Pi} = \mathbf{\Pi\Gamma}$ implies

$$\mathbf{\Gamma}(p)\boldsymbol{\psi}_z(p) = \boldsymbol{\psi}_z(p) = \boldsymbol{\psi}_z(p)\mathbf{\Gamma}(p). \tag{9.114}$$

When multiplied through by p_0 and transformed to space-time, these equations read

$$\nabla' \times \mathbf{\Psi}_z(x') = -i\partial_0' \mathbf{\Psi}_z(x') = \mathbf{\Psi}_z(x') \times \overleftarrow{\nabla}', \tag{9.115}$$

where ∂_0' denotes the partial with respect to x_0', ∇' the gradient with respect to $\mathbf{x}'$, and $\overleftarrow{\nabla}'$ indicates that ∇' acts to the left, i.e., on the column index. This states that *not only the columns, but also the rows of $\mathbf{\Psi}_z$ are solutions of Maxwell's equations. The three wavelets $\mathbf{\Psi}_{z,k}$ are thus coupled.* Note also that since $\mathbf{\Psi}_z(x') = \mathbf{\Psi}_{z-x'}(0)$, equation (9.115) can be rewritten as

$$\nabla \times \mathbf{\Psi}_z = -i\partial_0 \mathbf{\Psi}_z = \mathbf{\Psi}_z \times \overleftarrow{\nabla}, \tag{9.116}$$

where now ∂_0 and ∇ are the corresponding operators with respect to the *labels* $x_0 = \operatorname{Re} z_0$ and $\mathbf{x} = \operatorname{Re} \mathbf{z}$.

We will see in Section 7 that the reconstruction of $\mathbf{F}$ from $\tilde{\mathbf{F}}$ can be obtained by a simpler method than (9.108), using *scalar wavelets* Ψ_z associated with the wave equation instead of the matrix wavelets $\mathbf{\Psi}_z$ associated with Maxwell's equations. However, that presumes that we already know $\tilde{\mathbf{F}}(\mathbf{x}, is)$, and without this knowledge the simpler reconstruction becomes meaningless, since no new solutions can be obtained this way. *The use of matrix wavelets will be necessary in order to give a generalization of (9.108) through (9.110), where $\tilde{\mathbf{F}}(z)$ can be replaced with an unconstrained coefficient function.* In other words, we need matrix wavelets in space-time for exactly the same reason that $\mathbf{\Pi}(p)$ was needed in Fourier space (9.21): to eliminate the constraints in the coefficient function.

9.5 Computing the Wavelets and Reproducing Kernel

To obtain detailed information about the wavelets, we need them in explicit form. Since the wavelets are boundary values of the reproducing kernel, our computation also gives an explicit form for that kernel as a byproduct. Note first of all that

$$\mathbf{\Psi}_z(x') = \int_C d\tilde{p}\ \ \omega^2\, 2\theta(p\cdot y)\, e^{ip\cdot(z-\bar{z}')}\, \mathbf{\Pi}(p) = \mathbf{\Psi}_{iy}(x'-x). \tag{9.117}$$

That is, all the wavelets can be obtained from $\mathbf{\Psi}_{iy}$ by space-time translations. Therefore it suffices to compute $\mathbf{\Psi}_{iy}(x)$, for $y \in V'$. But

$$\mathbf{\Psi}_{iy}(x) = \int_C d\tilde{p}\ \ \omega^2\, 2\theta(p\cdot y)\, e^{-p\cdot(y-ix)}\, \mathbf{\Pi}(p) \tag{9.118}$$

is analytic in $w \equiv y - ix$, for $iw \in \mathcal{T}$. Thus we need only compute $\mathbf{\Psi}_{iy}(0)$, since then $\mathbf{\Psi}_{iy}(x)$ can be obtained by analytic continuation. Now

$$\mathbf{\Psi}_{iy}(0) = \int_C d\tilde{p}\ \ \omega^2\, 2\theta(p\cdot y)\, e^{-p\cdot y}\, \mathbf{\Pi}(p) = \mathbf{\Psi}_{-iy}(0), \tag{9.119}$$

since $\mathbf{\Pi}(-p) = \mathbf{\Pi}(p)$. Therefore we can assume that $y \in V_+'$. Then the integral extends only over C_+ and

$$\mathbf{\Psi}_{iy}(0) = \int_{C_+} d\tilde{p}\ \ 2\omega^2\, e^{-p\cdot y}\, \mathbf{\Pi}(p). \tag{9.120}$$

Now the matrix elements of $2\omega^2\,\mathbf{\Pi}(p)$ are given by (9.20):

$$2\,\omega^2\,\Pi_{mn}(p) = \delta_{mn}p_0^2 - p_m p_n + i\sum_{k=1}^{3}\varepsilon_{mnk}\,p_0 p_k\,. \tag{9.121}$$

(Recall that ε_{mnk} is the completely antisymmetric tensor with $\varepsilon_{123} = 1$.) To compute $\mathbf{\Psi}_{iy}(0)$, it is useful to write the coordinates of y in *contravariant* form:

$$\begin{cases} y^0 = y_0 \\ y^1 = -y_1 \\ y^2 = -y_2 \\ y^3 = -y_3 \end{cases} \quad\Rightarrow\quad p\cdot y = \sum_{\alpha=0}^{3} p_\alpha\, y^\alpha. \tag{9.122}$$

(This is related to the Lorentz metric; p is really in the *dual* $\mathbf{R}_4$ of the space-time $\mathbf{R}^4$, as explained in Sections 1.1 and 1.2. By writing the pairing beween $p \in \mathbf{R}_4$ and $y \in \mathbf{R}^4$ as $p \cdot y$, we have implicitly *identified* y with $y^* \in \mathbf{R}_4$, so that the pairing can be expressed as an inner product. The above "contravariant" form reverts y to $\mathbf{R}^4$. See the discussion below (1.85).)

To evaluate (9.120), let

$$S(y) \equiv \int_{C_+} d\tilde{p}\;\; e^{-p\cdot y}, \qquad y \in V'_+\,. \tag{9.123}$$

We will evaluate $S(y)$ below using symmetry arguments. $S(y)$ acts as a generating function for the matrix elements of $\mathbf{\Psi}_{iy}(0)$, as follows: Let ∂_α denote the partial derivative with respect to y^α for $\alpha = 0, 1, 2, 3$. Then

$$\int_{C_+} d\tilde{p}\;\; p_\alpha\, p_\beta\, e^{-p\cdot y} = \partial_\alpha\,\partial_\beta\, S(y) \equiv S_{\alpha\beta}(y), \quad \alpha,\beta = 0,1,2,3. \tag{9.124}$$

By (9.120) and (9.121), the matrix elements of $\mathbf{\Psi}_{iy}(0)$ are therefore

$$[\mathbf{\Psi}_{iy}(0)]_{m,n} = \delta_{mn}S_{00}(y) - S_{mn}(y) + i\sum_{k=1}^{3}\varepsilon_{mnk}\,S_{0k}(y), \quad m,n = 1,2,3. \tag{9.125}$$

It remains only to compute $S(y)$. For this, we use the fact that $S(y)$ is invariant under Lorentz transformations, since $p \cdot y$ and $d\tilde{p}$ are both invariant and C_+ is a homogeneous space for the proper Lorentz group. Since $y \in V'_+$, there exists a Lorentz transformation mapping y to $(\mathbf{0}, \lambda)$, where

$$\lambda(y) \equiv \sqrt{y^2} = \sqrt{y_0^2 - |\mathbf{y}|^2} > 0. \tag{9.126}$$

The invariance of S implies that $S(y) = S(\mathbf{0}, \lambda)$. Therefore

$$\begin{aligned} S(y) &= \int_{\mathbf{R}^3} \frac{d^3\mathbf{p}}{16\pi^3|\mathbf{p}|}\, e^{-\lambda|\mathbf{p}|} = \frac{1}{4\pi^2}\int_0^\infty \omega\, d\omega\, e^{-\omega\lambda} \\ &= \frac{1}{4\pi^2\lambda^2} = \frac{1}{4\pi^2 y^2}\,. \end{aligned} \tag{9.127}$$

Taking partials with respect to y^α and y^β gives

$$S_{\alpha\beta}(y) = \frac{4y_\alpha y_\beta - g_{\alpha\beta}\, y^2}{2\pi^2 (y^2)^3}, \qquad \alpha, \beta = 0, 1, 2, 3, \tag{9.128}$$

where $g_{\alpha\beta} = \mathrm{diag}(1, -1, -1, -1)$ is the Lorentz metric. It follows that

$$[\mathbf{\Psi}_{iy}(0)]_{m,n} = \frac{\delta_{mn}(y_0^2 + y_1^2 + y_2^2 + y_3^2) - 2y_m y_n + 2i \sum_{k=1}^{3} \varepsilon_{mnk} y_0 y_k}{\pi^2 (y^2)^3}. \tag{9.129}$$

Theorem 9.10. *The matrix elements* $[\mathbf{\Psi}_z(x')]_{m,n}$ *of* $\mathbf{\Psi}_z(x')$ *are*

$$\frac{\delta_{mn}(w_0^2 + w_1^2 + w_2^2 + w_3^2) - 2w_m w_n + 2i \sum_{k=1}^{3} \varepsilon_{mnk} w_0 w_k}{\pi^2 (w^2)^3}, \tag{9.130}$$

where

$$\begin{aligned} w &\equiv i(\bar{z} - x') = y + i(x - x') \\ w^2 &= w \cdot w = w_0^2 - w_1^2 - w_2^2 - w_3^2 . \end{aligned} \tag{9.131}$$

The reproducing kernel is given in terms of the wavelets by

$$\mathbf{K}(z' \,|\, z) = 2\theta(y' \cdot y)\, \mathbf{\Psi}_{z - \bar{z}'}(0). \tag{9.132}$$

The matrix elements of the wavelets $\mathbf{\Psi}_z(x')$ *and the kernel* $\mathbf{K}(z' \,|\, z)$ *are finite whenever* z' *and* z *belong to* $\mathcal{T}$, *since*

$$z^2 \neq 0 \quad \textit{for all} \quad z \in \mathcal{T}. \tag{9.133}$$

Proof: By analyticity, to find $\mathbf{\Psi}_z(x')$ we need only replace y by $w = y + i(x - x')$ in (9.129). (It can be shown that the analytic continuation is unique, i.e., that all paths from y to w give the same result. See Streater and Wightman (1964).) This proves (9.130). To prove (9.132), note that $\mathbf{K}(z' \,|\, z)$ vanishes when y' and y are in opposite halves of V', since $p \cdot y'$ and $p \cdot y$ have opposite signs in (9.112). If y' and y belong to the same half of V', then $p \cdot y'$ and $p \cdot y$ have the same sign, and we may replace $\theta(p \cdot y')\theta(p \cdot y)$ by $\theta(p \cdot y)$ in (9.112). But then

$$\mathbf{K}(z' \,|\, z) = \int_C d\tilde{p}\; 4\omega^2 \theta(p \cdot y) e^{ip \cdot (z' - \bar{z})} \mathbf{\Pi}(p) = 2\mathbf{\Psi}_{z - \bar{z}'}(0). \tag{9.134}$$

For general $z', z \in \mathcal{T}$, the kernel is therefore obtained by multiplying (9.134) by $\theta(y' \cdot y)$. Finally, to show that all the matrix elements are finite, it suffices to prove (9.133). Suppose that $z^2 = 0$ for some $z \in \mathcal{T}$. Then

$$x^2 - y^2 = 0, \qquad x \cdot y = x_0 y_0 - \mathbf{x} \cdot \mathbf{y} = 0.$$

The second equation implies

$$|x_0| \leq \frac{|\mathbf{x}|\, |\mathbf{y}|}{|y_0|} \leq |\mathbf{x}|,$$

since $y \in V'$. (The second inequality is strict unless $\mathbf{x} = \mathbf{0}$.) But then $x^2 = x_0^2 - |\mathbf{x}|^2 \le 0$, hence $y^2 > 0$ implies that $x^2 - y^2 < 0$. This contradicts the first equation. ■

Note that if z' and z are in the *same* half of $\mathcal{T}$, then $z - \bar{z}' = (x - x') + i(y + y')$ is also in that same half, since V'_+ and V'_- are each *convex*. Therefore $z - \bar{z}' \in \mathcal{T}$, and $\mathbf{\Psi}_{z-\bar{z}'}$ exists as a bounded map from $\mathbf{C}^3$ to $\mathcal{H}$. (This has been tacitly assumed for all $\mathbf{\Psi}_z$ with $z \in \mathcal{T}$, and will be proved below.) If z' and z are in *opposite* halves of $\mathcal{T}$, then $y' + y$ may not belong to V' (since $V' = V'_+ \cup V'_-$ is *not* convex); thus $z - \bar{z}'$ may not belong to $\mathcal{T}$ and therefore $\mathbf{\Psi}_{z-\bar{z}'}(0)$ need not exist. But then we are saved by the factor $\theta(y' \cdot y)$ in (9.132), since it vanishes!

In Section 4 we stated that the evaluation maps $\mathcal{E}_z$ are bounded, and that they become unbounded as $y \to 0$. The boundedness of $\mathcal{E}_z$ is of fundamental importance, since otherwise $\mathbf{\Psi}_z \equiv \mathcal{E}_z^*$ does not exist as a (bounded) map from $\mathbf{C}^3$ to $\mathcal{H}$ and we have no (normalizable) wavelets.† The statement that $\mathcal{E}_z : \mathcal{H} \to \mathbf{C}^3$ is bounded means that there is a constant N, (which turns out to depend on z) such that

$$\|\mathcal{E}_z \mathbf{F}\|^2 \equiv \|\tilde{\mathbf{F}}(z)\|^2_{\mathbf{C}^3} \le N(z)\|\mathbf{F}\|^2. \tag{9.135}$$

But in terms of an orthonormal basis $\{\mathbf{u}_k\}$ in $\mathbf{C}^3$ and the corresponding (vector) wavelets $\mathbf{\Psi}_{z,k} = \mathbf{\Psi}_z \mathbf{u}_k$, the Schwarz inequality gives

$$\begin{aligned} \|\tilde{\mathbf{F}}(z)\|^2_{\mathbf{C}^3} &= \sum_k |\mathbf{\Psi}^*_{z,k}\mathbf{F}|^2 \\ &\le \sum_k \|\mathbf{\Psi}_{z,k}\|^2 \, \|\mathbf{F}\|^2 \\ &= \|\mathbf{F}\|^2 \sum_k \mathbf{u}_k^* \mathbf{\Psi}_z^* \mathbf{\Psi}_z \mathbf{u}_k \\ &= \|\mathbf{F}\|^2 \operatorname{trace} \mathbf{\Psi}_z^* \mathbf{\Psi}_z \,, \end{aligned} \tag{9.136}$$

so (9.135) holds with

$$N(z) = \operatorname{trace} \mathbf{\Psi}_z^* \mathbf{\Psi}_z = \operatorname{trace} \mathbf{K}(z \,|\, z). \tag{9.137}$$

In fact, the proof shows that $N(z)$, as given by (9.137), is the *best* such bound.

Theorem 9.11. *For any $z \in \mathcal{T}$, the evaluation map $\mathcal{E}_z : \mathcal{H} \to \mathbf{C}^3$ is bounded. It becomes unbounded whenever z approaches the 7-dimensional boundary of $\mathcal{T}$,*

$$\partial\mathcal{T} = \{x + iy : y^2 \equiv y_0^2 - |\mathbf{y}|^2 = 0\}. \tag{9.138}$$

† If our solutions were scalar valued, as will be the case for acoustics in Chapter 11, then $\mathcal{E}_z$ would be a linear functional on $\mathcal{H}$ and Ψ_z would be its unique representative in $\mathcal{H}$, as guaranteed by the Riesz representation theorem. That theorem fails for *unbounded* linear functionals!

Therefore, the electromagnetic wavelets $\mathbf{\Psi}_z$ *exist as bounded linear maps from* $\mathbf{C}^3$ *to* $\mathcal{H}$, *for every* $z \in \mathcal{T}$. *They satisfy*

$$\|\mathbf{\Psi}_z^* \mathbf{F}\|^2 \le N(z) \|\mathbf{F}\|^2, \tag{9.139}$$

where $N(z)$ *is the trace of the reproducing kernel,*

$$N(z) = \text{trace}\, \mathbf{\Psi}_z^* \mathbf{\Psi}_z = \frac{1}{8\pi^2} \frac{3y_0^2 + |\mathbf{y}|^2}{(y_0^2 - |\mathbf{y}|^2)^3}. \tag{9.140}$$

For the wavelets labeled by the Euclidean region,

$$\mathbf{K}(\mathbf{x}, is \,|\, \mathbf{x}, is) = \frac{1}{8\pi^2 s^4} I, \tag{9.141}$$

where I *is the* 3×3 *identity matrix.*

Proof: By (9.132),

$$\mathbf{K}(z \,|\, z) = 2\theta(y^2) \mathbf{\Psi}_{2iy}(0) = 2\mathbf{\Psi}_{2iy}(0), \tag{9.142}$$

since $y^2 > 0$ in V'. Therefore by (9.129),

$$\begin{aligned} N(z) = 2\, \text{trace}\, \mathbf{\Psi}_{2iy}(0) &= \frac{1}{8\pi^2} \frac{3(y_0^2 + |\mathbf{y}|^2) - 2|\mathbf{y}|^2}{(y_0^2 - |\mathbf{y}|^2)^3} \\ &= \frac{1}{8\pi^2} \frac{3y_0^2 + |\mathbf{y}|^2}{(y_0^2 - |\mathbf{y}|^2)^3}, \end{aligned} \tag{9.143}$$

so (9.139) and (9.140) hold in view of (9.136). If $y^2 = 0$, then all the matrix elements of $\mathbf{K}(z \,|\, z)$ diverge, and (9.136) shows that $\mathbf{\Psi}_z$ is unbounded. Finally, (9.141) follows from (9.142) and (9.129) with $\mathbf{y} = \mathbf{0}$ and $y_0 = s$. ■

9.6 Atomic Composition of Electromagnetic Waves

The reproducing kernel computed in the last section can be used to *construct* electromagnetic waves according to local specifications, rather than merely to *reconstruct* known solutions from their analytic signals on E. This is especially interesting because the Fourier method for constructing solutions (Section 9.2) uses plane waves and is therefore completely unsuitable to deal with questions involving local properties of the fields. It will be shown in Section 9.7 that the wavelets $\mathbf{\Psi}_{x+iy}(x')$ are localized solutions of Maxwell's equations at the "initial" time $x_0' = x_0$. Hence we call the composition of waves from wavelets "atomic." (See Coifman and Rochberg (1980) for "atomic decompositions" of entire functions; see also Feichtinger and Gröchenig (1986, 1989 a, b), and Gröchenig (1991), for other atomic decompositions.)

Suppose $\tilde{\mathbf{F}}$ is the analytic signal of a solution $\mathbf{F} \in \mathcal{H}$ of Maxwell's equations. Then according to Theorem 9.8,

$$\int_E d\mu(z)\, |\tilde{\mathbf{F}}(z)|^2 = \|\mathbf{F}\|^2 < \infty. \tag{9.144}$$

Let $\mathcal{L}^2(E)$ be the set of *all* measurable functions $\mathbf{\Phi} : E \to \mathbf{C}^3$ for which the above integral converges. $\mathcal{L}^2(E)$ is a Hilbert space under the obvious inner product, obtained from (9.144) by polarization.† Define the map $R_E : \mathcal{H} \to \mathcal{L}^2(E)$ by

$$(R_E\, \mathbf{F})(z) \equiv \mathbf{\Psi}_z^* \mathbf{F} = \tilde{\mathbf{F}}(z), \quad z = (\mathbf{x}, is) \in E. \tag{9.145}$$

That is, $R_E\, \mathbf{F}$ is the restriction $\tilde{\mathbf{F}}\,|_E$ to E of the analytic signal $\tilde{\mathbf{F}}$ of $\mathbf{F}$. Then (9.144) implies that the *range* $\mathcal{A}_E$ of R_E is a closed subspace of $\mathcal{L}^2(E)$, and R_E maps $\mathcal{H}$ isometrically onto $\mathcal{A}_E$. The following theorem characterizes the range of R_E and gives the adjoint R_E^*.

Theorem 9.12.
(a) The range of R_E is the set $\mathcal{A}_E$ of all $\mathbf{\Phi} \in \mathcal{L}^2(E)$ satisfying the "consistency condition"

$$\mathbf{\Phi}(z') = \int_E d\mu(z)\, \mathbf{K}(z' \,|\, z)\, \mathbf{\Phi}(z), \tag{9.146}$$

pointwise in $z' \in E$.
(b) The adjoint operator $R_E^* : \mathcal{L}^2(E) \to \mathcal{H}$ is given by

$$R_E^* \mathbf{\Phi} = \int_E d\mu(z)\, \mathbf{\Psi}_z\, \mathbf{\Phi}(z), \tag{9.147}$$

where now the integral converges weakly in $\mathcal{H}$. That is,

$$R_E^* \mathbf{\Phi}(x') = \int_E d\mu(z)\, \mathbf{\Psi}_z(x')\, \mathbf{\Phi}(z) \quad a.e. \tag{9.148}$$

Proof: If $\mathbf{\Phi} \in \mathcal{A}_E$, then $\mathbf{\Phi}(z) = \tilde{\mathbf{F}}(z)$ for some $\mathbf{F} \in \mathcal{H}$, and (9.146) reduces to (9.110), which holds pointwise in $z' \in E$ by Theorem 9.9. On the other hand, given a function $\mathbf{\Phi} \in \mathcal{L}^2(E)$ that satisfies (9.146), let $\mathbf{F}$ denote the right-hand side of (9.146). Then for any $\mathbf{G} \in \mathcal{H}$,

$$\mathbf{G}^* \mathbf{F} = \int_E d\mu(z)\, \tilde{\mathbf{G}}(z)^* \mathbf{\Phi}(z) = \langle\, R_E\, \mathbf{G}, \mathbf{\Phi} \,\rangle_{\mathcal{L}^2}, \tag{9.149}$$

† We could identify E with $\mathbf{R}^4 \approx \{(\mathbf{x}, is) \in \mathbf{C}^4 : \mathbf{x} \in \mathbf{R}^3, s \in \mathbf{R}\}$ and $\mathcal{L}^2(E)$ with $L^2(\mathbf{R}^4)$, since the set $\mathbf{R}^4 \backslash E = \{(\mathbf{x}, 0) : \mathbf{x} \in \mathbf{R}^3\}$ has zero measure in $\mathbf{R}^4$. But this could cause confusion between the Euclidean region E and the *real* space-time $\mathbf{R}^4$.

where we have used $\mathbf{G}^*\mathbf{\Psi}_z = (\mathbf{\Psi}_z^*\mathbf{G})^* = \tilde{\mathbf{G}}(z)^*$. Hence the integral in (9.147) converges weakly in $\mathcal{H}$. The transform of $\mathbf{F}$ under R_E is

$$\begin{aligned}(R_E\,\mathbf{F})(z') = \mathbf{\Psi}_{z'}^*\mathbf{F} &= \int_E d\mu(z)\,\mathbf{\Psi}_{z'}^*\mathbf{\Psi}_z\mathbf{\Phi}(z)\\ &= \int_E d\mu(z)\,\mathbf{K}(z'\,|\,z)\,\mathbf{\Phi}(z) = \mathbf{\Phi}(z'),\end{aligned} \tag{9.150}$$

by (9.146). Hence $\mathbf{\Phi} \in \mathcal{A}_E$ as claimed, proving (a). Equation (9.149) states that $\langle\,\mathbf{G}, \mathbf{F}\,\rangle_{\mathcal{H}} = \langle\, R_E\,\mathbf{G}, \mathbf{\Phi}\,\rangle_{\mathcal{L}^2}$ for all $\mathbf{G} \in \mathcal{H}$. This shows that $\mathbf{F} = R_E^*\mathbf{\Phi}$, hence (b) is true. ■

According to (9.147), R_E^* constructs a solution $R_E^*\mathbf{\Phi} \in \mathcal{H}$ from an arbitrary coefficient function $\mathbf{\Phi} \in \mathcal{L}^2(E)$. When $\mathbf{\Phi}$ is actually the transform $R_E\,\mathbf{F}$ of a solution $\mathbf{F} \in \mathcal{H}$, then $\mathbf{\Phi}(z) = \mathbf{\Psi}_z^*\mathbf{F}$ and

$$R_E^*R_E\,\mathbf{F} = R_E^*\mathbf{\Phi} = \int_E d\mu(z)\,\mathbf{\Psi}_z\,\mathbf{\Psi}_z^*\mathbf{F} = \mathbf{F}, \tag{9.151}$$

by (9.108). Thus $R_E^*R_E = I$, the identity in $\mathcal{H}$. (This is equivalent to (9.144).) We now examine the opposite composition.

Theorem 9.13. *The orthogonal projection to $\mathcal{A}_E$ in $\mathcal{L}^2(E)$ is the composition $P \equiv R_E R_E^* : \mathcal{L}^2(E) \to \mathcal{L}^2(E)$, given by*

$$P\mathbf{\Phi}(z') \equiv \int_E d\mu(z)\,\mathbf{K}(z'\,|\,z)\,\mathbf{\Phi}(z). \tag{9.152}$$

Proof: By (9.147),

$$(R_E R_E^*\mathbf{\Phi})(z') \equiv \mathbf{\Psi}_{z'}^* R_E^*\mathbf{\Phi} = \int_E d\mu(z)\,\mathbf{\Psi}_{z'}^*\mathbf{\Psi}_z\mathbf{\Phi}(z) = P\mathbf{\Phi}(z'), \tag{9.153}$$

since $\mathbf{\Psi}_{z'}^*\mathbf{\Psi}_z = \mathbf{K}(z'\,|\,z)$. Hence $R_E R_E^* = P$. This also shows that $P^* = P$. Furthermore, $R_E^*R_E = I$ implies that $P^2 = P$; hence P is indeed the orthogonal projection to its range. It only remains to show that the range of P is $\mathcal{A}_E$. If $\mathbf{\Phi} = R_E\,\mathbf{F} \in \mathcal{A}_E$, then $R_E R_E^*\mathbf{\Phi} = R_E R_E^* R_E\,\mathbf{F} = R_E\,\mathbf{F} = \mathbf{\Phi}$. Conversely, any function in the range of P has the form $\mathbf{\Phi} = R_E R_E^*\mathbf{\Theta}$ for some $\mathbf{\Theta} \in \mathcal{L}^2(E)$; hence $\mathbf{\Phi} = R_E\mathbf{F}$ where $\mathbf{F} = R_E^*\mathbf{\Theta} \in \mathcal{H}$. ■

When the coefficient function $\mathbf{\Phi}$ in (9.147) is the transform of an actual solution, then R_E^* reconstructs that solution. However, this process does not appear to be too interesting, since we must have a complete knowledge of $\mathbf{F}$ to compute $\tilde{\mathbf{F}}(z)$. For example, if $z = (\mathbf{x}, is) \in E$, then (9.53) gives

$$\begin{aligned}\tilde{\mathbf{F}}(\mathbf{x}, is) &= \frac{1}{\pi i}\int_{-\infty}^{\infty} \frac{d\tau}{\tau - i}\,\mathbf{F}((\mathbf{x}, 0) + \tau(\mathbf{0}, s))\\ &= \frac{1}{\pi i}\int_{-\infty}^{\infty} \frac{d\tau}{\tau - i}\,\mathbf{F}(\mathbf{x}, \tau s).\end{aligned} \tag{9.154}$$

So we must know $\mathbf{F}(\mathbf{x}, t)$ for all $\mathbf{x}$ and all t in order to compute $\tilde{\mathbf{F}}$ on E. Therefore, no "initial-value problem" is solved by (9.108) and (9.109) when applied to $\mathbf{\Phi} = \tilde{\mathbf{F}} \in \mathcal{A}_E$. However, the option of applying (9.147) and (9.148) to *arbitrary* $\mathbf{\Phi} \in \mathcal{L}^2(E)$ is a very attractive one, since it is guaranteed to produce a solution without any assumptions on $\mathbf{\Phi}$ other than square-integrability. Moreover, the choice of $\mathbf{\Phi}$ gives some control over the local features of the field $R_E^* \mathbf{\Phi}$ at $t = 0$. It is appropriate to call R_E^* the *synthesizing operator* associated with the resolution of unity (9.107) (see Chapter 4). It can be used to build solutions in $\mathcal{H}$ from *unconstrained* functions $\mathbf{\Phi} \in \mathcal{L}^2(E)$. In fact, it is interesting to compare the wavelet construction formula

$$\mathbf{F}(x') = \int_E d\mu(z)\, \mathbf{\Psi}_z(x')\, \mathbf{\Phi}(z) \tag{9.155}$$

directly with its Fourier counterpart (9.21):

$$\mathbf{F}(x') = \int_C d\tilde{p}\; e^{ip\cdot x'}\, \mathbf{\Pi}(p)\, \mathbf{f}(p). \tag{9.156}$$

In both cases, the coefficient functions ($\mathbf{\Phi}$ and $\mathbf{f}$) are unconstrained (except for the square-integrability requirements). The building blocks in (9.155) are the matrix-valued wavelets parameterized by E, whereas those in (9.156) are the matrix-valued plane-wave solutions $e^{ip\cdot x'}\, \mathbf{\Pi}(p)$ parameterized by C. E plays the role of a *wavelet space*, analogous to the time-scale plane in one dimension, and replaces the light cone C as a parameter space for "elementary" solutions.

9.7 Interpretation of the Wavelet Parameters

Our goal in this section is twofold: (a) To reduce the wavelets $\mathbf{\Psi}_z$ to a sufficiently simple form that can actually be visualized, and (b) to use the ensuing picture to give a complete physical and geometric interpretation of the eight complex space-time parameters $z \in \mathcal{T}$ labeling $\mathbf{\Psi}_z$. That the wavelets can be visualized at all is quite remarkable, since $\mathbf{\Psi}_z(x')$ is a complex matrix-valued function of $x' \in \mathbf{R}^4$ and $z \in \mathcal{T}$. However, the symmetries of Maxwell's equations can be used to reduce the number of effective variables one by one, until all that remains is a single complex-valued function of two real variables whose real and imaginary parts can be graphed separately.

We begin by showing that the parameters $z \in \mathcal{T}$ can be eliminated entirely. Note that since $\mathbf{\Pi}(p)^* = \mathbf{\Pi}(p) = \mathbf{\Pi}(-p)$, the Hermitian adjoint of $\mathbf{\Psi}_z(x')$ is

$$\begin{aligned} \mathbf{\Psi}_z(x')^* &= \int_C d\tilde{p}\; 2\omega^2 \theta(p\cdot y) e^{-ip\cdot(x'-z)} \mathbf{\Pi}(p) \\ &= \int_C d\tilde{p}\; 2\omega^2 \theta(-p\cdot y) e^{ip\cdot(x'-z)} \mathbf{\Pi}(p) \\ &= \mathbf{\Psi}_{\bar{z}}(x'). \end{aligned} \tag{9.157}$$

(9.167) help us find $\mathbf{\Psi}_{\Lambda z}$ if we know $\mathbf{\Psi}_z$? Use the unitarity of U_Λ! For $z, w \in \mathcal{T}$, we have

$$\begin{aligned} \mathbf{\Psi}_w^* \mathbf{\Psi}_z &= \mathbf{\Psi}_w^* U_\Lambda^* U_\Lambda \mathbf{\Psi}_z \\ &= (U_\Lambda \mathbf{\Psi}_w)^* U_\Lambda \mathbf{\Psi}_z \\ &= (\mathbf{\Psi}_{\Lambda w} M_{\Lambda^{-1}}^*)^* (\mathbf{\Psi}_{\Lambda z} M_{\Lambda^{-1}}^*) \\ &= M_{\Lambda^{-1}} \mathbf{\Psi}_{\Lambda w}^* \mathbf{\Psi}_{\Lambda z} M_{\Lambda^{-1}}^* \end{aligned} \tag{9.168}$$

Using $M_\Lambda^{-1} = M_{\Lambda^{-1}}$, we finally obtain

$$\mathbf{K}(\Lambda w \,|\, \Lambda z) \equiv \mathbf{\Psi}_{\Lambda w}^* \mathbf{\Psi}_{\Lambda z} = M_\Lambda \mathbf{\Psi}_w^* \mathbf{\Psi}_z M_\Lambda^* = M_\Lambda \mathbf{K}(w \,|\, z) M_\Lambda^* \,. \tag{9.169}$$

This shows that the reproducing kernel transforms "covariantly." To get the kernel in the new coordinate frame, simply multiply by two 3×3 matrices on both sides to adjust the field values. (See also Jacobsen and Vergne (1977), where such invariances of reproducing kernels are used to *construct* unitary representations of the conformal group.) To find $\mathbf{\Psi}_{\Lambda z}(x')$ from $\mathbf{\Psi}_z(x')$, we now take the boundary values of $\mathbf{K}(\Lambda z' \,|\, \Lambda z)$ as $y' \to 0$, according to (9.68), and use (9.113):

$$\mathbf{\Psi}_{\Lambda z}(\Lambda x') = M_\Lambda \mathbf{\Psi}_z(x') M_\Lambda^* \,, \quad \Rightarrow \quad \mathbf{\Psi}_{\Lambda z}(x') = M_\Lambda \mathbf{\Psi}_z(\Lambda^{-1} x') M_\Lambda^* \,. \tag{9.170}$$

Returning to our problem of finding $\mathbf{\Psi}_{iy}$ with $y \in V'_+$, recall that we can write $y = \Lambda y'$, where $y' = (\mathbf{0}, \lambda)$ with

$$\lambda = \sqrt{y^2} = \sqrt{y_0^2 - |\mathbf{y}|^2} > 0 \quad \text{and} \quad \mathbf{v}_\Lambda = \frac{\mathbf{y}}{y_0} \,. \tag{9.171}$$

Then (9.170) gives

$$\mathbf{\Psi}_{iy}(x') = M_\Lambda \mathbf{\Psi}_{\mathbf{0},i\lambda}(\Lambda^{-1} x') M_\Lambda^* \,. \tag{9.172}$$

Since M_Λ is known (although we have not given its explicit form), (9.172) enables us to find $\mathbf{\Psi}_{iy}$ explicitly, once we know $\mathbf{\Psi}_{\mathbf{0},i\lambda}$ for $\lambda > 0$.

So far, we have used the properties of the wavelets under conjugations, translations and Lorentz transformations. We still have one parameter left, namely, $\lambda > 0$. This can now be eliminated using dilations. From (9.130) it follows that, for any $z \in \mathcal{T}$,

$$\mathbf{\Psi}_{az}(ax') = a^{-4} \mathbf{\Psi}_z(x'), \qquad a \neq 0. \tag{9.173}$$

Therefore

$$\mathbf{\Psi}_{\mathbf{0},i\lambda}(x) = \lambda^{-4} \mathbf{\Psi}_{\mathbf{0},i}(\lambda^{-1} x). \tag{9.174}$$

We have now arrived at a single wavelet from which all others can be obtained by dilations, Lorentz transformations, translations and conjugation. We therefore denote it without any label and call it a

$$\textit{reference wavelet:} \qquad \mathbf{\Psi} \equiv \mathbf{\Psi}_{\mathbf{0},i} \,. \tag{9.175}$$

Of course, this choice is completely arbitrary: Since we can go from $\boldsymbol{\Psi}$ to $\boldsymbol{\Psi}_z$, we can also go the other way. Therefore we could have chosen *any* of the $\boldsymbol{\Psi}_z$'s as the reference wavelet.† The wavelets with $\mathbf{y}=\mathbf{0}$, i.e., $z=x+i(\mathbf{0},s)$ have an especially simple form, reminiscent of the one-dimensional wavelets:

$$\boldsymbol{\Psi}_{\mathbf{x},t+is}(x') = s^{-4}\boldsymbol{\Psi}\left(\frac{x'-x}{s}\right) = s^{-4}\boldsymbol{\Psi}\left(\frac{\mathbf{x}'-\mathbf{x}}{s}, \frac{t'-t}{s}\right). \tag{9.176}$$

In particular, this gives all the wavelets $\boldsymbol{\Psi}_{\mathbf{0},is}$ labeled by E, which were used in the resolution of unity of Theorem 9.9.

The point of these symmetry arguments is not *only* that all wavelets can be obtained from $\boldsymbol{\Psi}$. We already *have* an expression for all the wavelets, namely (9.130). The main benefit of the symmetry arguments is that they give us a *geometrical* picture of the different wavelets, once we have a picture of a single one. If we can picture $\boldsymbol{\Psi}$, then we can picture $\boldsymbol{\Psi}_{x+iy}$ as a scaled, moving, translated version of $\boldsymbol{\Psi}$. Thus it all comes down to understanding $\boldsymbol{\Psi}$.

The matrix elements of $\boldsymbol{\Psi}(x)$ (we drop the prime, now that $z=(\mathbf{0},is)$) can be obtained form (9.130) by letting $\mathbf{w}=-i\mathbf{x}$ and $w_0=1-it$:

$$[\boldsymbol{\Psi}(\mathbf{x},t)]_{mn} = \frac{\delta_{mn}[(1-it)^2-r^2]+2x_m x_n+2(1-it)\sum_{k=1}^{3}\varepsilon_{mnk}x_k}{\pi^2[(1-it)^2+r^2]^3}, \tag{9.177}$$

where $r\equiv|\mathbf{x}|$. But $\boldsymbol{\Psi}(x)$ is still a complex matrix-valued function in $\mathbf{R}^4$, hence impossible to visualize directly. We now eliminate the polarization degrees of freedom. Returning to the Fourier representation of solutions, note that if $\mathbf{f}(p)$ already satisfies the constraint (9.12), then $\mathbf{\Pi}(p)\,\mathbf{f}(p)=\mathbf{f}(p)$ and (9.101) reduces to

$$\int_E d\mu(z)\,|\tilde{\mathbf{F}}(z)|^2 = \int_C \frac{d\tilde{p}}{\omega^2}\,|\mathbf{f}(p)|^2 = \|\mathbf{F}\|^2. \tag{9.178}$$

Define the *scalar* wavelets by

$$\Psi_z(x') \equiv \int_C d\tilde{p}\;2\omega^2\theta(p\cdot y)\,e^{ip\cdot(x'-\bar{z})}, \tag{9.179}$$

and the corresponding scalar kernel $K:\mathcal{T}\times\mathcal{T}\to\mathbf{C}$ by

$$K(z'\,|\,z) = \int_C d\tilde{p}\;4\omega^2\theta(p\cdot y')\,\theta(p\cdot y)\,e^{ip\cdot(z'-\bar{z})}. \tag{9.180}$$

† For this reason we do not use the term "mother wavelet." The wavelets are more like siblings, and actually have their genesis in the conformal group. The term "reference wavelet" is also consistent with "reference frame," a closely related concept.

Then (9.178), with essentially the same argument as in the proof of Theorem 9.9, now gives the relations

$$\begin{aligned}\tilde{\mathbf{F}}(z') &= \int_E d\mu(z)\, K(z' \,|\, z)\, \tilde{\mathbf{F}}(z) \quad \text{pointwise in } z' \in \mathcal{T},\\ \mathbf{F} &= \int_E d\mu(z)\, \Psi_z\, \tilde{\mathbf{F}}(z) \quad \text{weakly in } \mathcal{H},\\ \mathbf{F}(x') &= \int_E d\mu(z)\, \Psi_z(x')\, \tilde{\mathbf{F}}(z) \quad \text{a.e.}\end{aligned} \tag{9.181}$$

The first equation states that $K(z' \,|\, z)$ is still a reproducing kernel on the range $\mathcal{A}_E$ of $R_E : \mathcal{H} \to \mathcal{L}^2(E)$. The second and third equations state that an arbitrary solution $\mathbf{F} \in \mathcal{H}$ can be represented as a superposition of the scalar wavelets, with $\tilde{\mathbf{F}} = R_E\, \mathbf{F}$ as a (vector) coefficient function. Thus, when dealing with coefficient functions in the range $\mathcal{A}_E$ of R_E, it is unnecessary to use the matrix wavelets. The main advantage of the latter (and a very important one) is that they can be used even when the ceofficient function $\mathbf{\Phi}(\mathbf{x}, is)$ does *not* belong to the range $\mathcal{A}_E$ of R_E, since their kernel *projects* $\mathbf{\Phi}$ to $\mathcal{A}_E$, by Theorem 9.13.

The scalar wavelets and scalar kernel were introduced and studied in Kaiser (1992b, c, 1993). They cannot, of course, be solutions of Maxwell's equations precisely because they are scalars. But they do satisfy the wave equation, since their Fourier transforms are supported on the light cone. We will see them again in Chapter 11, in connection with acoustics. To see their relation to the matrix wavelets, note that $\mathbf{\Pi}(p)$ is a projection operator of rank 1, hence trace $\mathbf{\Pi}(p) = 1$. Taking the trace on both sides of (9.92) and (9.112) therefore gives

$$\Psi_z(x') = \text{trace}\, \mathbf{\Psi}_z(x'), \quad K(z' \,|\, z) = \text{trace}\, \mathbf{K}(z' \,|\, \bar{z}). \tag{9.182}$$

Taking the trace in (9.182) amounts, roughly, to averaging over polarizations. From (9.130) we obtain the simple expression

$$\Psi_z(x') = \frac{1}{\pi^2} \frac{3c^2 w_0^2 - \mathbf{w} \cdot \mathbf{w}}{[c^2 w_0^2 + \mathbf{w} \cdot \mathbf{w}]^3}, \quad \begin{cases} w_0 \equiv y_0 + i(x_0 - x_0') \\ \mathbf{w} \equiv \mathbf{y} + i(\mathbf{x} - \mathbf{x}'). \end{cases} \tag{9.183}$$

The trace of the reference wavelet is therefore

$$\text{trace}\, \mathbf{\Psi}(\mathbf{x}, t) = \frac{1}{\pi^2} \frac{3(1 - it)^2 - r^2}{[(1 - it)^2 + r^2]^3} \equiv \Psi(r, t), \quad r \equiv |\mathbf{x}|, \tag{9.184}$$

where we have set $c = 1$. Because it is *spherically symmetric*, the scalar reference wavelet $\Psi(r, t)$ can be easily plotted. Its real and imaginary parts are shown at the end of Chapter 11, along with those of other acoustic wavelets. (See the graphs with $\alpha = 3$.) These plots confirm that Ψ is a spherical wave converging towards the origin as t goes from $-\infty$ to 0, becoming localized in a sphere of radius $\sqrt{3}$ around the origin at $t = 0$, and then diverging away from the origin as t goes from 0 to ∞. Even though $\Psi(r, 0)$ does not have compact support (it decays as r^{-4}), it is seen to be very well localized in $|\mathbf{x}| \le \sqrt{3}$. Reinserting c, the scaling and translation properties tell us that at $t = 0$, $\mathbf{\Psi}_{\mathbf{x},is}$ is well localized in the sphere $|\mathbf{x}' - \mathbf{x}| \le \sqrt{3}\, c\, |s|$. Also shown are the real and imaginary parts of

$$\Psi(0,t) = \frac{3}{\pi^2(1-it)^4}, \tag{9.185}$$

which give the approximate *duration* of the pulse Ψ. An observer fixed at the origin sees most of the pulse during the time interval $-\sqrt{3} \le t \le \sqrt{3}$, although the decay in time is not as sharp as the decay in space at $t = 0$. By the scaling property (9.176), $\mathbf{\Psi}_{\mathbf{x},is}$ *has a duration of order* $2\sqrt{3}\,|s|$. These pulses can therefore be made arbitrarily short.

Recall that in the one-dimensional case, wavelets must satisfy the admissibility condition $\int dt\, \psi(t) = 0$, which guarantees that ψ has a certain amount of oscillation and means that the wavelet transform $\psi_{s,t}^* f$ represents *detail* in f at the scale s (Section 3.4). The plots in Chapter 11 confirm that Ψ does indeed have this property in the time direction, but they also show that it does *not* have it in the radial direction. In the radial direction, Ψ acts more like an *averaging function* than a wavelet! This means that in the spatial directions, the analytic signals $\tilde{\mathbf{F}}(x + iy)$ are *blurred versions* of the solutions $\mathbf{F}(x)$ rather than their "detail." This is to be expected because $\tilde{\mathbf{F}}$ is an analytic continuation of $\mathbf{F}$. In fact, since y is the "distance" vector from $\mathbf{R}^4$, *y controls the blurring: y_0 controls the overall scale, and $\mathbf{y}$ controls the direction.* This can be seen directly from the coefficient functions of the wavelets in the Fourier domain. Consider the factor

$$R_y(p) \equiv 2\omega^2 \theta(p \cdot y) e^{-p \cdot y} \tag{9.186}$$

occurring in the definitions of $\mathbf{\Psi}_z$ and Ψ_z. Suppose that $y \in V_+'$. Then $R_y(p) = 0$ for $p \in C_-$, and for $p \in C_+$ we have

$$p \cdot y \ge \omega(y_0 - |\mathbf{y}|/c) > 0, \tag{9.187}$$

with equality if and only if either $\mathbf{y} = \mathbf{0}$ or

$$\mathbf{p} = \omega \frac{\mathbf{y}}{|\mathbf{y}|}, \qquad \mathbf{y} \ne \mathbf{0}. \tag{9.188}$$

If $\mathbf{y} = \mathbf{0}$, $\mathbf{p}$ cannot be parallel to $\mathbf{y}$ (except for the trivial case $p = 0$, which was excluded from the beginning). Then all directions $\mathbf{p} \ne \mathbf{0}$ get equal weights, which accounts for the spherical symmetry of $\Psi(r,t)$. If $\mathbf{y} \ne \mathbf{0}$, then $R_y(p)$ decays most slowly in the direction where $\mathbf{p}$ is parallel to $\mathbf{y}$. Along this direction,

$$R_y(p) = 2\omega^2 e^{-\omega(y_0 - |\mathbf{y}|)} \equiv \rho_y(\omega). \tag{9.189}$$

To get a rough interpretation of R_y, regard $\rho_y(\omega)$ as a one-dimensional weight function on the positive frequency axis. Then it is easily computed that the mean frequency and the standard deviation ("bandwidth") are

$$\bar{\omega}(y) = \frac{3}{y_0 - |\mathbf{y}|/c}\,, \qquad \sigma(y) = \frac{\sqrt{3}}{y_0 - |\mathbf{y}|/c}\,, \tag{9.190}$$

respectively. A similar argument applies when $y \in V'_-$, showing that $R_y(p)$ favors the directions in C_- where $\mathbf{p}$ is parallel to $\mathbf{y}$. The factor $\theta(p \cdot y)$ in (9.186) provides *stability*, guaranteeing that Fourier components in the *opposite* direction (which would give divergent integrals) do not contribute. Therefore we interpret $R_y(p)$ as a *ray filter* on the light cone C, favoring the Fourier components along the ray

$$\text{Ray}\,(y) \equiv \{p \in C : \text{sign}\, p_0 = \text{sign}\, y_0,\ \mathbf{p} = \omega\mathbf{y}/|\mathbf{y}|\}, \quad \mathbf{y} \neq \mathbf{0}. \tag{9.191}$$

Now that we have a reasonable interpretation of $\mathbf{\Psi}_{\mathbf{0},is}$ with $s > 0$, let us return to the wavelets $\mathbf{\Psi}_{iy}$ with arbitrary $y \in V'_+$. By (9.172), such wavelets can be obtained by applying a Lorentz transformation to $\mathbf{\Psi}_{\mathbf{0},is}$. To obtain $y = (\mathbf{y}, y_0)$, we must apply a Lorentz transformation with velocity

$$\mathbf{v}(y) = \frac{\mathbf{y}}{y_0} \tag{9.192}$$

to the four-vector $(\mathbf{0}, s)$. To find s, we apply the constraint (9.164):

$$c^2 y_0^2 - |\mathbf{y}|^2 = c^2 s^2 \ \Rightarrow\ s = \sqrt{y_0^2 - |\mathbf{y}|^2/c^2} = y_0\sqrt{1 - v^2/c^2}\,, \tag{9.193}$$

where $v \equiv |\mathbf{v}|$ and we have reinserted the speed of light for clarity. The physical picture of $\mathbf{\Psi}_{iy}$ is, therefore, as follows: $\mathbf{\Psi}_{iy}$ is a *Doppler-shifted* version of the spherical wavelet $\mathbf{\Psi}_{\mathbf{0},is}$. Its center moves with the velocity $\mathbf{v} = \mathbf{y}/y_0$, so its spherical wave fronts are not concentric. They are compressed in the direction of $\mathbf{y}$ and rarified in the opposite direction. This may also be checked directly using (9.183).

To summarize: *All eight parameters* $z = (\mathbf{x}, x_0) + i(\mathbf{y}, y_0) \in \mathcal{T}$ *have a direct physical interpretation. The real space-time point* $(\mathbf{x}, x_0)$ *is the focus of the wavelet in space-time, i.e., the place and time when the wavelet is localized;* $\mathbf{y}/y_0$ *is the velocity of its center; the quantity*

$$R(y) \equiv \sqrt{3\,y^2} \tag{9.194}$$

is its radius at the time of localization, in the reference frame where the center is at rest; and the sign of y_0 *is its helicity* (Section 9.2). Since all these parameters, as well as the wavelets that they label, were derived simply by extending the electromagnetic field to complex space-time, it would appear that the causal tube $\mathcal{T}$ is a natural arena in which to study electromagnetics. There are reasons to believe that it may also be a preferred arena for other physical theories (Kaiser [1977a, b, 1978a, 1981, 1987, 1990a, 1993, 1994b,c]).

9.8 Conclusions

The construction of the electromagnetic wavelets has been completely unique in the following sense: (a) The inner product (9.31) on solutions is uniquely determined, up to a constant factor, by the requirement that it be Lorentz-invariant. (b) The analytic extensions of the positive- and negative-frequency parts of $\mathbf{F}$ to $\mathcal{T}_+$ and $\mathcal{T}_-$, respectively, are certainly unique, hence so is $\tilde{\mathbf{F}}(z)$. (c) The evaluation maps $\mathcal{E}_z\mathbf{F} = \tilde{\mathbf{F}}(z)$ are unique, hence so are their adjoints $\mathbf{\Psi}_z \equiv \mathcal{E}_z^*$ with respect to the invariant inner product. On the other hand, the choice of the Euclidean space-time region E as a parameter space for expanding solutions is rather arbitrary. But because of the nice transformation properties of the wavelets under the conformal group, this arbitrariness can be turned to good advantage, as explained in the below.

Although we have used only the symmetries of the electromagnetic wavelets under translations, dilations and Lorentz transformations, their symmetries under the full conformal group $\mathcal{C}$, including reflections and special conformal transformations, will be seen to be extremely useful in the next chapter, where we apply them to the analysis of electromagnetic scattering and radar. The Euclidean region E, chosen for constructing the resolution of unity, may be regarded as the group of space translations and scalings acting on real space-time by $g_{\mathbf{x},is}x' = sx' + (\mathbf{x}, 0)$. As such, it is a subgroup of $\mathcal{C}$. E is invariant under space rotations but not under time translations, Lorentz transformations or special conformal transformations. This noninvariance can be exploited by applying any of the latter transformations to the resolution of unity over E and using the transformation properties of the wavelets. The general idea is that when $g \in \mathcal{C}$ is applied to (9.107), then another such resolution of unity is obtained in which E is replaced by its image gE under g. If $gE = E$, nothing new results. If g is a time translation, then the wavelets parameterized by gE are all localized at some time $t \neq 0$ rather than $t = 0$. If g is a Lorentz transformation, then gE is "tilted" and all the wavelets with $z \in gE$ have centers that move with a uniform nonzero velocity rather than being stationary. Finally, if g is a special conformal transformation, then gE is a curved submanifold of $\mathcal{T}$ and the wavelets parameterized by gE have centers with varying velocities. This is consistent with results obtained by Page (1936) and Hill (1945, 1947, 1951), who showed that special conformal transformations can be interpreted as mapping to an *accelerating* reference frame. The idea of generating new resolutions of unity by applying conformal transformations to the resolution of unity over E is another example of "deformations of frames," explained in the case of the Weyl–Heisenberg group in Section 5.4 and in Kaiser (1994a).

Chapter 10

Applications to Radar and Scattering

Summary: In this chapter we propose an application of electromagnetic wavelets to radar signal analysis and electromagnetic scattering. The goal in radar, as well as in sonar and other remote sensing, is to obtain information about objects (e.g., the location and velocity of an airplane) by analyzing waves (electromagnetic or acoustic) reflected from these objects, much as visual information is obtained by analyzing reflected electromagnetic waves in the visible spectrum. The location of an object can be obtained by measuring the *time delay* τ between an outgoing signal and its echo. Furthermore, the motion of the object produces a *Doppler effect* in the echo amounting to a *time scaling*, where the scale factor s is in one-to-one correspondence with the object's velocity. The *wideband ambiguity function* is the correlation between the echo and an arbitrarily time-delayed and scaled version of the outgoing signal. It is a maximum when the time delay and scaling factor best match those of the echo. This allows a determination of s and τ, which then give the approximate location and velocity of the object. The wideband ambiguity function is, in fact, nothing but the continuous wavelet transform of the echo, with the outgoing signal as a mother wavelet! When the outgoing signal occupies a narrow frequency band around a high carrier frequency, the Doppler effect can be approximated by a uniform frequency shift. The wideband ambiguity function then reduces to the *narrow band ambiguity function*, which depends on the time delay and the frequency shift. This is essentially the windowed Fourier transform of the echo, with the outgoing signal as a basic window. The ideas of Chapters 2 and 3 therefore apply naturally to the analysis of *scalar-valued radar and sonar signals in one (time) dimension.* But radar and sonar signals are actually waves in *space* as well as time. Therefore they are subject to the physical laws of propagation and scattering, unlike the unconstrained time signals analyzed in Chapters 2 and 3. Since the analytic signals of such waves are their wavelet transforms, it is natural to interpret the analytic signals as *generalized, multi-dimensional wideband ambiguity functions.* This idea forms the basis of Section 10.2.

Prerequisites: Chapters 2, 3, and 9.

10.1 Ambiguity Functions for Time Signals

Suppose we want to find the location and velocity of an object, such as an airplane. One way to do this approximately is to send out an electromagnetic wave in the direction of the object and observe the echo reflected to the source. As explained below, the comparison between the outgoing signal and its echo allows an approximate determination of the distance *(range)* R of the object and its *radial velocity* v along the line-of-sight. This is the problem of radar in its most elementary form. For a thorough treatment, see the classical books of Woodward (1953), Cook and Bernfeld (1967), and Rihaczek (1969). The wavelet-based analysis presented here has been initiated by Naparst (1988), Auslander and Gertner (1990), and Miller (1991), but some of the ideas date back to Swick (1966, 1969). Although we speak mainly of radar, the considerations of this section apply to sonar as well. The main difference is that the outgoing and returning signals represent acoustic waves satisfying the wave equation instead of electromagnetic waves satisfying Maxwell's equations. In Section 10.2 we outline some ideas for applying electromagnetic wavelets to radar.

The outgoing radar signal begins as a real-valued function of time, i.e., $\psi : \mathbf{R} \to \mathbf{R}$, representing the voltage fed into a transmitting antenna. The antenna converts $\psi(t)$ to a full electromagnetic wave and beams this wave in the desired direction. (In the "search" phase, all directions of interest are scanned; once objects are detected, they can be "tracked" by the method described here.) The outgoing electromagnetic wave can be represented as a complex vector-valued function $\mathbf{F}(\mathbf{x}, t) = \mathbf{B}(\mathbf{x}, t) + i\mathbf{E}(\mathbf{x}, t)$ on space-time, i.e., $\mathbf{F} : \mathbf{R}^4 \to \mathbf{C}^3$, as explained in Chapter 9. (Of course, only the values of ψ and $\mathbf{F}$ in bounded regions of $\mathbf{R}$ and $\mathbf{R}^4$, respectively, are of practical interest.) The exact relation between $\mathbf{F}(\mathbf{x}, t)$ and $\psi(t)$ does not concern us here. For now, the important thing is that the returning echo, also a full electromagnetic wave, is converted by the same apparatus (the antenna, now in a receiving mode) to a real time signal $f(t)$, which again represents a voltage. We will say that $\psi(t)$ and $f(t)$ are *proxies* for the outgoing electromagnetic wave $\mathbf{F}(\mathbf{x}, t)$ and its echo, respectively.

Although it might appear that a detailed knowledge of the whole process (transmission, propagation, reflection and reception) might be necessary in order to gain range and velocity information by comparing f and ψ, that process can, in fact, be largely ignored. The situation is much like a telephone conversation: a question $\psi(t)$ is asked and a response $f(t)$ is received. To extract the desired information, we need only two facts, both of which follow from the laws of propagation and reflection of electromagnetic waves:

(a) Electromagnetic waves propagate at the constant speed c, the speed of light ($\approx 3 \times 10^8$ m/sec). Consequently, if the reflecting object is at rest and the reflection occurs at a distance R from the radar site, then the proxy f suffers a *delay* equal to the time $2R/c$ necessary for the wave to make the round trip. Thus

$$f(t) = a\psi(t - d), \quad \text{where} \quad d \equiv \frac{2R}{c} \quad \text{if} \quad v = 0. \tag{10.1}$$

Here a is a constant representing the reflectivity of the object and the attenuation (loss of amplitude) suffered by the signal.

(b) An electomagnetic wave reflected by an object moving at a radial velocity v suffers a *Doppler effect*: the incident wave is *scaled* by the factor $s(v) = (c+v)/(c-v)$ upon reflection. To see this, assume for the moment that the object being tracked is small, essentially a point, and that the outgoing signal is sufficiently short that during the time interval when the reflection occurs, the object's velocity v is approximately constant. Then the range varies with time as

$$R(t) = R_0 + vt. \tag{10.2}$$

Since the range varies during the observation interval, so does the delay in the echo. Let $d(t)$ be the delay of the echo arriving at time t. Then

$$c\,d(t) = 2R(t - d(t)/2) = 2R_0 + 2vt - vd(t), \tag{10.3}$$

which gives

$$d(t) = 2\,\frac{R_0 + vt}{c+v}, \tag{10.4}$$

generalizing (10.1). When this is substituted into (10.1), we get

$$f(t) = a\psi\left(t - \frac{2v}{c+v}\,t - \frac{2R_0}{c+v}\right) = a\psi\left(\frac{t-\tau_0}{s_0}\right), \tag{10.5}$$

where

$$s_0 \equiv \frac{c+v}{c-v}, \quad \tau_0 \equiv \frac{2R_0}{c-v}, \quad \Rightarrow \quad v = \frac{s_0 - 1}{s_0+1}\,c, \quad R_0 = \frac{c\tau_0}{s_0+1}. \tag{10.6}$$

All material objects move with speeds less than c, hence $s_0 > 0$ always. If $v > 0$ (i.e., the object is moving away), then f is a stretched version of ψ. (A simple intuitive explanation is that when $v > 0$, successive wave fronts take longer to catch up with the object, hence the separation of wave fronts in the echo is larger.) Similarly, when $v < 0$, the reflected signal is a compressed version of ψ. The value of a clearly depends on the amount of amplification performed on the echo. We choose $a = s_0^{-1/2}$, so that f has the same "energy" as ψ, i.e., $\|f\|^2 = \|\psi\|^2$.

For *any* $\tau \in \mathbf{R}$ and any $s > 0$, introduce the notation

$$\psi_{s,\tau}(t) \equiv s^{-1/2}\psi\left(\frac{t-\tau}{s}\right), \tag{10.7}$$

already familiar from wavelet theory, so that

$$f(t) = \psi_{s_0,\tau_0}(t). \tag{10.8}$$

In the frequency domain, this gives

$$\hat{f}(\omega) = s_0^{1/2}\,e^{-2\pi\,i\omega\tau_0}\,\hat{\psi}(s_0\omega). \tag{10.9}$$

If $v > 0$, then $s_0 > 1$ and all frequency components are scaled down. If $v < 0$, then $s_0 < 1$ and all frequency components are scaled up. This is in accordance with everyday experience: the pitch of the horn of an oncoming car rises and that of a receding car drops.

The objective is to "predict" the trajectory of the object, and this is accomplished (within the limitations of the above model) by finding R_0 and v. To this end, we consider the whole *family* of scaled and translated versions of ψ:

$$\{\psi_{s,\tau} : s > 0, \tau \in \mathbf{R}\}. \tag{10.10}$$

We regard $\psi_{s,\tau}$ as a *test signal* with which to compare f. A given return $f(t)$ is *matched* with $\psi_{s,\tau}$ by taking its inner product with $\psi_{s,\tau}$:

$$\tilde{f}(s,\tau) \equiv \langle\, \psi_{s,\tau}\, , f\, \rangle \equiv \psi^*_{s,\tau} f = \int_{-\infty}^{\infty} dt\ s^{-1/2} \psi\left(\frac{t-\tau}{s}\right) f(t). \tag{10.11}$$

(Recall that ψ is assumed to be real. If it were complex, then $\bar{\psi}$ would have to be used in (10.11).) $\tilde{f}(s,\tau)$ is called the *wideband ambiguity function* of f (see Auslander and Gertner [1990], Miller [1991]). If the simple "model" (10.8) for f holds, then

$$\tilde{f}(s,\tau) = \langle\, \psi_{s,\tau}\, , \psi_{s_0,\tau_0}\, \rangle. \tag{10.12}$$

By the Schwarz inequality,

$$|\tilde{f}(s,\tau)| \le \|\psi_{s,\tau}\|\, \|\psi_{s_0,\tau_0}\| = \|\psi\|^2, \tag{10.13}$$

with equality if and only if $s = s_0$ and $\tau = \tau_0$. Thus, *we need only maximize* $|\tilde{f}(s,\tau)|$ *in order to find* s_0 *and* τ_0, assuming (10.8) holds. (Note that since f and ψ are real, it actually suffices to maximize $\tilde{f}(s,\tau)$ instead of $|\tilde{f}(s,\tau)|$.)

But in general, (10.8) may fail for various reasons. The reflecting object may not be rigid, with different parts having different radial velocities (consider a cloud). Or the object, though rigid, may be changing its "aspect" (orientation with respect to the radar site) during the reflection interval, so that its different parts have different radial velocities. Or, there may be many reflecting objects, each with its own range and velocity. We will model all such situations by assuming that there is a *distribution* of reflectors in the time-scale plane, described by a density $D(s,\tau)$. Then (10.8) is replaced by

$$f(t) = \iint \frac{ds_0\, d\tau_0}{s_0^2}\, \psi_{s_0,\tau_0}(t)\, D(s_0,\tau_0). \tag{10.14}$$

(We have included the factor s_0^{-2} in the measure for later convenience. This simply amounts to a choice of normalization for D, since the factor can always be absorbed into D. It will actually prove to be insignificant, since only values $s_0 \approx 1$ will be seen to be of interest in radar.) $D(s_0,\tau_0)$ is sometimes called the *reflectivity distribution.* In this generalized context, the objective is to find $D(s,\tau)$. This, then, is the *inverse problem* to be solved: knowing ψ and given f, find D.

As the above notation suggests, wavelet analysis is a natural tool here. $\tilde{f}(s,\tau)$ is just the wavelet transform of f with ψ as the mother wavelet. If ψ is admissible,[†] i.e., if

$$C \equiv \int_{-\infty}^{\infty} \frac{d\omega}{|\omega|}\, |\hat{\psi}(\omega)|^2 < \infty, \tag{10.15}$$

then f can be reconstructed by

$$f(t) = C^{-1} \iint \frac{ds\, d\tau}{s^2}\, \psi_{s,\tau}(t) \tilde{f}(s,\tau) \quad \text{a.e..} \tag{10.16}$$

Comparison of (10.14) and (10.16) suggests that *one possible solution of the inverse problem is*

$$D(s,\tau) = C^{-1} \tilde{f}(s,\tau), \tag{10.17}$$

i.e., D is the wavelet transform of $C^{-1} f$. However, this may not be true in general. (10.17) would imply that D is in the *range* of the wavelet transform with respect to ψ:

$$D \in \mathcal{F}_\psi \equiv \{\tilde{g} : \tilde{g}(s,\tau) \equiv \psi_{s,\tau}^* g : g \in L^2(\mathbf{R})\}. \tag{10.18}$$

But we have seen in Chapters 3 and 4 that (a) $\mathcal{F}_\psi$ is a closed subspace of $L^2(ds\, d\tau/s^2)$, and (b) a function $D(s,\tau)$ can belong to $\mathcal{F}_\psi$ if and only if it satisfies the *consistency condition* with respect to the reproducing kernel K associated with ψ. For the reader's convenience, we recall the exact statement here.

Theorem 10.1. *Let $D \in L^2(ds\, d\tau/s^2)$, i.e.,*

$$\iint \frac{ds\, d\tau}{s^2}\, |D(s,\tau)|^2 < \infty. \tag{10.19}$$

Then D belongs to $\mathcal{F}_\psi$ if and only if it satisfies the integral equation

$$D(s,\tau) = \iint \frac{ds_0\, d\tau_0}{s_0^2}\, K(s,\tau \,|\, s_0,\tau_0)\, D(s_0,\tau_0), \tag{10.20}$$

where K is the reproducing kernel associated with ψ:

$$K(s,\tau \,|\, s_0,\tau_0) \equiv C^{-1} \langle\, \psi_{s,\tau}\, ,\, \psi_{s_0,\tau_0} \,\rangle.$$

[†] It turns out that the "DC component" of ψ, i.e., $\hat{\psi}(0) = \int_{-\infty}^{\infty} dt\, \psi(t)$, has no effect on the radiated wave. That is, $\psi(t) - \hat{\psi}(0)$ has the same effect as $\psi(t)$, so we may as well assume that $\hat{\psi}(0) = 0$. This is related to the fact that solutions in $\mathcal{H}$ have no DC component; see (9.32). If ψ has rapid decay in the time domain (which is always the case in practice), then $\hat{\psi}(\omega)$ is smooth near $\omega = 0$ and $\hat{\psi}(0) = 0$ implies that ψ is admissible. Hence in this case, *the physics of radiation goes hand-in-hand with the mathematics of wavelet analysis!*

If (10.20) holds, then $D(s,\tau) = \tilde{g}(s,\tau)$*, where* $g \in L^2(\mathbf{R})$ *is given by*

$$g(t) = C^{-1} \iint \frac{ds\, d\tau}{s^2}\, \psi_{s,\tau}(t) D(s,\tau) \quad a.e. \tag{10.21}$$

Incidentally, (10.20) also gives a necessary and suffficient condition for any given function $D(s,\tau)$ to be the wideband ambiguity function of *some* time signal. (For example, $\tilde{f}(s,\tau)$ must satisfy (10.20).) In radar terminology, $K(s,\tau \,|\, s_0, \tau_0)$ is the (wideband) *self-ambiguity function* of ψ, since it matches translated and dilated versions of ψ against one another. (By contrast, $\tilde{f}(s,\tau)$ is called the wideband *cross*-ambiguity function.) Thus *(10.20) gives a relation that must be satisfied between all cross-ambiguity functions and the self-ambiguity function.*

Returning to the inverse problem (10.14), we see that (10.17) is far from a unique solution. There is no fundamental reason why the reflectivity D should satisfy the consistency condition (10.20). As explained in Chapter 4, the orthogonal projection P_ψ to $\mathcal{F}_\psi$ in $L^2(ds\, d\tau/s^2)$ is precisely the integral operator in (10.20):

$$(P_\psi D)(s,\tau) \equiv \iint \frac{ds_0\, d\tau_0}{s_0^2}\, K(s,\tau \,|\, s_0,\tau_0)\, D(s_0,\tau_0).$$

Equation (10.20) merely states that D belongs to $\mathcal{F}_\psi$ if and only if $P_\psi D = D$. A *general* element $D \in L^2(ds\, d\tau/s^2)$ can be expressed uniquely as

$$D(s,\tau) = D_\parallel(s,\tau) + D_\perp(s,\tau), \tag{10.22}$$

where $D_\parallel$ belongs to $\mathcal{F}_\psi$ and $D_\perp$ belongs to the orthogonal complement $\mathcal{F}_\psi^\perp$ of $\mathcal{F}_\psi$ in $L^2(ds\, d\tau/s^2)$. Then (10.14) implies that

$$\begin{aligned} C^{-1}\tilde{f}(s,\tau) = C^{-1}\psi_{s,\tau}^* f &= \iint \frac{ds_0\, d\tau_0}{s_0^2}\, K(s,\tau \,|\, s_0,\tau_0)\, D(s_0,\tau_0) \\ &= (P_\psi D)(s,\tau) = D_\parallel(s,\tau). \end{aligned} \tag{10.23}$$

This is the correct replacement of (10.17), since it does not assume that $D \in \mathcal{F}_\psi$. Equations (10.22) and (10.23) combine to give the following (partial) solution to the inverse problem:

Theorem 10.2. *The most general solution* $D(s,\tau)$ *of (10.14) in* $L^2(ds\, d\tau/s^2)$ *is given by*

$$D(s,\tau) = C^{-1}\tilde{f}(s,\tau) + h(s,\tau), \tag{10.24}$$

where $h(s,\tau)$ *is any function in* $\mathcal{F}_\psi^\perp$*, i.e.,*

$$\iint \frac{ds_0\, d\tau_0}{s_0^2}\, K(s,\tau \,|\, s_0,\tau_0)\, h(s_0,\tau_0) = 0. \tag{10.25}$$

The inverse problem has, therefore, no unique solution if only one outgoing signal is used. We are asking too much: using a function of one independent variable, we want to determine a function of two independent variables. It has therefore been suggested that a *set* of waveforms $\psi^1, \psi^2, \psi^3, \ldots$ be used instead. The first proposal of this kind, made in the narrow-band régime, is due to Bernfeld (1984, 1992), who suggested using a family of *chirps*. In the wideband regime, similar ideas have been investigated by Maas (1992) and Tewfik and Hosur (1994). Equation (10.23) then generalizes to

$$P_{\psi^k} D = C_k^{-1} \tilde{f}^{\,k}, \qquad k = 1, 2, 3, \ldots, \tag{10.26}$$

where $\tilde{f}^{\,k}(s, \tau) \equiv \langle \psi^k_{s,\tau}, f^k \rangle$ is the ambiguity function of the echo $f^k(t)$ of $\psi^k(t)$ and P_{ψ^k} is the orthogonal projection to the range of the wavelet transform associated with ψ^k. However, such schemes implicitly assume that the distribution function $D(s, \tau)$ has an *objective* significance in the sense that the *same* $D(s, \tau)$ gives all echos f^k from their incident signals ψ^k. There is no *a priori* reason why this should be so. *Any* finite-energy function and, in particular, any echo $f(t)$, can be written in the form (10.14) for some coefficient function $D(s, \tau)$, namely, $D = C^{-1}\tilde{f}$. The claim that there is a *universal* coefficient function giving all echos amounts to a *physical* statement, which must be examined in the light of the laws governing the propagation and scattering of electromagnetic waves. I am not aware of any such analyses in the wideband regime.

The term "ambiguity function," as used in the radar literature (see Woodward [1953], Cook and Bernfeld [1967], and Barton [1988]), usually refers to the *narrow-band ambiguity functions*, which are related to the wideband versions as follows.† The latter can be expressed in the frequency domain by applying Parseval's identity to its definition (10.11):

$$\tilde{f}(s, \tau) = \langle \hat{\psi}_{s,\tau}, \hat{f} \rangle = \int_{-\infty}^{\infty} d\omega \, \sqrt{s}\, e^{2\pi i \omega \tau}\, \overline{\hat{\psi}}(s\omega)\, \hat{f}(\omega). \tag{10.27}$$

All objects of interest in radar travel with speeds much smaller than light. Thus $|v|/c \ll 1$ and

$$s \equiv \frac{c+v}{c-v} \approx 1 + \frac{2v}{c}, \qquad \tau \equiv \frac{2R_0}{c-v} \approx \frac{2R_0}{c}. \tag{10.28}$$

If the maximum expected speed is $0 < v_{\max} \ll c$ and the maximum range of the radar is $R_{\max}$, then we have, to first order in c^{-1}:

$$|s-1| \le \frac{2v_{\max}}{c}, \qquad |\tau| \le \frac{2R_{\max}}{c}. \tag{10.29}$$

---

† Various schemes exist for going from the wideband to the narrow-band regimes. The procedure given here is meant to be conceptually as simple as possible, and no attempt is made to be mathematically rigorous or invoke subtle group-theoretical considerations. For more detailed analyses, see Cook and Bernfeld (1967), Auslander and Gertner (1990), and Kalnins and Miller (1992).

Therefore, in practice only $\tau \approx 0$ and $s \approx 1$ are of interest, and we may assume that $D(s, \tau)$ is small outside of a neighborhood of $(1, 0)$. (The factor s_0^{-2} in (10.14) is therefore merely of academic interest in radar.) Suppose now that $\hat{\psi}(\omega)$ is concentrated mainly in the double frequency band

$$0 < \alpha \leq |\omega| \leq \gamma, \quad \text{where} \quad \frac{\gamma - \alpha}{\gamma + \alpha} \ll 1. \tag{10.30}$$

The *bandwidth* β and the *central frequency* ω_c are defined by

$$\beta \equiv \gamma - \alpha\,, \qquad \omega_c \equiv \frac{\alpha + \gamma}{2}\,, \tag{10.31}$$

and (10.30) means that $\beta \ll 2\omega_c$. A signal ψ satisfying this condition will be called a *narrow-band signal.* It may be regarded as a relatively slowly-varying "message" imprinted or coded onto a high *carrier frequency* ω_c, like a speech waveform carried by a radio wave. The integral in (10.27) extends only over the two frequency intervals

$$\omega_c - \frac{\beta}{2} = \alpha \leq |s\omega| \leq \gamma = \omega_c + \frac{\beta}{2}\,. \tag{10.32}$$

Now

$$s\omega \approx \left(1 + \frac{2v}{c}\right)\omega = \omega + \frac{2v\omega}{c}\,. \tag{10.33}$$

If ψ is a narrow-band signal, the second term can be approximated by letting

$$\omega \to \begin{cases} \omega_c & \text{if } \omega > 0 \\ -\omega_c & \text{if } \omega < 0. \end{cases} \tag{10.34}$$

Thus

$$s\omega \approx \begin{cases} \omega + \phi & \text{if } \omega > 0 \\ \omega - \phi & \text{if } \omega < 0 \end{cases}, \quad \text{where} \quad \phi \equiv \phi(v) \equiv \frac{2\omega_c v}{c} = \frac{2v}{\lambda_c}\,, \tag{10.35}$$

where λ_c is the carrier wavelength. That is, *for narrow-band signals, the effect of Doppler scaling can be approximated as a "Doppler shift" that translates all positive-frequency components by* $\phi(v)$ *and all negative-frequency components by* $-\phi(v)$. We can then express the ambiguity function in terms of ϕ and τ instead of s and τ. Because positive and negative frequencies undergo opposite Doppler shifts, this so-called "narrow-band approximation" is made considerably simpler by ignoring the negative-frequency band. This can be done by the replacement

$$\hat{\psi}(\omega) \to 2\theta(\omega)\hat{\psi}(\omega) = \begin{cases} 2\hat{\psi}(\omega) & \text{if } \omega > 0 \\ 0 & \text{if } \omega < 0. \end{cases} \tag{10.36}$$

The inverse Fourier transform of the right-hand side is the *Gabor analytic signal* of $\psi(t)$. It is necessarily complex-valued, and its real part is $\psi(t)$ (see Chapter 9). Since $\hat{\psi}(-\omega) = \overline{\hat{\psi}(\omega)}$, no information is lost in the process. Furthermore, since the remaining spectrum consists of a narrow frequency band centered around ω_c, the numerical analysis will be much simpler if the analytic signal is *demodulated*, i.e., if the high carrier frequency is removed, so that only the "message" remains. This amounts to translating the spectrum to the left by ω_c. Thus we define a new (complex) signal $\Psi(t)$ by

$$\Psi(t) = 2\,e^{-2\pi\, i\omega_c t} \int_0^\infty d\omega\, e^{2\pi\, i\omega t}\, \hat{\psi}(\omega). \tag{10.37}$$

The echo $f(t)$ has a band structure similar to that of $\psi(t)$, since the Doppler shift is very small compared to the carrier frequency. Thus we perform the same operations on it: filter out the negative-frequency band, then demodulate. This gives the complex signal

$$F(t) = 2\,e^{-2\pi\, i\omega_c t} \int_0^\infty d\omega\, e^{2\pi\, i\omega t}\, \hat{f}(\omega). \tag{10.38}$$

Ψ and F carry the same information as do ψ and f, but have the advantage that they oscillate much more slowly and undergo uniform Doppler shifts (rather than Doppler scalings, which act oppositely on the negative frequencies). If $f = \psi_{s,\tau}$, then for $\omega > 0$ (10.35) gives

$$\begin{aligned} \hat{f}(\omega) &= \sqrt{s}\; e^{-2\pi\, i\omega\tau}\, \hat{\psi}(s\omega) \\ &\approx e^{-2\pi\, i\omega\tau}\, \hat{\psi}(\omega + \phi). \end{aligned} \tag{10.39}$$

Hence

$$\begin{aligned} F(t) &\approx 2\,e^{-2\pi\, i\omega_c t} \int_0^\infty d\omega\; e^{2\pi\, i\omega(t-\tau)}\, \hat{\psi}(\omega + \phi) \\ &= 2\,e^{-2\pi\, i\omega_c t}\; e^{-2\pi\, i\phi(t-\tau)} \int_\phi^\infty d\omega\; e^{2\pi\, i\omega(t-\tau)}\, \hat{\psi}(\omega) \\ &= 2\,e^{-2\pi\, i\omega_c t}\; e^{-2\pi\, i\phi(t-\tau)} \int_0^\infty d\omega\; e^{2\pi\, i\omega(t-\tau)}\, \hat{\psi}(\omega), \end{aligned} \tag{10.40}$$

since $\hat{\psi}(\omega)$ vanishes for $|\omega| \le |\phi|$. It follows that

$$F(t) \approx e^{-2\pi\, i\theta}\, \Psi_{\phi,\tau}(t), \tag{10.41}$$

where

$$\Psi_{\phi,\tau}(t) = e^{-2\pi\, i\phi t}\, \Psi(t-\tau), \qquad \theta \equiv \theta(\phi,\tau) = (\omega_c - \phi)\tau. \tag{10.42}$$

The phase factor $e^{-2\pi\, i\theta}$ is time-independent, hence it does not affect the analysis and can be ignored. Therefore, the echo from a point reflector can be represented by $\Psi_{\phi,\tau}$ in the narrow-band approximation. For a given return $f(t)$, we define the *narrow-band cross-ambiguity function* in terms of $F(t)$ by†

$$\tilde{F}(\phi,\tau) \equiv \Psi^*_{\phi,\tau} F = \int_{-\infty}^{\infty} dt\; e^{2\pi\, i\phi t}\, \overline{\Psi}(t-\tau)\, F(t). \tag{10.43}$$

If $f = \psi_{s_0,\tau_0}$, then $F = e^{-2\pi\, i\theta_0}\, \Psi_{\phi_0,\tau_0}$, where

$$s_0 = 1 + \frac{2v_0}{c}, \quad \phi_0 = \frac{2v_0}{\lambda_c}, \quad \theta_0 = (\omega_c - \phi_0)\tau_0\,. \tag{10.44}$$

Therefore

$$|\tilde{F}(\phi,\tau)| = |\langle\, \Psi_{\phi,\tau}\,, \Psi_{\phi_0,\tau_0}\,\rangle| \le \|\Psi\|^2, \tag{10.45}$$

with equality if and only if $\tau = \tau_0$ and $\phi = \phi_0$ (i.e., $v_0 = v$). This gives a method for determining the range and velocity of a point reflector. As in the wideband case, a general echo is modeled by assuming a distribution of reflectors, now described as a function of the delay and Doppler shift:

$$F(t) = \iint d\phi\, d\tau\; \Psi_{\phi,\tau}(t)\, D_{\text{NB}}(\phi,\tau). \tag{10.46}$$

(The narrow-band distribution $D_{\text{NB}}(\phi,\tau)$ can be related to the wideband distribution $D(s,\tau)$ by noting that it contains the phase factor $e^{-2\pi\, i\theta}$ and that $d\phi = \omega_c\, ds$.) Equation (10.46) poses an inverse problem: given F, find D_{NB}. Just as wavelet analysis was used to analyze and solve the wideband inverse problem, so can time-frequency analysis be used to do the same for the narrow-band problem. This is based on the observation that $\tilde{F}(\phi,\tau)$ *is just the windowed Fourier transform of* F *with respect to the complex window* $\Psi(t)$ (Chapter 2). The corresponding *self-ambiguity function*

$$K(\phi,\tau\,|\,\phi_0,\tau_0) \equiv C_\Psi^{-1}\langle\, \Psi_{\phi,\tau}\,, \Psi_{\phi_0,\tau_0}\,\rangle, \qquad C_\Psi \equiv \|\Psi\|^2, \tag{10.47}$$

is the reproducing kernel associated with the frame $\{\Psi_{\phi,\tau}\}$. All the observations made earlier about the existence and uniqueness of solutions to the wideband inverse problem now apply to the narrow-band case as well, since the mathematical structure in both cases is essentially identical, as explained in Chapter 4. In particular, the most general solution of (10.46) is

$$D_{\text{NB}}(\phi,\tau) = C_\Psi^{-1}\tilde{F}(\phi,\tau) + H(\phi,\tau), \tag{10.48}$$

where $H(\phi,\tau)$ is any function in the orthogonal complement $\mathcal{F}_\Psi^\perp$ of the range $\mathcal{F}_\Psi$ of the windowed Fourier transform (with respect to Ψ) in $L^2(\mathbf{R}^2)$. (See

† In many texts, the narrow-band ambiguity function is defined as $|\tilde{F}(\phi,\tau)|^2$ rather than $\tilde{F}(\phi,\tau)$. The wideband ambiguity functions are real when the signals are real, therefore they can be maximized directly to find the matching delay and scale. The narrow-band versions, on the other hand, are necessarily complex and therefore it is their modulus that must be maximized.

Chapters 2 and 4.) Again, the solution is not unique and it is necessary to use a variety of outgoing signals to obtain uniqueness. Here, too, the question of the "objectivity" of D_{NB} (its independence of Ψ) must be addressed (see Schatzberg and Deveney [1994]).

The narrow-band ambiguity function was first defined by Woodward (1953) in his seminal monograph on radar. Wideband ambiguity functions were studied by Swick (1966, 1969). The renewed interest in the wideband functions (Auslander and Gertner [1990], Miller [1991], Kalnins and Miller [1992], Maas [1992]) seems to have been inspired by the development of wavelet analysis. One might argue that since radar signals are usually narrow-band, there is no point in considering the wideband ambiguity functions. However, there are some good reasons to do so.

(a) Since acoustic waves satisfy the wave equation, they also propagate at a constant speed. Therefore, ambiguity functions play as fundamental a role in *sonar* as they do in radar. However, whereas radar usually involves narrow-band signals, this is no longer true for sonar since the frequencies involved are much lower and the speed of sound (in water, say) is much smaller than the speed of light. Acoustic wavelets similar to the electromagnetic wavelets are constructed in Chapter 11.

(b) *Even for narrow-band signals, the formalism associated with the wideband ambiguity functions is conceptually simpler and more intuitive than the one associated with the narrow-band version.* The relation of the velocity to the scale factor is direct and clear, whereas its relation to the Doppler frequency is more convoluted, depending as it does on transforming to the frequency domain. Also, the elimination of negative frequencies and subsequent demodulation make $\tilde{F}(\phi, \tau)$ a less intuitive and direct object than $\tilde{f}(s, \tau)$.

(c) Accelerations are simple to describe in the time domain but very complicated in the frequency domain. Therefore an extension of the ambiguity function concept to include accelerations is much more likely to exist in the time-scale representation than in the time-frequency representation.

Yet, most treatments of radar use the narrow-band ambiguity functions without even referring to the wideband versions. One reason is practical: the demodulated signals oscillate much more slowly than the original ones, hence they are easier to handle. (Lower sampling rates can be used, for example.) Another reason may be that wavelet analysis is still new and little-known. A third reason (suggested to me by M. Bernfeld, private communication) could be that before the advent of fast digital computers, ambiguity functions were routinely computed by analog methods using filter banks, which are better able to handle windowed Fourier transforms than wavelet transforms since the former require modulations while the latter require scaling. Given all this, a case can be made that modern radar analysis ought to explore the wideband regime, even when all the signals are narrow-band.

10.2 The Scattering of Electromagnetic Wavelets

When the detailed physical process of propagation and scattering is ignored, the net effect of a point object on the *proxy* $\psi(t)$ representing the outgoing electromagnetic wave is the transformation

$$\psi(t) \mapsto \psi_{s,\tau}(t) \equiv s^{-1/2}\psi\left(\frac{t-\tau}{s}\right), \tag{10.49}$$

where τ is the delay and $s = (c+v)/(c-v)$ is the scaling factor induced by the velocity of the object. In the narrow-band regime, this reduces to

$$\Psi(t) \mapsto \Psi_{\phi,\tau}(t) \equiv e^{-2\pi\, i\phi t}\,\Psi(t-\tau), \tag{10.50}$$

where $\phi = 2v/\lambda_c$ is the Doppler frequency shift induced on the positive-frequency band by the velocity and we have ignored the phase factor $e^{-2\pi\, i\theta(\phi,\tau)}$ in (10.41). $\Psi(t)$ is the demodulated Gabor analytic signal of $\psi(t)$.

In each of these cases, we have a set of *operations* acting on signals: in the first case, it is translations and scalings, while in the second it is translations and modulations. For *any* finite-energy signal $f(t)$, define $U_{s,\tau}f$ and $W_{\phi,\tau}f$ by

$$(U_{s,\tau}f)(t) \equiv s^{-1/2}f\left(\frac{t-\tau}{s}\right), \quad (W_{\phi,\tau}f)(t) \equiv e^{-2\pi\, i\phi t}\,f(t-\tau), \tag{10.51}$$

so that, in particular,

$$\psi_{s,\tau} = U_{s,\tau}\psi, \qquad \Psi_{\phi,\tau} = W_{\phi,\tau}\Psi. \tag{10.52}$$

$U_{s,\tau}$ and $W_{\phi,\tau}$ are invertible operators on $L^2(\mathbf{R})$ which *preserve the energy*:

$$\|U_{s,\tau}f\|^2 = \|f\|^2 \quad \text{and} \quad \|W_{\phi,\tau}f\|^2 = \|f\|^2. \tag{10.53}$$

By polarization, it follows that they preserve inner products, hence they are *unitary*:

$$U^*_{s,\tau} = U^{-1}_{s,\tau}, \qquad W^*_{\phi,\tau} = W^{-1}_{\phi,\tau}. \tag{10.54}$$

Furthermore, it is easily verified that in each case, applying two of the operators consecutively is equivalent to applying a single one, uniquely determined by the two. Namely,

$$U_{s',\tau'}U_{s,\tau} = U_{s's,\,s'\tau+\tau'}, \qquad W_{\phi',\tau'}W_{\phi,\tau} = e^{2\pi\, i\phi\tau'}\,W_{\phi'+\phi,\tau'+\tau}. \tag{10.55}$$

Since $U_{1,0}$ is the identity operator, $U_{1/s,\,-\tau/s}$ is seen to be the inverse of $U_{s,\tau}$. Hence the operators $U_{s,\tau}$ form a *group*: the inverses and products of U's are

U's. In fact, the index pairs (s, τ) themselves form a group, called the *affine group* or $ax + b$ group, which we denote by $\mathcal{A}$. It consists of the half-plane $\{(s, \tau) \in \mathbf{R}^2 : s > 0\}$, acting on *time* variable by *affine transformations*:

$$t \mapsto (s, \tau)t \equiv st + \tau. \tag{10.56}$$

We write $g = (s, \tau)$ for brevity, and (10.56) defines it as a map $g : \mathbf{R} \to \mathbf{R}$. Two consecutive transformations give

$$(s', \tau')(s, \tau)t = (s', \tau')(st + \tau) = (s's)t + (s'\tau + \tau'), \tag{10.57}$$

which implies the multiplication rule in $\mathcal{A}$,

$$(s', \tau')(s, \tau) = (s's, s'\tau + \tau'). \tag{10.58}$$

Equation (10.55) therefore shows that the U's satisfy the same the group law as the elements of $\mathcal{A}$:

$$U_{g'}U_g = U_{g'g}\,, \qquad g', g \in \mathcal{A}. \tag{10.59}$$

That is, the action of (s, τ) on the time $t \in \mathbf{R}$ is now *represented* by an *induced* action of $U_{s,\tau}$ on signals $f \in L^2(\mathbf{R})$. One therefore says that the correspondence $U : g \mapsto U_g$ (mapping the action on time to the action on signals) is a *representation* of $\mathcal{A}$ on $L^2(\mathbf{R})$. Furthermore, since each U_g is a unitary operator, this representation is said to be *unitary*. Another important property of the representation is that, according to (10.52), all the proxy wavelets are obtained from the mother wavelet by applying U's:

$$\psi_g = U_g \psi.$$

Consequently,

$$U_{g'}\psi_g = U_{g'}U_g\psi = U_{g'g}\psi = \psi_{g'g}\,. \tag{10.60}$$

That is, *affine transformations, as represented on $L^2(\mathbf{R})$ by U, map wavelets to wavelets simply by transforming their labels.* Note that since $(s, \tau)^{-1}t = (t - \tau)/s$, U_g acts on functions by acting *inversely* on their argument:

$$(U_g f)(t) = s^{-1/2} f(g^{-1}t), \qquad g \equiv (s, \tau). \tag{10.61}$$

(For example, to translate a function to the right by τ, we translate its argument t to the *left* by τ.)

The phase factor in the second equation of (10.55) spoils the group law for the W's. For example, the inverse of $W_{\phi,\tau}$ does not belong to the set of W's unless $\phi\tau$ is an integer, since the phase factor prevents $W_{-\phi,-\tau}$ (the only possible candidate) from being the inverse. From this point of view, the W's are more complicated than the U's. This can be attributed to the repeated "surgery" performed on signals in going from the wideband to the narrow-band regimes. We will not be concerned here with these fine points, except to note that this further supports the idea that the wideband regime is conceptually simpler than its narrow-band approximation.†

The operators $U_{s,\tau}$ and $W_{\phi,\tau}$ act on *unconstrained functions*, i.e., on signals whose only requirement is to have a finite energy. However, they owe their significance to the fact that these signals are "proxies" for electromagnetic waves that *are* constrained (by Maxwell's equations) and undergo physical scattering. In fact, *$U_{s,\tau}$ and $W_{\phi,\tau}$ are the effects on the proxies of the propagation and scattering suffered by the corresponding waves: τ is the delay due to the constant propagation speed, while s and ϕ are the effects of "elementary" scattering in the wideband and narrow-band regimes, respectively.* In this section we attempt to make the physics *explicit* by dealing with the electromagnetic waves themselves rather than their proxies. Thus we consider solutions $\mathbf{F}(\mathbf{x}, t)$ of Maxwell's equations in the Hilbert space $\mathcal{H}$ (Section 9.2) rather than finite-energy time signals $\psi(t)$ in $L^2(\mathbf{R})$. Suppose we know the electric and magnetic fields at $t = 0$. Then there is a unique solution $\mathbf{F} \in \mathcal{H}$ with those initial values. To find the fields at a later time $t > 0$, we simply evaluate $\mathbf{F}$ at t. Therefore, the propagation of the waves is already built into $\mathcal{H}$. It can be represented simply by *space-time translations* of solutions, generalizing the time delay suffered by the proxies: $\mathbf{F}(x) \mapsto \mathbf{F}(x - b)$, $b \in \mathbf{R}^4$. (Causality requires that $\tau > 0$; in space-time, this corresponds to $b \in V'_+$, the future cone; see Chapter 9. However, we must consider all $b \in \mathbf{R}^4$ in order to have a group.)

What about scattering? That is a difficult problem, and we merely propose a rough *model* that remains to be developed and tested in future work. No attempt is made here to derive this model from basic theory, although that presents a very interesting challenge. We take our clue from the effect of scattering on the proxies. Since the wideband regime is both more general and simpler, it will be our guide. The scaling transformation representing the Doppler effect on the proxies was derived by assuming that the scatterer is small (essentially a point) and the signal is short. In this sense, it represents an "elementary" scattering event.

More general scattering is modeled by superposing such elementary events, as in (10.14). But the Doppler effect is actually a *space-time* phenomenon rather than just a *time* phenomenon. It results from the coordinate transformation

† The set of W's can be made into a group by including arbitrary phase factors $e^{2\pi i\lambda}$ (thus also enlarging the labels to (λ, ϕ, τ)). This expanded set of labels forms a group known as the *Weyl–Heisenberg group* $\mathcal{W}$, which is related to $\mathcal{A}$ by a process called "contraction." See Kalnins and Miller (1992). The narrow-band approximation is very similar to the procedure for going from relativistic to nonrelativistic quantum mechanics, with ω_c playing the role of the rest energy mc^2 divided by Planck's constant $\hbar$. See Kaiser (1990a), Section 4.6.

of space-time $\mathbf{R}^4$ representing a "boost" from any reference frame (Cartesian coordinate system in space-time $\mathbf{R}^4$) to another reference frame, in motion relative to the first. If the relative velocity is uniform, such a transformation is represented by a 4×4 matrix Λ called a *Lorentz transformation.* Because Λ preserves the indefinite Lorentzian scalar product in $\mathbf{R}^4$, it has the structure of a rotation through an imaginary angle. (This hyperbolic "rotation" is in the two-dimensional plane in $\mathbf{R}^4$ determined by the time axis together with the velocity vector; see Chapter 9.) This suggests that elementary scattering should be represented by Lorentz transformations. But these transformations do not form a group: the combination of two Lorentz transformation along different directions results in a *rotation* as well as a third Lorentz transformation. If we want to have a group of operations (and this is indeed a thing to be desired!), we must include rotations. Lorentz transformations and rotations together do form a group called the *proper Lorentz group* $\mathcal{L}_0$. Rotations are represented in space-time as linear maps $g : \mathbf{R}^4 \to \mathbf{R}^4$ defined by

$$x = (\mathbf{x}, t) \mapsto gx = (R\mathbf{x}, t), \tag{10.62}$$

where R is the 3×3 matrix implementing the rotation. Thus R is an orthogonal matrix with determinant 1. *Reflections* in $\mathbf{R}^3$ are represented similarly, but now R has determinant -1.

Clearly we want to include reflections in our group since they are the simplest scattering events. The group consisting of all Lorentz transformations, rotations and space reflections will be denoted by $\mathcal{L}$. It is called the *orthochronous* Lorentz group (see Streater and Wightman [1964]). Like $\mathcal{L}_0$, $\mathcal{L}$ is a six-parameter group. Elements are specified by giving three continuous parameters representing a velocity, and three more representing a rotation. Reflections are specified by the discrete parameter "yes" or "no," once a plane for a possible rotation has been selected. ("Yes" means reflect in that plane.) We denote arbitrary elements of $\mathcal{L}$ by Λ and call them "Lorentz transformations," even if they include rotations and space reflections.

Therefore, the minimum set of operations needed to model propagation *and* elementary scattering consists of all space-time translations (four parameters) and orthochronous Lorentz transformations (six parameters). Together, these form the 10-parameter *Poincaré group*, which we denote by $\mathcal{P}$. As the name implies, $\mathcal{P}$ *is* a group. A Lorentz transformation Λ followed by a translation $x \mapsto x + b$ gives

$$g \equiv (\Lambda, b) : x \mapsto gx \equiv \Lambda x + b, \tag{10.63}$$

generalizing the action (10.56) of the affine group on the time axis. If $g' = (\Lambda', b')$ is another such transformation, then

$$g'gx = \Lambda' gx + b' = \Lambda'(\Lambda x + b) + b' = (\Lambda'\Lambda)x + (\Lambda' b + b'), \tag{10.64}$$

so

$$g'g = (\Lambda', b')(\Lambda, b) = (\Lambda'\Lambda,\ \Lambda' b + b'). \tag{10.65}$$

This is the law of group multiplication in $\mathcal{P}$, which generalizes the law (10.58) for the affine group.

The Poincaré group is a central object of interest in relativistic quantum mechanics. It might be argued that it is unnecessary to consider relativistic effects in radar since all velocities of interest are much smaller than c. This is true, but it misses the point: the electromagnetic wavelets were constructed using the symmetries of Maxwell's equations, which form the *conformal group* $\mathcal{C}$. Consequently, these wavelets map into one another under conformal transformations, much as the proxy wavelets $\psi_{s,\tau}$ map into one another under affine transformations by (10.60). But $\mathcal{C}$ contains $\mathcal{P}$ as a subgroup, so the wavelets also map into one another under Poincaré transformations. *If relativistic effects are neglected, the electromagnetic wavelets no longer transform to one another under boosts, and the theory becomes much more involved from a mathematical and conceptual point of view.* The nonrelativistic approximation is quite analogous to the narrow-band approximation, and it leads to a similar mutilation of the group laws. Indeed, to make full use of the symmetries of Maxwell's equations, we should include *all* conformal transformations. Aside from the Poincaré group, $\mathcal{C}$ contains the uniform dilations $x \mapsto ax$ $(a \neq 0)$ and the so-called *special conformal transformations*, which form a four-parameter subgroup. A simple way to understand these transformations is to begin with the *space-time inversion*

$$J : x \mapsto \frac{x}{x \cdot x}, \qquad \text{i.e.,} \qquad J(\mathbf{x}, t) \equiv \frac{(\mathbf{x}, t)}{c^2t^2 - |\mathbf{x}|^2}. \tag{10.66}$$

If translations are denoted by $T_b x = x + b$, then special conformal transformations are compositions $C_b \equiv J T_b J$, i.e.,

$$C_b x = J\left(\frac{x}{x \cdot x} + b\right) = \frac{\frac{x}{x \cdot x} + b}{\left(\frac{x}{x \cdot x} + b\right)^2} = \frac{x + (x \cdot x) b}{1 + 2b \cdot x + (b \cdot b)(x \cdot x)}, \qquad b \in \mathbf{R}^4. \tag{10.67}$$

Since C_b is nonlinear, it maps flat subsets in $\mathbf{R}^4$ into curved ones. Depending on the choice of b, this capacity can be used in different ways. Let $\mathbf{R}^3_0 \equiv \{(\mathbf{x}, 0) : \mathbf{x} \in \mathbf{R}^3\}$ be the configuration space at time zero. If $b = (\mathbf{b}, 0)$, then C_b maps $\mathbf{R}^3_0$ into itself. Thus C_b maps flat surfaces in *space* at time zero to curved surfaces in the same space. This will be used to study the reflection of wavelets from curved surfaces. For some different choices of b, C_b can be interpreted as *boosting to an accelerating reference frame* (Page [1936], Hill [1945, 1947, 1951]), just as the (linear) Lorentz transformations boost to uniformly moving reference frames. This could be used to estimate the acceleration of an object, just as Lorentz transformations are used to estimate the velocity.

Thus we have arrived at the conformal group $\mathcal{C}$ as a model combining propagation and elementary scattering. In order for this model to be useful, we need to know how conformal transformations affect electromagnetic waves. That is, we need a *representation* of $\mathcal{C}$ on $\mathcal{H}$. Furthermore, we want the action of $\mathcal{C}$ on $\mathcal{H}$ to preserve the norms $\|\mathbf{F}\|$ of solutions, which means it must be *unitary.* A unitary representation of $\mathcal{C}$ analogous to the above representation of the affine group can be constructed as follows. Every conformal transformation is a map $g : \mathbf{R}^4 \to \mathbf{R}^4$, not necessarily linear because g may include a special conformal transformation like (10.67). g can be extended uniquely to a map on the complex space-time domain $\mathcal{T}$, i.e.,

$$g : \mathcal{T} \to \mathcal{T}. \tag{10.68}$$

For example, a Poincaré transformation acts on $z = x + iy \in \mathcal{T}$ by

$$(\Lambda, b)z = \Lambda z + b = (\Lambda x + b) + i\Lambda y, \tag{10.69}$$

and a special conformal transformations looks just like (10.67) but with x replaced by z. In fact, the action of g on $\mathcal{T}$ is *nicer* than its action on $\mathbf{R}^4$, since the latter is singular due to the inversion J, which maps x to infinity when x is light-like ($x \cdot x = 0$). The inversion in $\mathcal{T}$ is $Jz = z/z \cdot z$, and it is nonsingular since z^2 cannot vanish on $\mathcal{T}$ (see Theorem 9.10). Thus *all conformal transformations are nonsingular in $\mathcal{T}$*. We define the action of $\mathcal{C}$ on solutions $\mathbf{F} \in \mathcal{H}$ by first defining it on the space of analytic signals of solutions, i.e., on

$$\tilde{\mathcal{H}} \equiv \{\tilde{\mathbf{F}}(z) : \mathbf{F} \in \mathcal{H}\}. \tag{10.70}$$

Namely, the operator U_g representing $g \in \mathcal{C}$ on $\tilde{\mathcal{H}}$ is

$$(U_g \tilde{\mathbf{F}})(z) = M_g(z)\tilde{\mathbf{F}}(g^{-1}z), \tag{10.71}$$

where $M_g(z)$ is a 3×3 matrix representing the effect of g on the electric and magnetic fields at z (see Section 9.7). $M_g(z)$ generalizes the factor $s^{-1/2}$ that was necessary in order to make scaling operations on proxies unitary. Indeed, when g is the scaling transformation $z \mapsto az$ with $a \neq 0$, it can be easily seen from the definition of $\|\mathbf{F}\|^2$ that unitarity implies

$$(U_g \tilde{\mathbf{F}})(z) = a^{-2}\tilde{\mathbf{F}}(z/a), \tag{10.72}$$

so $M_g(z) = a^{-2}I$, where I is the identity matrix. Note that M_g is independent of z in this case. In fact, M_g is independent of z whenever g is linear, which means that g is a combination of a Poincaré transformation and scaling. Translations do not affect the polarization, so $M_g = I$. But a space rotation should be accompanied by a uniform rotation of the fields, so M_g is a real orthogonal matrix. A Lorentz transformation results in a scaling and mixing of the electric and magnetic fields (Jackson [1975]), so M_g must be complex (in order to mix $\mathbf{E}$

and $\mathbf{B}$). Finally, when g is a special conformal transformation, $M_g(z)$ depends on z due to the nonlinearily of g. The explicit form of $M_g(z)$ is known, although we will not need to use it here; see Rühl (1972), and Jacobsen and Vergne (1977). The properties of $M_g(z)$ ensure that $g \mapsto U_g$ is a unitary representation of $\mathcal{C}$ on $\tilde{\mathcal{H}}$. Finally, the action of $\mathcal{C}$ on $\mathcal{H}$ is obtained by taking boundary values as $y \to 0$ simultaneously from the past and future cones, as explained in Chapter 9. We write it formally as

$$(U_g \mathbf{F})(x) = M_g(x)\mathbf{F}(g^{-1}x), \tag{10.73}$$

although $M_g(x)$ (unlike $M_g(z)$) may have singularities due to the singular nature of special conformal transformations on $\mathbf{R}^4$. (The singular set has zero measure; thus (10.73) still defines a unitary representation.) That makes the representation $g \mapsto U_g$ unitary on $\mathcal{H}$ as well. This means that from an "objective" point of view, there is absolutely no difference between the original Cartesian coordinate system in $\mathbf{R}^4$ and its image under g: To an observer using the transformed system, $U_g\mathbf{F}$ looks just as $\mathbf{F}$ does to an observer using the original system. This is an *extension* of special relativity theory that is valid when only free electromagnetic waves are considered. See Cunningham (1910), Bateman (1910), Page (1936), and Hill (1945, 1947, 1951).

We can now derive the action of $\mathcal{C}$ on the wavelets. (This was already done in Section 9.7 for the special case of Lorentz transformations.) Setting $g \to g^{-1}$ in (10.71) and using $\tilde{\mathbf{F}}(z) = \mathbf{\Psi}_z^*\mathbf{F}$, we have

$$\mathbf{\Psi}_z^* U_{g^{-1}} \mathbf{F} = M_{g^{-1}}(z)\mathbf{\Psi}_{gz}^*\mathbf{F}. \tag{10.74}$$

Dropping $\mathbf{F}$ and taking adjoints gives

$$U_{g^{-1}}^* \mathbf{\Psi}_z = \mathbf{\Psi}_{gz} M_{g^{-1}}(z)^*. \tag{10.75}$$

But U_g is unitary, so $U_{g^{-1}}^* = U_g$. This gives the simple action

$$U_g \mathbf{\Psi}_z = \mathbf{\Psi}_{gz} M_{g^{-1}}(z)^*, \tag{10.76}$$

which generalizes (10.60) from proxies to physical wavelets. Note that the polarization matrix M operates on the wavelets from the *right*. Aside from it, U_g acts on $\mathbf{\Psi}_z$ simply by transforming the *label* from z to gz. That means that U_g changes not only the point of focus x of $\mathbf{\Psi}_{x+iy}$, but also its center velocity $\mathbf{v} = \mathbf{y}/s$, scale $|s|$ and helicity (sign of s). For example, if $g = \Lambda$ is a pure Lorentz transformation, then $gz = \Lambda x + i\Lambda y$, so $\mathbf{\Psi}_{gz}$ is focused at Λx and has velocity and scale determined by Λy. Thus U_g "boosts" $\mathbf{\Psi}_z$ by transforming the coordinates of its point of focus and simultaneously boosting its velocity/scale vector. Translations, on the other hand, act only on x since $z + b = (x + b) + iy$.

In order for a scattering event to be "elementary," both the wave and the scatterer must be elementary in some sense. In the case of the proxies, the elementary waves were the wavelets $\psi_{s,\tau}(t)$. If the mother wavelet $\psi(t)$ is emitted at $t = 0$, then causality requires that $\psi(t) = 0$ for $t < 0$. This implies that $\psi_{s,\tau}(t) = 0$ for $t < \tau$. We would like to say that the wavelets $\mathbf{\Psi}_z$ are elementary electromagnetic waves. But recall that the wavelets with stationary centers, $\mathbf{\Psi}_{\mathbf{x},is}(\mathbf{x}', t')$, are *converging* to $\mathbf{x}$ when $t' < t$ and a *diverging* from $\mathbf{x}$ when $t' > t$. For arbitrary $z = x + iy \in \mathcal{T}$, $\mathbf{\Psi}_z(x')$ is a converging wave when $x' - x$ belongs to the past cone V'_- and a diverging wave when $x' - x \in V'_+$. We will think of $\mathbf{\Psi}_z$ as being *emitted*† at time $t' = t$ from its support region near $\mathbf{x}' = \mathbf{x}$. For this to make sense, it is necessary to ignore the converging part. That, along with some other necessary operations, will be done below.

Now that we have defined elementary waves, we need to define elementary scatterers. For the $\psi_{s,\tau}$'s, these were point reflectors. That was enough because the only parameters were time and scale. The $\mathbf{\Psi}_z$'s have many more parameters: space-time and velocity-scale on the side of the independent variables, and polarization on the side of the dependent variables. In order not to lose the potential information carried by these parameters, an elementary scatterer must have more structure than a point. Since we are presently interested in radar, our scatterers will be idealized as "mirrors" in various states. (More general situations may be contemplated by assuming that the scatterers reflect a part of the incident wavelet and absorb, transmit or refract the rest. Here we consider reflection only.) We begin by defining an elementary reflector in a standard state, which will be called the *standard reflector*, and then transform it to arbitrary states using conformal transformations. The standard reflector is a *flat, round mirror at rest at the origin* $\mathbf{x} = \mathbf{0}$ *at time* $t = 0$*, facing the positive* x_1*-axis.* It is assumed to have a certain radius and exist only during a certain time interval around $t = 0$, so it occupies a compact space-time region R centered at the origin $x = 0$ in $\mathbf{R}^4$. This region is cylindrical in the time direction since the mirror is assumed to be at rest during its entire interval of existence. A general elementary reflector is obtained by applying any conformal transformation g to R. We will order the different operations in g as follows: first *scale* R by s (assume $s > 0$), then *curve* it by a special conformal transformation C_b with $b = (\mathbf{b}, 0)$, *rotate* it to an arbitrary orientation, and *boost* it to an arbitrary velocity and acceleration (using Lorentz and special conformal transformations); finally, *translate* it to an arbitrary point b in space-time. The resulting scaled, curved, rotated, moving mirror at b will be denoted by gR. All of its parameters are conveniently vested in g.

We now ask: What happens when an elementary wave encounters an ele-

---

† The natural splitting of electromagnetic wavelets into absorbed and emitted parts is discussed in Chapter 11.

mentary scatterer? As with all wave phenomena, the rigorous details are difficult and complicated, and so far have not been worked out. Fortunately, it is possible to give a rough intuitive account, since we are completely in the space-time domain. (Scattering phenomena are often analyzed for harmonic waves, i.e., in the frequency domain, where much of common-sense intuition is lost. An even more serious problem is that nonlinear operations, such as special conformal transformations, cannot be transferred to the frequency domain since the Fourier variables are related to space-time by *duality*, which is a linear concept. See Section 1.4.) The simplest situation occurs when the reflector is in the standard state, since we then have a pulse reflected from a flat mirror. If we can guess how an arbitrary wavelet $\mathbf{\Psi}_z$ is reflected from R, we can apply conformal transformations to this encounter to see how wavelets are reflected from an arbitrary gR. *This method derives its legitimacy from the invariance of Maxwell's equations under* $\mathcal{C}$.

Suppose, therefore, that an emitted wavelet $\mathbf{\Psi}_z$ encounters R.† We claim that $z = x + iy$ must satisfy certain conditions in order that a substantial reflection take place. First of all, if $t \equiv x_0 > 0$, very little reflection can take place because the standard mirror exists only for a time interval around $t = 0$. Thus, *causality* requires (approximately) that $t < 0$ for reflection. But even when $t < 0$, the reflection is still weak if $x_1 < 0$, since then most of the wavelet originates behind the mirror. This gives two conditions on x. There are also conditions on y. The duration of $\mathbf{\Psi}_z$ is of the order $|s| \pm |\mathbf{y}|/c$. If the "lifetime" of the mirror is small compared to this, then most of the wavelet will not be reflected even if the conditions on x are satisfied. Altogether, we assume that there is a certain set of conditions on $z = x+iy$ that are necessary in order for $\mathbf{\Psi}_z$ to produce a substantial reflection from R. We say that z is *admissible* if all these conditions are satisfied, *inadmissible* otherwise. (Note that this admissibility is relative to the *standard* reflector R!) We assume that the strength of the reflection can be modeled by a function $P_r(z)$ with $0 \le P_r(z) \le 1$, such that $P_r(z) \approx 1$ when z is admissible and $P_r(z) \approx 0$ when it is inadmissible. We call $P_r(z)$ the *reflection coefficient function.* Actually, $P_r(z) = 1$ only if the reflection is perfect. Imperfect reflection (due to absorption or transmission, for example) can be modeled with $0 < P_r(z) < 1$.

---

† Actually, since $\mathbf{\Psi}_z$ is a *matrix* solution of Maxwell's equations, this means (if taken literally) that *three* waves are emitted. The return $\mathbf{F}$ then also consists of three waves, i.e., it is a matrix as well. This has a natural interpretation: The matrix elements of $\mathbf{F}$ give the *correlations* between the outgoing and incoming triplets. We can avoid this multiplicity by emitting any column of $\mathbf{\Psi}_z$ instead or, more generally, a linear combination of the columns given by $\mathbf{\Psi}_z\mathbf{u} \in \mathcal{H}$, where $\mathbf{u} \in \mathbf{C}^3$. Then $\mathbf{F}$ is also an ordinary solution. But it seems natural to work with $\mathbf{\Psi}_z$ as a whole.

To model the reflection of Ψ_z in R, define the linear map $\rho : \mathbf{R}^4 \to \mathbf{R}^4$ by

$$\rho(x_1, x_2, x_3, t) \equiv (-x_1, x_2, x_3, t). \tag{10.77}$$

ρ is the geometrical reflection in the hyperplane $\{x_1 = 0\}$ in $\mathbf{R}^4$. Note that ρ belongs to the Lorentz group $\mathcal{L} \subset \mathcal{C}$. We expect that the reflection of Ψ_{x+iy} in R seems to originate from $\tilde{x} = \rho x$, and its center moves with the velocity $\tilde{v} = \tilde{y}/\tilde{s}$, where $\tilde{y} = \rho y$. This is consistent with the assumption that the reflected wave is itself a wavelet parameterized by $\tilde{z} \equiv \rho x + i\rho y = \rho z$. Since $\rho = \rho^{-1} \in \mathcal{L}$, (10.76) gives

$$U_\rho \Psi_z = \Psi_{\rho z} M_\rho^* \tag{10.78}$$

for the reflected wave when $P_r(z) = 1$. For *general* $z \in \mathcal{T}$, we multiply $U_\rho \Psi_z$ by the reflection coefficient $P_r(z)$. Thus *our basic model for the reflection of an arbitrary wavelet* Ψ_z *by the standard reflector* R *is given by*

$$S\Psi_z \equiv P_r(z) U_\rho \Psi_z\,. \tag{10.79}$$

This model might be extended to include *transmission* as follows: The transmitted wave is modeled by $P_t(z)\Psi_z$, where $P_t(z)$ is a "transmission coefficient function" similar to $P_r(z)$. The total effect of the scattering is then modeled by $P_r(z)U_\rho\Psi_z + P_t(z)\Psi_z$. (It may even be possible to model refraction in a similar geometric way.) If no absorption occurs, then $P_r(z)$ and $P_t(z)$ are such that the scattering preserves the energy. Since we are presently interested in radar, only reflection will be considered here.

A *general* elementary scattering event is an encounter between a general wavelet Ψ_z and a general elementary reflector gR. We assume that the reflection is obtained by transforming everything to the standard reference frame (where the reflector is a flat mirror at rest near the origin), viewing the reflection there, and transforming back to the original frame. This gives

$$\begin{aligned}\Psi_z &\to \Psi_{g^{-1}z} M_g^*(z) \to P_r(g^{-1}z)\Psi_{\rho g^{-1}z} M_g^*(z) M_\rho^* \\ &\to P_r(g^{-1}z)\Psi_{g\rho g^{-1}z} M_g(z)^* M_\rho^* M_{g^{-1}}(\rho g^{-1} z)^*\,.\end{aligned} \tag{10.80}$$

The right-hand side of (10.80) is seen to be (by the group property of the U's)

$$P_r(g^{-1}z) U_g U_\rho U_{g^{-1}} \Psi_z = P_r(g^{-1}z) U_{g\rho g^{-1}} \Psi_z\,. \tag{10.81}$$

Let

$$\rho_g \equiv g\rho g^{-1} = \rho_g^{-1}\,. \tag{10.82}$$

Here ρ_g represents the geometrical reflection in the (not necessarily linear) three-dimensional hypersurface $\{gx : x_1 = 0\}$ in $\mathbf{R}^4$, i.e., the image under g of $\{x : x_1 = 0\}$. Thus *our model for the elementary reflection of* Ψ_z *from* gR *is given by*

$$S_g \Psi_z \equiv P_r(g^{-1}z) U_{\rho_g} \Psi_z = P_r(g^{-1}z) \Psi_{\rho_g z} M_{\rho_g}(z)^*\,. \tag{10.83}$$

Equation (10.83) generalizes the model (10.8) for proxies.

We can now extend the method of ambiguity functions to the present setting. Suppose we want to find the state of an unknown reflector. We assume that the reflector is elementary, and "illuminate" it with $\mathbf{\Psi}_z$. The parameters $z \in \mathcal{T}$ are fixed, determined by the "state" of the radar. For example, if $z = (\mathbf{x}, is) \in E$, then the radar is at rest at $\mathbf{x}$ and $\mathbf{\Psi}_z$ is emitted at $t = 0$. But this need not be assumed, and $z \in \mathcal{T}$ can be arbitrary. Then the return is given by $\mathbf{F} = S_g \mathbf{\Psi}_z$ as in (10.83). The problem is to obtain information about the state g of the reflector, given only $\mathbf{F}$. For simplicity, we assume for the moment that $\mathbf{F}(\mathbf{x}, t)$ can be measured everywhere in space (at some fixed time t), which is, of course, quite unrealistic. Define the *conformal cross-ambiguity function* as the correlation between $\mathbf{F}$ and $U_{\rho_h} \mathbf{\Psi}_z$, with $h \in \mathcal{C}$ arbitrary:

$$\mathbf{A}_z(h) \equiv (U_{\rho_h} \mathbf{\Psi}_z)^* \mathbf{F} = \mathbf{\Psi}_z^* U_{\rho_h}^* \mathbf{F}. \tag{10.84}$$

That is, $\mathbf{A}_z(h)$ *matches* the return (which is assumed to have the form $S_g \mathbf{\Psi}_z$, with g unknown) with $U_{\rho_h} \mathbf{\Psi}_z$. If indeed $\mathbf{F} = S_g \mathbf{\Psi}_z$, then

$$\mathbf{A}_z(h) = P_r(g^{-1}z)(U_{\rho_h} \mathbf{\Psi}_z)^* (U_{\rho_g} \mathbf{\Psi}_z). \tag{10.85}$$

By (10.76), (10.82), and (10.83),

$$\begin{aligned} \mathbf{A}_z(h) &= P_r(g^{-1}z) M_{\rho_h}(z)\, \mathbf{\Psi}_{\rho_h z}^* \mathbf{\Psi}_{\rho_g z} M_{\rho_g}(z)^* \\ &= P_r(g^{-1}z) M_{\rho_h}(z)\, \mathbf{K}(\rho_h z \mid \rho_g z)\, M_{\rho_g}(z)^*, \end{aligned} \tag{10.86}$$

where

$$\mathbf{K}(z \mid w) \equiv \mathbf{\Psi}_z^* \mathbf{\Psi}_w \tag{10.87}$$

is the reproducing kernel for the electromagnetic wavelets, now playing the role of *conformal self-ambiguity function.* Recall that for the proxies, we used Schwarz's inequality at this point to find the parameters of the reflector (10.13). Since $\mathbf{A}_z(h)$ is a matrix, this argument must now be modified. Let $\mathbf{u}_1, \mathbf{u}_2, \mathbf{u}_3$ be any orthonormal basis in $\mathbf{C}^3$, and define the vectors $\mathbf{\Phi}_{h,z,n}, \mathbf{\Phi}_{g,z,n} \in \mathcal{H}$ by

$$\mathbf{\Phi}_{h,z,n} \equiv U_{\rho_h} \mathbf{\Psi}_z \mathbf{u}_n, \qquad \mathbf{\Phi}_{g,z,n} \equiv U_{\rho_g} \mathbf{\Psi}_z \mathbf{u}_n. \tag{10.88}$$

The *trace* of $\mathbf{A}_z(h)$ is defined as

$$\begin{aligned} \operatorname{trace} \mathbf{A}_z(h) &\equiv \sum_n \mathbf{u}_n^* \, \mathbf{A}_z(h)\, \mathbf{u}_n \\ &= P_r(g^{-1}z) \sum_n \mathbf{\Phi}_{h,z,n}^* \mathbf{\Phi}_{g,z,n} \\ &= P_r(g^{-1}z) \sum_n \langle\, \mathbf{\Phi}_{h,z,n}\,, \mathbf{\Phi}_{g,z,n} \,\rangle. \end{aligned} \tag{10.89}$$

Applying Schwarz's inequality and using the unitarity of U_{ρ_h} and U_{ρ_g}, we obtain

$$\begin{aligned}|\text{trace}\,\mathbf{A}_z(h)| &\leq P_r(g^{-1}z)\sum_n |\langle\, \mathbf{\Phi}_{h,z,n}\,,\,\mathbf{\Phi}_{g,z,n}\,\rangle| \\ &\leq P_r(g^{-1}z)\sum_n \|\mathbf{\Phi}_{h,z,n}\|\,\|\mathbf{\Phi}_{g,z,n}\| \\ &= P_r(g^{-1}z)\sum_n \|\mathbf{\Psi}_z\mathbf{u}_n\|^2 \\ &= P_r(g^{-1}z)\sum_n \mathbf{u}_n^*\mathbf{\Psi}_z^*\mathbf{\Psi}_z\mathbf{u}_n \\ &= P_r(g^{-1}z)\ \text{trace}\,\mathbf{\Psi}_z^*\mathbf{\Psi}_z = P_r(g^{-1}z)\,N(z),\end{aligned} \tag{10.90}$$

where $N(z)$ is given by (9.140). Equality is attained in (10.90) if and only if $\mathbf{\Phi}_{h,z,n} = \mathbf{\Phi}_{g,z,n}$ for all n, i.e., if and only if

$$U_{\rho_h}\mathbf{\Psi}_z = U_{\rho_g}\mathbf{\Psi}_z\,. \tag{10.91}$$

The right-hand side of the last equation in (10.90) is independent of h, and so represents the absolute maximum of $|\text{trace}\,\mathbf{A}_z(h)|$. It can be shown that (10.91) holds if and only if

$$\rho_h z = \rho_g z. \tag{10.92}$$

Equation (10.92) states that the reflection of z in hR is the same as the reflection of z in gR, which is in accord with intuition. By (10.82) and (10.92),

$$\rho_h\rho_g z = z. \tag{10.93}$$

This means that $\rho_h\rho_g$ belongs to the subgroup $K_z \equiv \{k \in \mathcal{C} : kz = z\}$, which is called the *isotropy group* at z and is isomorphic to the maximal compact subgroup $K = U(1) \times SU(2) \times SU(2)$ of $\mathcal{C}$. This gives a partial determination of the state g of the reflector, and it represents the most we can expect to learn from the reflection of a single wavelet.

It may happen that a reflector that looks elementary at one resolution (choice of $s \equiv y_0$ in the outgoing wavelet) is "resolved" to show fine-structure at a higher resolution (smaller value of s). Since conformal ambiguity functions depend on many more physical parameters than do the wideband (affine) ambiguity functions of Section 10.1, they can be expected to provide more detailed imaging information. A *complex* reflector is one that can be built by patching together elementary reflectors gR in different states $g \in \mathcal{C}$. There may be a finite or infinite number of gR's, and they may or may not coexist at the same time. We model this "building" process as follows: There is a *left-invariant Haar measure* on $\mathcal{C}$, which we denote by $d\nu(g)$. This measure is unique, up to a constant factor, and it is characterized by its invariance under left translations by elements of $\mathcal{C}$: $d\nu(g_0 g) = d\nu(g)$, for any fixed $g_0 \in \mathcal{C}$. $d\nu(g)$ is, in fact, precisely the generalization to $\mathcal{C}$ of the measure $d\nu(s,\tau) \equiv s^{-2} ds\, d\tau$ used on the affine group in Section 10.1. We model the complex reflector by choosing a density function $w(g)$, telling us which reflectors gR are present and in what strength. The complex reflector is then represented by the *measure*

$$d\sigma(g) = w(g) d\nu(g). \tag{10.94}$$

The measure $d\sigma(g)$ is a "blueprint" that tells us how to construct the complex reflector: Apply each g in the support of $w(g)$ to the standard reflector R (giving the scaled, rotated, boosted and translated reflector gR), then assemble these pieces. Suppose this reflector is illuminated with a single wavelet $\mathbf{\Psi}_z$. Then the return can be modeled as a superposition of (10.83), with $d\sigma(g)$ providing the weight:

$$S\mathbf{\Psi}_z = \int d\sigma(g) S_g \mathbf{\Psi}_z = \int d\sigma(g) P_r(g^{-1}z)\, U_{\rho_g} \mathbf{\Psi}_z\,. \tag{10.95}$$

Note that $d\sigma$ is generally concentrated on very small, lower-dimensional subsets of $\mathcal{C}$. If the reflector consists of a finite number of elementary reflectors $g_1 R, g_2 R, \ldots$, then $w(g)$ is a sum of delta-like functions and the integral in (10.95) becomes a sum.

Equation (10.95) is a generalization to electromagnetic waves of (10.14) for proxies, with the correspondence

$$\begin{aligned} &\psi \to \mathbf{\Psi}_z & &f \to S_g \mathbf{\Psi}_z \\ &(s,\tau) \in \mathcal{A} \to g \in \mathcal{C} & &s^{-2} ds\, d\tau \to d\nu(g) \\ &\psi_{s,\tau} \to U_{\rho_g} \mathbf{\Psi}_z & &D(s,\tau) \to w(g) P_r(g^{-1}z). \end{aligned} \tag{10.96}$$

In particular,

$$D_z(g) \equiv w(g) P_r(g^{-1}z) \tag{10.97}$$

is interpreted as the "reflectivity distribution" of the complex reflector with respect to the incident wavelet $\mathbf{\Psi}_z$. $w(g)$ tells us whether gR is present, and in what strength, while $P_r(g^{-1}z)$ gives the (predictable) reflection coefficient of gR with respect to $\mathbf{\Psi}_z$, *if* gR is present. Just as $D(s,\tau)$ was defined on the affine group $\mathcal{A}$, which parameterized the possible states (range and radial velocity) of the point reflector for the proxies, so is $D_z(g)$ defined on the conformal group $\mathcal{C}$, which parameterizes the states of the elementary reflectors gR. Note that $D_z(g) = 0$ if $g^{-1}z$ is not admissible. For $g^{-1}z$ to be admissible, the time component of $g^{-1}x$ must be negative, which means that g translates to the future. This corresponds to $D(s,\tau) = 0$ for $\tau < 0$, and represents *causality*. The delayed and Doppler-scaled proxies $\psi_{s,\tau}$ in (10.14) are now replaced by $U_{\rho_g} \mathbf{\Psi}_z$, as indicated in (10.96).

Equation (10.95) represents an *inverse problem*: given $S\mathbf{\Psi}_z$, find $D_z(g)$. Similar considerations can be applied toward the solution of this problem as were applied to the corresponding problem for proxies in Section 10.1, since the underlying mathematics is similar, although more involved. For example, the reproducing kernel $\mathbf{K}(z' \mid z)$ gives the orthogonal projection to the range of the analytic-signal transform $\mathbf{F} \to \tilde{\mathbf{F}}$, as explained in Chapter 9.

As an example of complex reflectors, consider the following scheme. Suppose we expect the return to be a result of reflection from one or more airplanes with more or less known characteristics. Then we can first build a "model airplane" P in a standard state, analogous to the standard reflector R. P is to represent a complex reflector at rest near the origin $\mathbf{x} = \mathbf{0}$ at time $t = 0$ in a standard orientation. Hence it is built from patches gR that do not include any time translations, boosts or accelerations. Recall that the special conformal transformations C_b with $b = (\mathbf{b}, 0)$ leave the configuration space $\mathbf{R}_0^3 \equiv \{t = 0\}$ invariant but curve flat objects in $\mathbf{R}_0^3$. For example, letting $x = (0, x_2, x_3, 0)$ be the position vector in $R_0 \equiv \{x \in R : t = 0\}$ and choosing $b = (a, 0, 0, 0)$, (10.67) gives

$$C_b(0, x_2, x_3, 0) = \frac{(-ar^2, x_2, x_3, 0)}{1 + a^2r^2} \equiv (\mathbf{r}', 0), \tag{10.98}$$

where $r^2 \equiv x_2^2 + x_3^2$. Hence the image of the flat mirror R_0 at $t = 0$ is a curved disc C_bR_0, also at $t = 0$, convex if $a < 0$ and concave if $a > 0$. Choosing $b = (\mathbf{b}, 0)$ with $b_2 \neq 0$ or $b_3 \neq 0$ gives lop-sided instead of uniform curving. Thus by scaling, curving, rotating and translating R in space, we can build a "model airplane" P in a standard state. This can be done ahead of time, to the known specifications. Possibly, only certain characteristic parts of the actual reflector are included in P, either for the sake of economy or because of ignorance about the actual object. Then, in a field situation, this model can be *propelled as a whole* to an arbitrary state by applying a conformal transformation $h \in \mathcal{C}$, this time including time translations, boosts and accelerations. Thus hP is a version of P flying with an arbitrary velocity, at an arbitrary location, with an arbitrary orientation, etc. If the weight function for P is $w(g)$, then that for hP is $w(h^{-1}g)$. If hP is illuminated with $\mathbf{\Psi}_z$, the expected return is

$$\int d\nu(g)\, w(h^{-1}g) P_r(g^{-1}z)\, U_{\rho_g} \mathbf{\Psi}_z\,, \tag{10.99}$$

where now the reflectivity distribution $w(h^{-1}g)P_r(g^{-1}z)$ is presumed known. This can be matched with the actual return to estimate the state h of the plane. If several such planes are expected, we need only apply several conformal transformations $h_1, h_2, \ldots$ to P.

Note that when imaging a complex object like hP, the scale parameter in the return may carry valuable information about the *local curvature* (for example, *edges*) of the object. Usually this kind of information is obtained in the frequency domain where (as we have mentioned before) it may be more complicated.

Modified from spherical waves to nondiffracting beams, they have been proposed for possible ultrasound applications (Zou, Lu, and Greenleaf [1993a, b]).

Most of the results in this section are straightforward extensions of ideas developed in Chapter 9 and Section 10.2, and the explanations and proofs will be rather brief. I have tried to make the presentation self-contained so that readers interested exclusively in acoustics can get the general idea without referring to the previous material, doing so only when they decide to learn the details. At a minimum, the reader should read Section 9.3 on the analytic-signal transform before attempting to read this section.

As explained in Section 9.2, solutions of the wave equation can be obtained by means of Fourier analysis in the form

$$\begin{aligned} F(x) &= \int_{\mathbf{R}^3} \frac{d^3\mathbf{p}}{16\pi^3\omega} \left[e^{i(\omega t - \mathbf{p}\cdot\mathbf{x})} f(\mathbf{p},\omega) + e^{i(-\omega t - \mathbf{p}\cdot\mathbf{x})} f(\mathbf{p},-\omega) \right] \\ &= \int_C d\tilde{p}\; e^{ip\cdot x} f(p), \end{aligned} \tag{11.1}$$

where

$$\begin{aligned} & x = (\mathbf{x}, t) = \text{space-time four-vector} \\ & p = (\mathbf{p}, p_0) = \text{wavenumber-frequency four-vector} \\ & p_0 = \pm|\mathbf{p}| \quad \text{by the wave equation} \\ & \omega = |\mathbf{p}| = \text{absolute value of frequency} \\ & p\cdot x = p_0 t - \mathbf{p}\cdot\mathbf{x} = \text{Lorentz-invariant scalar product} \\ & C = \{p = (\mathbf{p}, p_0) \in \mathbf{R}^4 : p^2 \equiv p_0 - |\mathbf{p}|^2 = 0\} = C_+ \cup C_- = \text{light cone} \\ & d\tilde{p} \;=\; \frac{d^3\mathbf{p}}{16\pi^3\omega} = \text{Lorentz-invariant measure on } C. \end{aligned} \tag{11.2}$$

In order to define the wavelets, we need a Hilbert space of solutions. To this end, fix $\alpha \ge 1$ and define

$$\begin{aligned} \|F\|^2 &\equiv \int_{\mathbf{R}^3} \frac{d^3\mathbf{p}}{16\pi^3\omega^\alpha} \left[|f(\mathbf{p},\omega)|^2 + |f(\mathbf{p},-\omega)|^2 \right] \\ &= \int_C \frac{d\tilde{p}}{\omega^{\alpha-1}} \, |f(p)|^2. \end{aligned} \tag{11.3}$$

Let $\mathcal{H}_\alpha$ be the set of all solutions (11.1) for which the above integral converges, i.e.,

$$\mathcal{H}_\alpha = \{F : \|F\|^2 < \infty\}. \tag{11.4}$$

Then $\mathcal{H}_\alpha$ is a Hilbert space under the inner product obtained by polarizing (11.3):

$$F^*G \equiv \langle F, G \rangle \equiv \int_C \frac{d\tilde{p}}{\omega^{\alpha-1}} \, \overline{f}(p)\, g(p), \qquad F^*F = \|F\|^2. \tag{11.5}$$

Recall that a similar Hilbert space was defined in the one-dimensional case in Section 9.3. The meaning of the parameter α is roughly as follows: For large values of α, $f(p)$ *must* decay rapidly near $\omega = 0$ in order for F to belong to $\mathcal{H}_\alpha$. This means that $F(x)$ necessarily has many zero moments in the time direction, and therefore it must have a high degree of *oscillation* along the time axis. On the other hand, a large value of α also means that $f(p)$ *may* grow or slowly decay as $\omega \to \infty$, and this would tend to make $F(x)$ *rough.* That is, a large value of α simultaneously forces oscillation and permits roughness. The wavelets we are about to construct will be seen to have the oscillation (as they must) but not the roughness. They decay exponentially in frequency.

To construct the wavelets, extend $F(x)$ from the real space-time $\mathbf{R}^4$ to the causal tube $\mathcal{T} = \{x+iy \in \mathbf{C}^4 : y^2 > 0\}$ by applying the analytic-signal transform (Section 9.3):

$$\begin{aligned} \tilde{F}(x+iy) &\equiv \frac{1}{\pi i} \int_{-\infty}^{\infty} \frac{d\tau}{\tau - i} F(x+\tau y) \\ &= \int_C d\tilde{p}\ 2\theta(p\cdot y) e^{ip\cdot(x+iy)} f(p), \end{aligned} \tag{11.6}$$

where θ is the unit step function. By multiplying and dividing the integrand in the last equation by $\omega^{\alpha-1}$, we can express the integral as an inner product of F with another solution Ψ_{x+iy}, which will be the acoustic wavelet parameterized by $z = x + iy$:

$$\begin{aligned} \tilde{F}(z) &= \int_C \frac{d\tilde{p}}{\omega^{\alpha-1}}\ 2\omega^{\alpha-1}\, \theta(p\cdot y) e^{ip\cdot z} f(p) \\ &= \int_C \frac{d\tilde{p}}{\omega^{\alpha-1}}\, \overline{\psi}_z(p) f(p), \end{aligned} \tag{11.7}$$

where

$$\psi_z(p) \equiv 2\omega^{\alpha-1}\, \theta(p\cdot y) e^{-ip\cdot\bar{z}}. \tag{11.8}$$

ψ_z is the Fourier coefficient function for the solution

$$\Psi_z(x') \equiv \int_C d\tilde{p}\ e^{ip\cdot x'} \psi_z(p) = \int_C d\tilde{p}\ 2\omega^{\alpha-1}\, \theta(p\cdot y) e^{ip\cdot(x'-\bar{z})}, \tag{11.9}$$

which we call an *acoustic wavelet of order* α. For $\alpha = 3$ we recognize the scalar wavelets of Section 9.7. By the definition (11.5) of the inner product,

$$\tilde{F}(z) = \Psi_z^* F = \langle \Psi_z, F \rangle \quad \text{for all } F \in \mathcal{H}_\alpha \text{ and } z \in \mathcal{T}. \tag{11.10}$$

For the acoustic wavelets to be useful, we must be able to express every solution $F(x)$ as a superposition of Ψ_z's. That is, we need a resolution of unity. We follow a procedure similar to that used for the electromagnetic wavelets. The restriction of the analytic signal $\tilde{F}(z)$ to the Euclidean region $E = \{(\mathbf{x}, is) : \mathbf{x} \in$

$\mathbf{R}^3, s \neq 0\}$ in $\mathcal{T}$ is

$$\begin{aligned}\tilde{F}(\mathbf{x}, is) &= \int_C d\tilde{p}\; 2\theta(p_0 s) e^{-p_0 s} e^{-i\mathbf{p}\cdot\mathbf{x}} f(p) \\ &= \int_{\mathbf{R}^3} \frac{d^3\mathbf{p}}{8\pi^3\omega} e^{-i\mathbf{p}\cdot\mathbf{x}} \left[\theta(s) e^{-\omega s} f(\mathbf{p}, \omega) + \theta(-s) e^{\omega s} f(\mathbf{p}, -\omega)\right].\end{aligned} \tag{11.11}$$

Therefore by Plancherel's theorem,

$$\begin{aligned}&\int_{\mathbf{R}^3} d^3\mathbf{x}\, |\tilde{F}(\mathbf{x}, is)|^2 \\ &\quad= \int_{\mathbf{R}^3} \frac{d^3\mathbf{p}}{8\pi^3\omega^2} \left[\theta(s) e^{-2\omega s} |f(\mathbf{p}, \omega)|^2 + \theta(-s) e^{2\omega s} |f(\mathbf{p}, -\omega)|^2\right].\end{aligned} \tag{11.12}$$

Note that for $\alpha > 2$,

$$\int_{-\infty}^{\infty} ds\; |s|^{\alpha-3}\, \theta(\pm s) e^{\mp 2\omega s} = \int_0^{\infty} ds\, s^{\alpha-3} e^{-2\omega s} = \frac{\Gamma(\alpha-2)}{(2\omega)^{\alpha-2}}. \tag{11.13}$$

In order to obtain $\|F\|^2$ on the right-hand side of (11.12), we integrate both sides with respect to s with the weight factor $|s|^{\alpha-3}$, assuming that $\alpha > 2$:

$$\begin{aligned}&\int_E d^3\mathbf{x}\; ds\, |s|^{\alpha-3}\, |\tilde{F}(\mathbf{x}, is)|^2 \\ &\qquad= \int_{\mathbf{R}^3} \frac{d^3\mathbf{p}}{8\pi^3\omega^2} \frac{\Gamma(\alpha-2)}{(2\omega)^{\alpha-2}} \left[|f(\mathbf{p}, \omega)|^2 + |f(\mathbf{p}, -\omega)|^2\right] \\ &\qquad= \frac{\Gamma(\alpha-2)}{2^{\alpha-3}} \|F\|^2.\end{aligned} \tag{11.14}$$

Let $\tilde{\mathcal{H}}_\alpha$ be the space of all analytic signals of solutions in $\mathcal{H}_\alpha$, i.e.,

$$\tilde{\mathcal{H}}_\alpha \equiv \{\tilde{F} : F \in \mathcal{H}_\alpha\}. \tag{11.15}$$

Equation (11.14) suggests that we define an inner product on $\tilde{\mathcal{H}}_\alpha$ by

$$\tilde{F}^* \tilde{G} \equiv \langle\, \tilde{F}, \tilde{G}\,\rangle \equiv \int_E d\mu_\alpha(z)\, \overline{\tilde{F}(z)}\, \tilde{G}(z), \tag{11.16}$$

where $d\mu_\alpha(z)$ is the measure on the Euclidean region defined by

$$d\mu_\alpha(\mathbf{x}, is) = \frac{2^{\alpha-3}}{\Gamma(\alpha-2)}\, d^3\mathbf{x}\; ds\, |s|^{\alpha-3}. \tag{11.17}$$

With this definition, we have the following result. The heart of the theorem is (11.18), but we also state some important consequences for emphasis. See Chapter 9 for parallel results and terminology in the case of electromagnetics.

Theorem 11.1. *If $\alpha > 2$, then*

(a) The inner product defined on $\tilde{\mathcal{H}}_\alpha$ by (11.16) satisfies

$$\tilde{F}^*\tilde{G} = F^*G, \quad i.e., \quad \int_E d\mu_\alpha(z)\, F^*\Psi_z\Psi_z^*G = F^*G. \tag{11.18}$$

In particular, $\|\tilde{F}\| = \|F\|$ and $\tilde{\mathcal{H}}_\alpha$ is a closed subspace of $L^2(d\mu_\alpha)$.

(b) The acoustic wavelets of order α give the following resolution of unity in $\mathcal{H}_\alpha$:

$$\int_E d\mu_\alpha(z)\, \Psi_z\Psi_z^* = I, \text{ weakly in } \mathcal{H}_\alpha\,. \tag{11.19}$$

(This means that (11.18) holds for all $F, G \in \mathcal{H}_\alpha$.)

(c) Every solution $G \in \mathcal{H}_\alpha$ can be written

$$G = \int_E d\mu_\alpha(z)\, \Psi_z\Psi_z^*G = \int_E d\mu_\alpha(z)\, \Psi_z\, \tilde{G}(z), \text{ weakly in } \mathcal{H}_\alpha\,. \tag{11.20}$$

(This means that (11.18) holds for all $F \in \mathcal{H}_\alpha$.) Therefore,

$$G(x') = \int_E d\mu_\alpha(z)\, \Psi_z(x')\, \tilde{G}(z) \quad a.e. \tag{11.21}$$

(d) The orthogonal projection in $L^2(d\mu_\alpha)$ to the closed subspace $\tilde{\mathcal{H}}_\alpha$ is given by

$$(P\Phi)(z') = \int_E d\mu_\alpha(z)\, K(z'\,|\,z)\,\Phi(z), \quad \Phi \in L^2(d\mu_\alpha), \tag{11.22}$$

where

$$\begin{aligned} K(z'\,|\,z) \equiv \Psi_{z'}^*\Psi_z &= 4\int_C d\tilde{p}\ \omega^{\alpha-1}\,\theta(p\cdot y')\theta(p\cdot y)\, e^{ip\cdot(z'-\bar{z})} \\ &= 4\theta(y\cdot y')\int_C d\tilde{p}\ \omega^{\alpha-1}\,\theta(p\cdot y)\, e^{ip\cdot(z'-\bar{z})} \end{aligned} \tag{11.23}$$

is the associated reproducing kernel.

For $\alpha \le 2$, (11.13) shows that the expression for the inner product in $\tilde{\mathcal{H}}_\alpha$ diverges and Theorem 11.1 fails. We therefore say that the Ψ_z's are *admissible* if and only if $\alpha > 2$. Only admissible wavelets give resolutions of unity over E.

11.2 Emission, Absorption, and Currents

In order for the inner products Ψ_z^*F to converge for all $F \in \mathcal{H}_\alpha$, Ψ_z must itself belong to $\mathcal{H}_\alpha$. The norm of Ψ_z is

$$\begin{aligned} \|\Psi_{x+iy}\|^2 &= \int_C \frac{d\tilde{p}}{\omega^{\alpha-1}}\, 4\omega^{2\alpha-2}\theta(p\cdot y)e^{-2p\cdot y} \\ &= 4\int_C d\tilde{p}\ \omega^{\alpha-1}\theta(p\cdot y)e^{-2p\cdot y}, \end{aligned} \tag{11.24}$$

where we used $\theta(p \cdot y)^2 = \theta(p \cdot y)$ a.e. Clearly $\|\Psi_{x-iy}\|^2 = \|\Psi_{x+iy}\|^2$, so we can assume that y belongs to the future cone V'_+. Then

$$\|\Psi_z\|^2 = 4 \int_{C_+} d\tilde{p}\ \omega^{\alpha-1} e^{-2p \cdot y} = 2^{1-\alpha} E_\alpha(y), \tag{11.25}$$

where

$$E_\alpha(y) \equiv \int_{C_+} d\tilde{p}\ \omega^{\alpha-1} e^{-p \cdot y}. \tag{11.26}$$

Since

$$p \cdot y = \omega s - \mathbf{p} \cdot \mathbf{y} \geq \omega s - \omega |\mathbf{y}| \tag{11.27}$$

and $s - |\mathbf{y}| > 0$ for $y \in V'_+$, the integrand in (11.26) decays exponentially in the radial (ω) direction and the integral converges. (It decays least rapidly along the direction where $\mathbf{p}$ is parallel to $\mathbf{y}$, which means that Ψ_z favors these Fourier components; see the discussion of *ray filters* at the end of Chapter 9.) This proves that $\Psi_z \in \mathcal{H}_\alpha$, for any $\alpha \geq 1$ and all $z \in \mathcal{T}$.

We now compute $E_\alpha(z)$ when α is a positive integer. The resulting expression will be valid for arbitrary $\alpha \geq 1$, and it will be used to find an explicit form for $\Psi_z(x')$. According to (9.127),

$$S(y) \equiv \int_{C_+} d\tilde{p}\ e^{-p \cdot y} = \frac{1}{4\pi^2 y^2} = \frac{1}{4\pi^2} \frac{1}{s^2 - |\mathbf{y}|^2}. \tag{11.28}$$

We split $S(y)$ into partial fractions, assuming $u \equiv |\mathbf{y}| \neq 0$:

$$S(y) = \frac{1}{8\pi^2 u} \left[\frac{1}{s-u} - \frac{1}{s+u} \right]. \tag{11.29}$$

Then

$$\begin{aligned} E_\alpha(y) &= (-\partial_s)^{\alpha-1} S(y) \\ &= \frac{1}{8\pi^2 u} (-\partial_s)^{\alpha-1} \left[\frac{1}{s-u} - \frac{1}{s+u} \right] \\ &= \frac{\Gamma(\alpha)}{8\pi^2 u} \left[\frac{1}{(s-u)^\alpha} - \frac{1}{(s+u)^\alpha} \right], \quad u \equiv |\mathbf{y}| \neq 0. \end{aligned} \tag{11.30}$$

To get $E_\alpha(y)$ when $\mathbf{y} = \mathbf{0}$, take the limit $u \to 0^+$ by applying L'Hospital's rule:

$$E_\alpha(\mathbf{0}, s) = \frac{\Gamma(\alpha+1)}{4\pi^2 s^{\alpha+1}}, \quad s > 0. \tag{11.31}$$

For general $y \in V'$, we need only replace s by $|s|$ in (11.30)–(11.31). Although these formulas were derived on the assumption that $\alpha \in \mathbf{N}$, a direct (but more tedious) computation shows that they hold for all $\alpha \geq 1$. In view of (11.26), this proves that $\Psi_z \in \mathcal{H}_\alpha$ for $\alpha \geq 1$.[†]

Equation (11.30) can also be used to compute the wavelets. Recall that it suffices to find $\Psi_{iy}(x)$ since

$$\Psi_{x+iy}(x') = \Psi_{iy}(x' - x). \tag{11.32}$$

But

$$\Psi_{iy}(x) = 2\int_{C_+} d\tilde{p}\ \omega^{\alpha-1} e^{-p\cdot(y-ix)} = 2E_\alpha(y - ix). \tag{11.33}$$

Theorem 9.10 tells us that

$$(y - ix)^2 = -(x + iy)^2 \neq 0 \quad \text{for} \quad y \in V'. \tag{11.34}$$

As a result, it can be shown that $E_\alpha(y)$ has a unique analytic continuation to $E_\alpha(y - ix)$. This continuation can be obtained formally by substituting

$$s \to s - it, \qquad u = (\mathbf{y}^2)^{1/2} \to ((\mathbf{y} - i\mathbf{x})^2)^{1/2} \tag{11.35}$$

in (11.30). We are using the notation

$$(\mathbf{y} - i\mathbf{x})^2 \equiv (\mathbf{y} - i\mathbf{x}) \cdot (\mathbf{y} - i\mathbf{x}) = \mathbf{y}^2 - \mathbf{x}^2 - 2i\mathbf{x} \cdot \mathbf{y} \tag{11.36}$$

because $|\mathbf{y} - i\mathbf{x}|^2$ could be confused with the Hermitian norm in $\mathbf{C}^3$. We find the explicit form of the reference wavelet for $\mathcal{H}_\alpha$, i.e., of

$$\begin{aligned}\Psi(r, t) &\equiv \Psi_{\mathbf{0},i}(\mathbf{x}, t)\\ &= 2\int_{C_+} d\tilde{p}\ \omega^{\alpha-1} e^{-\omega(1-it)-i\mathbf{p}\cdot\mathbf{x}}, \quad r \equiv |\mathbf{x}|.\end{aligned} \tag{11.37}$$

As the notation implies, Ψ is spherically symmetric. The substitution (11.35) is now

$$s \to 1 - it, \qquad u \to \left(-\mathbf{x}^2\right)^{1/2} \equiv ir, \tag{11.38}$$

giving

$$\begin{aligned}\Psi(r, t) &= \frac{\Gamma(\alpha)}{4\pi^2\, ir}\left[\frac{1}{(1 - it - ir)^\alpha} - \frac{1}{(1 - it + ir)^\alpha}\right], \quad r \neq 0\\ \Psi(0, t) &= \frac{\Gamma(\alpha + 1)}{2\pi^2(1 - it)^{\alpha+1}}\,.\end{aligned} \tag{11.39}$$

[†] Since Ψ_z depends on α, we should have indicated this somehow, e.g., by writing Ψ_z^α; to avoid cluttering the notation, we usually leave out the superscript, inserting it only when the dependence on α needs to be emphasized. Also, the integral for $\|\Psi_z\|^2$ actually converges for all $\alpha > 0$. Although only wavelets with $\alpha > 2$ give resolutions of unity, it may well be that wavelets with $0 < \alpha \leq 2$ turn out to have some significance, and therefore they should not be ruled out.

We choose the branch cut for the square-root function in (11.38) along the positive real axis, and the branch cut of $g(w) = w^\alpha$ along the negative real axis. Note that Ψ is independent of the choice of branches.

Equation (11.39) can be written as

$$\Psi(r,t) = \Psi^-(r,t) + \Psi^+(r,t), \tag{11.40}$$

where

$$\begin{aligned} \Psi^-(r,t) &\equiv \frac{\Gamma(\alpha)}{4\pi^2\, ir}\, \frac{1}{(1-it-ir)^\alpha} \\ \Psi^+(r,t) &\equiv \Psi^-(-r,t). \end{aligned} \tag{11.41}$$

$\Psi^\pm(r,t)$ are singular at $r=0$, and the singularity cancels in $\Psi(r,t)$. Aside from that singularity, $\Psi^-(r,t)$ is concentrated near the set

$$\{(\mathbf{x},t) \in \mathbf{R}^4 : t = -r = -|\mathbf{x}| < 0\} = \partial V'_-\,, \tag{11.42}$$

i.e., near *the boundary of the past cone* in space-time. (Remember that we have set $c=1$.) This is the set of events from which a light signal can be emitted in order to reach the origin $\mathbf{x} = \mathbf{0}$ at $t=0$. Thus

- $\Psi^-(r,t)$ *represents an incoming wavelet that gets absorbed near the origin* $x=0$ *in space-time.*

Similarly, $\Psi^+(r,t)$ is concentrated near

$$\{(\mathbf{x},t) \in \mathbf{R}^4 : t = r = |\mathbf{x}| > 0\} = \partial V'_+\,, \tag{11.43}$$

the boundary of the *future* cone. This is the set of all events that can be reached by light emitted from the origin.

- $\Psi^+(r,t)$ *represents an outgoing wavelet that is emitted near* $x=0$.

But these are just the kind of wavelets we had in mind when constructing the model of radar signal emission and scattering in Section 10.2! Upon scaling, boosting, and translating, $\Psi^-(r,t)$ generates a family of wavelets $\Psi_z^-(x')$ that are interpreted as being absorbed at $x' = x$ and whose center moves with velocity $\mathbf{y}/y_0$. Since Ψ_z^- is not spherically symmetric unless $\mathbf{y} = \mathbf{0}$, its spatial localization at the time of absorption is not uniform. It can be expected to be of the order of $y_0 \pm |\mathbf{y}|$, depending on the direction. (This has to do with the rates of decay of $\theta(p\cdot y)\, e^{-p\cdot y}$ along the different directions in $\mathbf{p}$. See "ray filters" in Chapter 9.) Similarly, $\Psi^+(r,t)$ generates a family of wavelets $\Psi_z^+(x')$ that are interpreted as being emitted at x and traveling with $\mathbf{v} = \mathbf{y}/y_0$.

The wavelets $\Psi_z^\pm$ are not global solutions of the wave equation since they have a singularity at the origin. In fact, using

$$\nabla^2 \frac{1}{r} = -4\pi\delta(\mathbf{x}), \tag{11.44}$$

it can be easily checked that $\Psi^-(r,t)$ and $\Psi^+(r,t)$ are *fundamental solutions* of the wave equation in the following sense:

$$\begin{aligned} -\partial_t^2\Psi^-(r,t)+\nabla^2\Psi^-(r,t) &= -\frac{\Gamma(\alpha)}{i\pi(1-it)^\alpha}\,\delta(\mathbf{x}) \\ -\partial_t^2\Psi^+(r,t)+\nabla^2\Psi^+(r,t) &= \frac{\Gamma(\alpha)}{i\pi(1-it)^\alpha}\,\delta(\mathbf{x}). \end{aligned} \tag{11.45}$$

The second equation is an analytic continuation of the first equation to the other branch of $ir=(-\mathbf{x}^2)^{1/2}$. These remarkable equations have a simple, intuitive interpretation. In view of the above picture, the second equation states that *to emit of Ψ^+, we need to produce an " elementary current"* at the origin given by

$$J(\mathbf{x},t)=\frac{\Gamma(\alpha)}{i\pi(1-it)^\alpha}\,\delta(\mathbf{x})\equiv -i\phi(t)\delta(\mathbf{x}), \tag{11.46}$$

and the first equation states that *the absorption of Ψ^- creates the same current with the opposite sign.* The sum of the two partial wavelets satisfies the wave equation because the two currents cancel. By "elementary current" I mean a rapidly decaying function of one variable, in this case

$$\phi(t)\equiv\frac{\Gamma(\alpha)}{\pi(1-it)^\alpha}, \tag{11.47}$$

concentrated on a line in space-time, in this case, the time axis. Thus $\phi(t)$ is just an old-fashioned *time signal*, detected or transmitted by an "aperture" (e.g., an antenna, a microphone, or loudspeaker) at rest at the origin! This brings us back to the "proxies" of Section 10.2. Evidently, equations (11.45) give a precise relation between the *waves* $\Psi^\pm$ and the *signal* ϕ! Note the simple relation between ϕ and the value of the wave at the origin, as given in (11.39):

$$\phi'(t)=2\pi i\,\Psi(0,t). \tag{11.48}$$

This relation is not an accident. $\phi(t)$ is the coefficient function of $\delta(\mathbf{x})$ in (11.46) because applying ∇^2 to Ψ^- changes the factor $(ir)^{-1}$ in (11.41) to $4\pi i\delta(\mathbf{x})$. Another way of eliminating this factor is to return to (11.39) and take the limit $r\to 0^+$. That involves using L'Hôpital's rule, which amounts to a differentiation with respect to r. But as far as Ψ^- is concerned, differentiation with respect to r has the same effect as differentiation with respect to t.

According to (11.48), the signal ϕ at the origin acts like a *potential* for Ψ. Using the properties of the acoustic wavelets under translations and dilations, it can be shown easily that equations (11.45) imply

$$\begin{aligned} \Box'\,\Psi^-_{\mathbf{x},t+is}(\mathbf{x}',t') &= i\phi_{s,t}(t')\,\delta(\mathbf{x}'-\mathbf{x})\equiv -J_{\mathbf{x},t+is}(x') \\ \Box'\,\Psi^+_{\mathbf{x},t+is}(\mathbf{x}',t') &= -i\phi_{s,t}(t')\,\delta(\mathbf{x}'-\mathbf{x})\equiv J_{\mathbf{x},t+is}(x'), \end{aligned} \tag{11.49}$$

where

$$\square' \equiv -\partial_{t'}^2 + \nabla_{\mathbf{x}'}^2 \quad \text{and} \quad \phi_{s,t}(t') \equiv |s|^{-\alpha}\phi\left(\frac{t'-t}{s}\right). \tag{11.50}$$

By applying a Lorentz transformation to (11.49) we obtain a similar equation for a general $z \in \mathcal{T}$ (with $\mathbf{v} \equiv \mathbf{y}/s$ not necessarily zero), relating $\square'\,\Psi_z^{\pm}(x')$ to a *moving* source $J_z(x')$. J_{x+iy} is supported on the line in $\mathbf{R}^4$ parameterized by

$$x'(\tau) = x + \tau y.$$

This extends the above relation between waves and signals to the whole family of wavelets $\Psi_z^{\pm}$.

Equations (11.49) suggest a practical method for analyzing and synthesizing with acoustic wavelets: The analysis is based on the recognition of the received signal $i\phi_{s,t}$ at $\mathbf{x} \in \mathbf{R}^3$ as the "signature" of $\Psi^-_{\mathbf{x},t+is}$, and the synthesis is based on using $-i\phi_{s,t}$ as an input at $\mathbf{x}$ to *generate* the outgoing wavelet $\Psi^+_{\mathbf{x},t+is}$. This sets up a direct link between the analysis of time signals (such as a line voltage or the pressure at the bridge of a violin) and the physics of the waves carrying these signals, giving strong support to the idea of a marriage between signal analysis and physics, as expressed in Section 9.1. In fact, if we use ϕ as a mother wavelet for the "proxies," then *equations (11.49) give a fundamental relation between the physical wavelets* $\Psi_z^{\pm}$ *and the proxy wavelets* $\phi_{s,t}$: The latter act as sources for the former! By taking linear superpositions, this relation extends to one between arbitrary waves and signals. Consequently, *signal information can be transmitted directly in the time-scale domain, provided we use the proxy wavelets* $\phi_{s,t}$ *compatible with the physical wavelets* Ψ_z.

The signal $\phi(t)$ is centered at $t = 0$ and has a degree of oscillation controlled by α. High values of α mean rapid oscillation and rapid decay in time, with envelope

$$|\phi(t)| = \frac{\Gamma(\alpha)}{\pi\,(1+t^2)^{\alpha/2}}. \tag{11.51}$$

For $\alpha \in \mathbf{N}$, it is easy to compute the Fourier transform of ϕ by contour integration:

$$\begin{aligned}
\hat{\phi}(\xi) &= \frac{\Gamma(\alpha)}{\pi}\int_{-\infty}^{\infty}\frac{dt}{(1-it)^{\alpha}}\,e^{-i\xi t} \\
&= \frac{1}{\pi}\,(-\partial_u)^{\alpha-1}\int_{-\infty}^{\infty}\frac{dt}{u-it}\,e^{-i\xi t}\,\Big|_{u=1} \\
&= \frac{i}{\pi}(-\partial_u)^{\alpha-1}\left[(-2\pi i)\theta(\xi)e^{-\xi u}\right]\Big|_{u=1} \\
&= 2\theta(\xi)\xi^{\alpha-1}e^{-\xi},
\end{aligned} \tag{11.52}$$

where we have closed the contour in the upper-half plane if $\xi < 0$ and in the lower-half plane if $\xi > 0$. But comparison of (11.52) with (11.08) shows that

$\hat{\phi}(\xi)$ *is the Fourier representative* $\psi_{\mathbf{0},i}$ *of the reference wavelet* $\Psi = \Psi_{\mathbf{0},i}$, i.e.,

$$\hat{\phi}(p_0) = 2\theta(p_0)|p_0|^{\alpha-1}e^{-p_0} = \psi_{\mathbf{0},i}(\mathbf{p}, p_0).$$

Referring back to (9.47), this shows that

$$C_\beta \equiv \int_{-\infty}^{\infty} d\xi \ |\xi|^{\beta-1}|\hat{\phi}(\xi)|^2 = 2^{4-2\alpha-\beta}\Gamma(2\alpha+\beta-2) < \infty$$

whenever $2\alpha + \beta > 2$. Therefore ϕ is admissible in the sense of (9.47), and it can be used as a mother wavelet for the proxy Hilbert space

$$\mathcal{H}_\beta^{\text{proxy}} = \{f : \mathbf{R} \to \mathbf{C} : \int_{-\infty}^{\infty} \frac{d\xi}{|\xi|^\beta} \ |\hat{f}(\xi)|^2 < \infty\}, \quad \beta > 2 - 2\alpha.$$

We can now return to Section 10.2 and fix a gap (one of many) in our discussion of electromagnetic scattering there. We postulated that the electromagnetic wavelets $\mathbf{\Psi}_z$ can be split up into an absorbed part $\mathbf{\Psi}_z^-$ and an emitted part $\mathbf{\Psi}_z^+$. This can now be established. Since the matrix elements of the electromagnetic wavelets were obtained by differentiating $S(y)$, the same splitting (11.29) that resulted in the splitting of Ψ_z also gives rise to a splitting of $\mathbf{\Psi}_z$. The partial wavelets $\mathbf{\Psi}_z^\pm$ are fundamental solutions of Maxwell's equations, the right-hand sides now representing elementary *electrical* charges and currents. (In electromagnetics, the charge density and current combine into a four-vector transforming naturally under Lorentz transformations.) Superpositions like (10.95) and (10.99) may perhaps be interpreted as modeling absorption and reemission of wavelets by scatterers. In that case, the "elementary reflectors" of Section 10.2 ought to be reexamined in terms of elementary electrical charges and currents.

11.3 Nonunitarity and the Rest Frame of the Medium

We noted that the scalar wavelets (9.179) associated with electromagnetics are a special case of (11.9) with $\alpha = 3$. The choice $\alpha = 3$ was dictated by the requirement that the inner product in $\mathcal{H}$ be invariant under the conformal group, which implies that the representation of $\mathcal{C}$ on $\mathcal{H}$ is unitary. It is important for Lorentz transformations to be represented by unitary operators in *free-space* electromagnetics since there is no preferred reference frame in free space. If the operators representing Lorentz transformations are not unitary, they can change the norm ("energy") of a solution. We can then choose solutions minimizing the norm, and that defines a preferred reference frame. Let us now see how the norm of Ψ_z depends on the speed $v = |\mathbf{v}| = |\mathbf{y}/s| = u/s < 1$. By (11.25) and (11.30),

$$\|\Psi_z\|^2 = \frac{\Gamma(\alpha)}{2\pi^2(2s)^{\alpha+1}\,v}\left[\frac{1}{(1-v)^\alpha} - \frac{1}{(1+v)^\alpha}\right], \quad z \in \mathcal{T}_+ . \tag{11.53}$$

Now

$$\lambda(y) \equiv \sqrt{y^2} = s\sqrt{1-v^2} \tag{11.54}$$

is invariant under Lorentz transformations. Therefore we eliminate s in favor of λ:

$$\begin{aligned} \|\Psi_z\|^2 &= \frac{\Gamma(\alpha)(1-v^2)^{(\alpha+1)/2}}{2\pi^2(2\lambda)^{\alpha+1}\,v}\left[\frac{1}{(1-v)^\alpha} - \frac{1}{(1+v)^\alpha}\right] \\ &= \frac{\Gamma(\alpha)\sqrt{1-v^2}}{2\pi^2(2\lambda)^{\alpha+1}\,v}\left[\left(\frac{1+v}{1-v}\right)^\alpha - \left(\frac{1-v}{1+v}\right)^\alpha\right]. \end{aligned} \tag{11.55}$$

For the statement of the following theorem, we reinsert c by dimensional analysis. The measure $d\tilde{p}$ has units $[l^{-1}t^{-1}]$, where $[t]$ and $[l]$ are the units of time and length, respectively; hence $\|\Psi_z\|^2$ has units $[l^{-1}t^{-\alpha}]$. Since λ has units of length, we multiply (11.55) by c^α, and replace v by v/c.

Proposition 11.2. *The unique value $\alpha \geq 1$ for which $\|\Psi_z\|^2$ is invariant under Lorentz transformations is $\alpha = 1$. For $\alpha > 1$, $\|\Psi_z\|^2$ is a monotonically increasing function of the speed $v(y) = |\mathbf{y}/y_0|$ of the center of the wavelet, given by*

$$\|\Psi_z\|^2 = \frac{\Gamma(\alpha)\,c^\alpha}{\pi^2(2\lambda)^{\alpha+1}}\,\frac{\sinh(\alpha\theta)}{\sinh\theta}, \tag{11.56}$$

where θ is defined by

$$\frac{v}{c} = \tanh\theta. \tag{11.57}$$

Proof: We have mentioned that v is the hyperbolic tangent of a "rotation angle" in the geometry of Minkowskian space-time. This suggests the change of variables (11.57) in (11.55), which indeed results in the reduction to (11.56). Since λ is Lorentz-invariant, only θ depends on v in (11.56). Hence the expression is independent of v if and only if $\alpha = 1$. For $\alpha > 1$, let

$$f(\theta) = \frac{\sinh(\alpha\theta)}{\sinh\theta}. \tag{11.58}$$

We must show that $f(\theta)$ is a monotonically increasing function of θ. The derivative of f is

$$f'(\theta) = \frac{\alpha\sinh\theta\,\cosh(\alpha\theta) - \cosh\theta\sinh(\alpha\theta)}{\sinh^2\theta}. \tag{11.59}$$

Therefore it suffices to prove that

$$\tanh(\alpha\theta) < \alpha\tanh\theta, \qquad \theta > 0,\ \alpha > 1. \tag{11.60}$$

Begin with the fact that

$$\cosh(\alpha\theta) > \cosh\theta, \qquad \theta > 0, \tag{11.61}$$

which follows from the monotonicity of the hyperbolic cosine in $[0, \infty)$. Thus

$$\operatorname{sech}^2(\alpha\theta') < \operatorname{sech}^2\theta', \qquad \theta' > 0. \tag{11.62}$$

Integrating both sides over $0 \leq \theta' \leq \theta$ gives (11.60). ■

Equation (11.56) can be expanded in powers of v/c. The first two terms are

$$\|\Psi_z\|^2 = \frac{\Gamma(\alpha+1)\,c^\alpha}{\pi^2(2\lambda)^{\alpha+1}}\left[1+\frac{\alpha^2-1}{6}\frac{v^2}{c^2}+\cdots\right]. \tag{11.63}$$

This is similar to the formula for the nonrelativistic approximation to the energy of a relativistic particle, whose first two terms are the rest energy mc^2 and the kinetic energy $mv^2/2$. The wavelets Ψ_z behave much like particles: their parameters include position and velocity, the classical phase space parameters of particles. The time and scale parameters, however, are *extensions* of the classical concept. Classical particles are perfectly localized (i.e., they have vanishing scale) and always "focused," making a time parameter superfluous. (The time parameter used to give the evolution of classical particles corresponds to t' in $\Psi_{x+iy}(x')$, and *not* to the time component t of x!) The wavelets are therefore a bridge between the particle concept and the wave concept. It is this particle-like quality that makes it possible to model their scattering in classical, geometric terms.

Suppose that $\alpha > 1$. Fix any $z = (\mathbf{x}, is) \in E$, and consider the whole family of wavelets that can be obtained from Ψ_z by applying Lorentz transformations, i.e., viewing Ψ_z from reference frames in all possible states of uniform motion. According to Proposition 11.2, Ψ_z has a lower energy than any of these other wavelets, since all the other wavelets have the same value of $\sqrt{(y')^2} = \lambda = |s|$ but only Ψ_z has $v = 0$. This defines a unique reference frame, which we interpret to be the frame in which the medium is at rest. Since Lorentz transformations do not preserve the norm of Ψ_z, the representation of $\mathcal{C}$ on $\mathcal{H}_\alpha$ cannot be unitary when $\alpha > 1$. (Recall that the electromagnetic representation of $\mathcal{C}$ in Chapter 9 was unitary, and the associated scalar wavelets had $\alpha = 3$. This is not a contradiction since those wavelets did not *belong* to $\mathcal{H}$ but were, rather, auxiliary objects.) The fact that $\alpha = 1$ is the unique choice giving a unitary representation of Lorentz transformations is well known to physicists (Bargmann and Wigner [1948]). For electromagnetics, the unique unitary representation is the one in Chapter 9, with $\alpha = 3$. It may well be that electromagnetic representations of $\mathcal{C}$ with $\alpha \neq 3$ (which can be constructed as for acoustics) can be used to model electromagnetic phenomena in *media*. If such models turn out to correspond to reality, then a physical interpretation of α in relation to the medium may be possible.

11.4 Examples of Acoustic Wavelets

Figures 11.1–11.8 show various aspects of $\Psi(r,t)$ and $\Psi^{\pm}(r,t)$ for $\alpha = 3, 10, 15$, and 50.

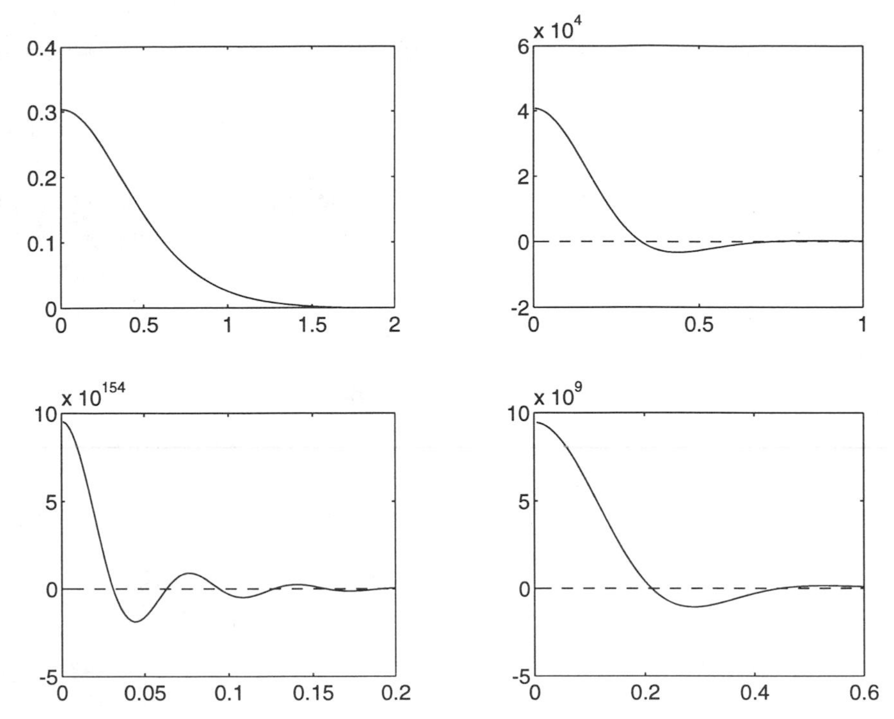

Figure 11.1. *Clockwise, from top left*: $\Psi(r, 0)$ with $\alpha = 3, 10, 15$, and 50, showing excellent localization at $t = 0$. (Note that $\Psi(r, 0)$ is real.)

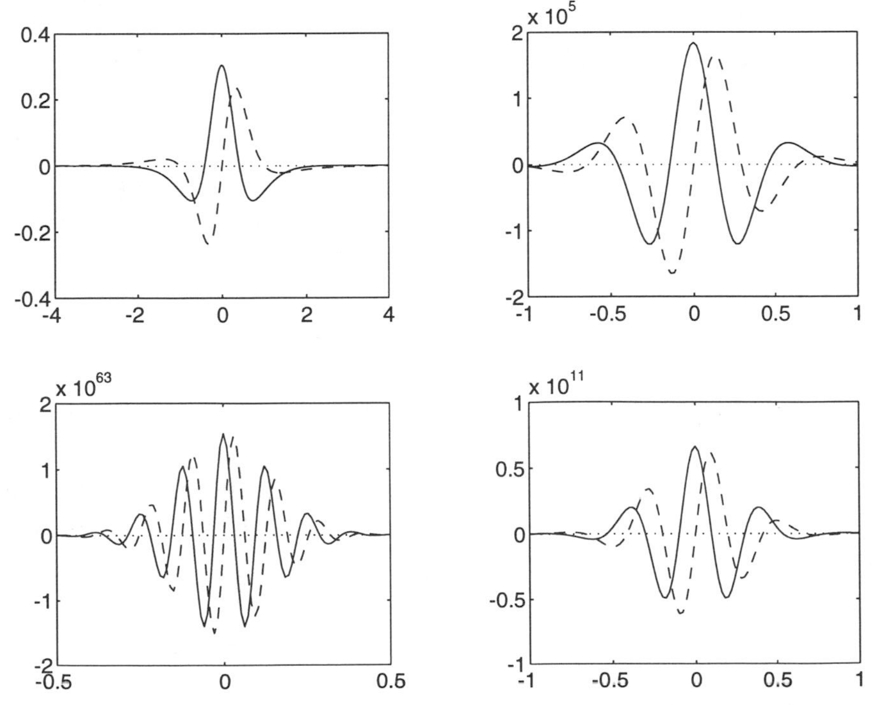

Figure 11.2. *Clockwise, from top left*: The real part (*solid*) and imaginary part (*dashed*) of $\Psi(0, t)$ with $\alpha = 3, 10, 15$, and 50, showing increasing oscillation and decreasing duration.

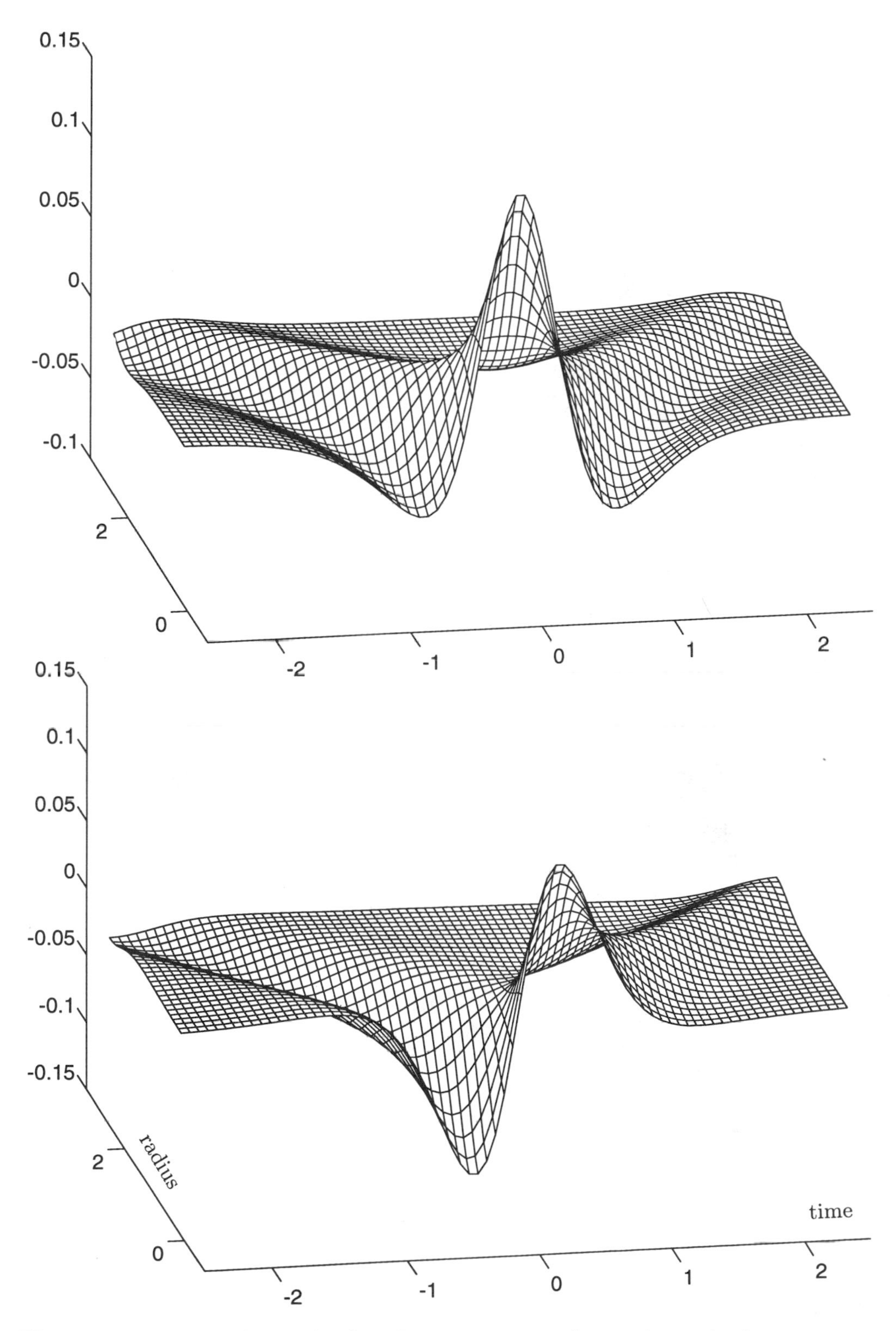

Figure 11.3. The real part (*top*) and imaginary part (*bottom*) of $\Psi(r, t)$ with $\alpha = 3$.

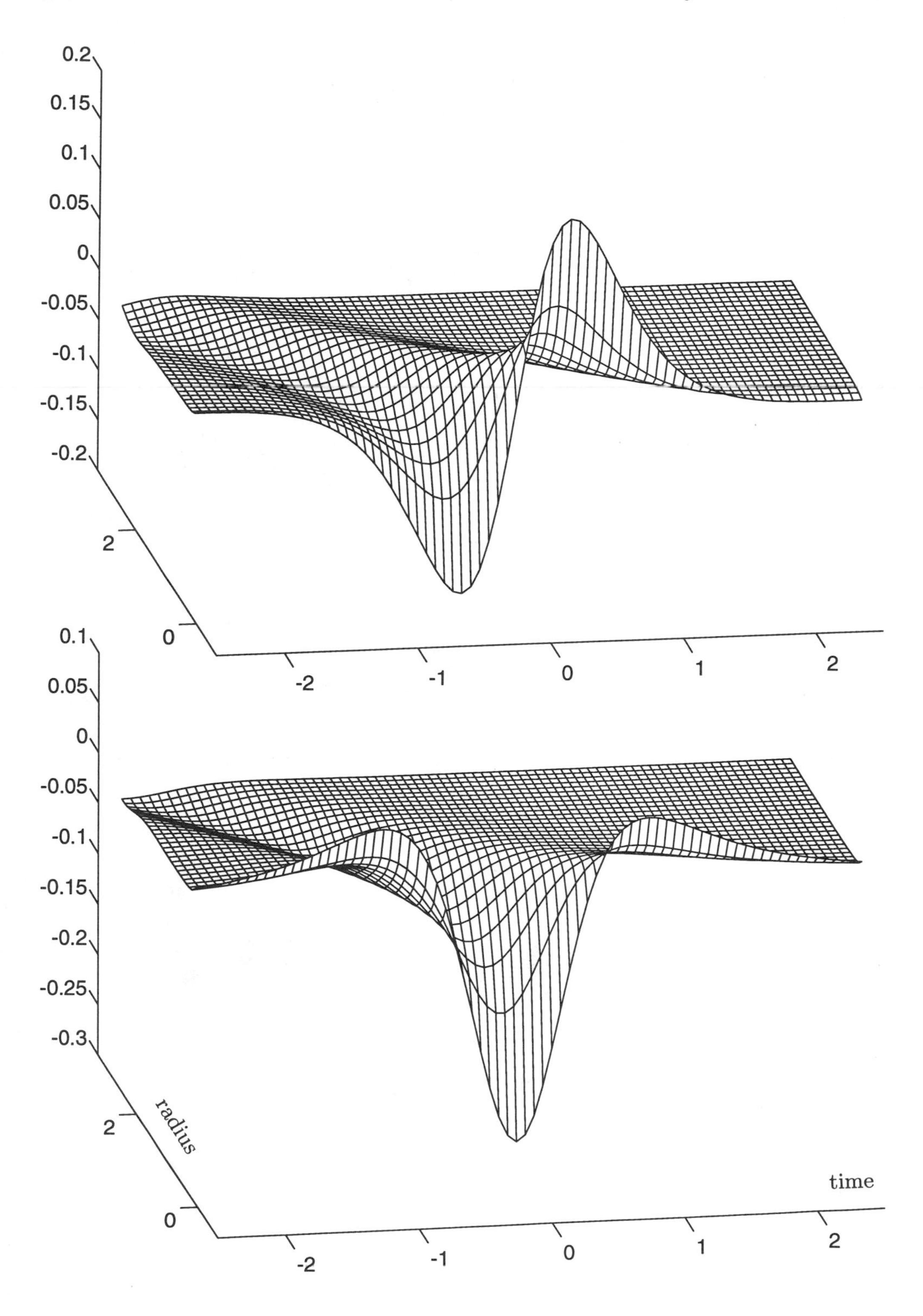

Figure 11.4. The real part (*top*) and imaginary part (*bottom*) of the absorbed (or detected) wavelet $\Psi^-(r,t)$ with $\alpha = 3$.

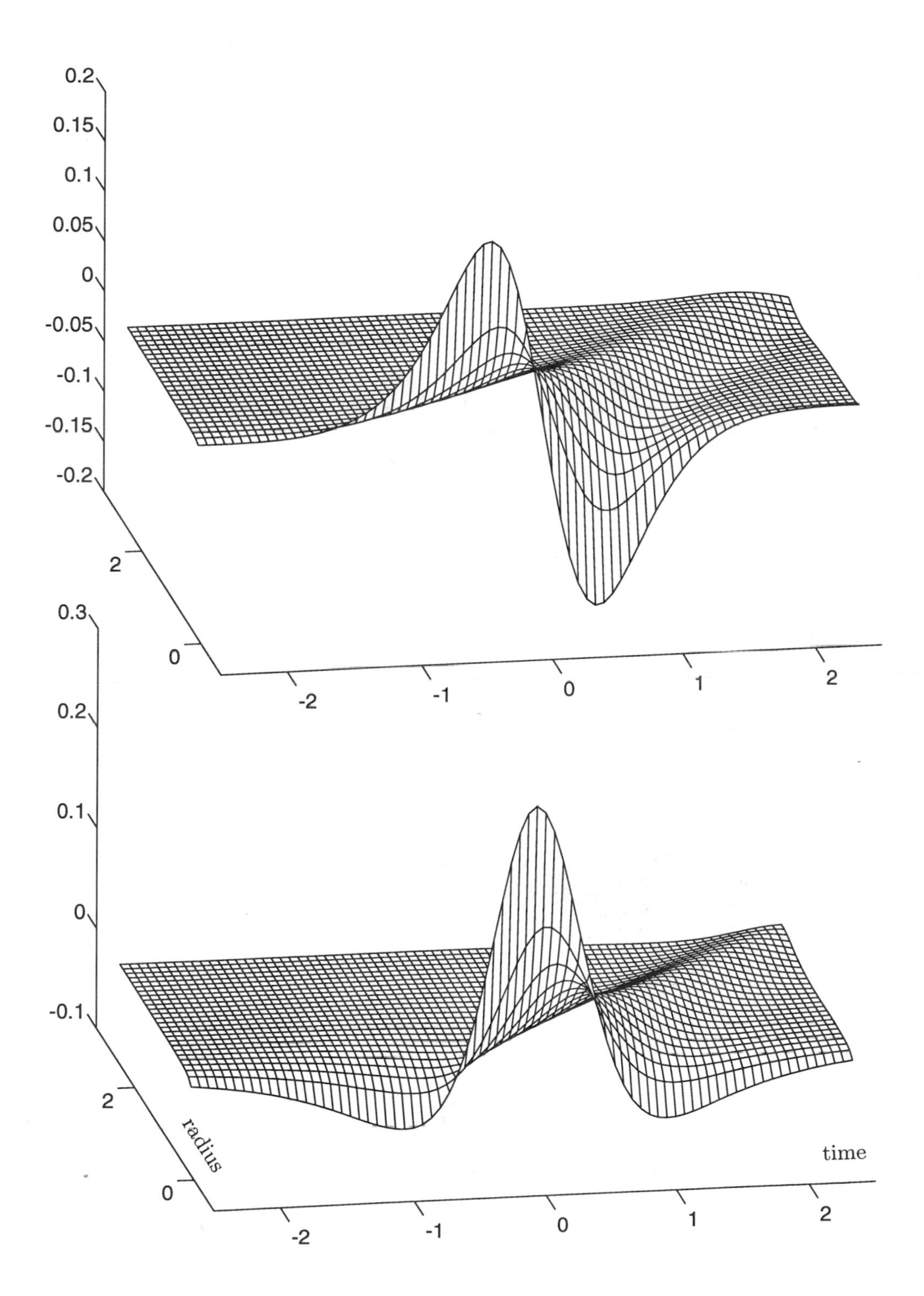

Figure 11.5. The real part (*top*) and imaginary part (*bottom*) of the emitted wavelet $\Psi^{+}(r, t)$ with $\alpha = 3$.

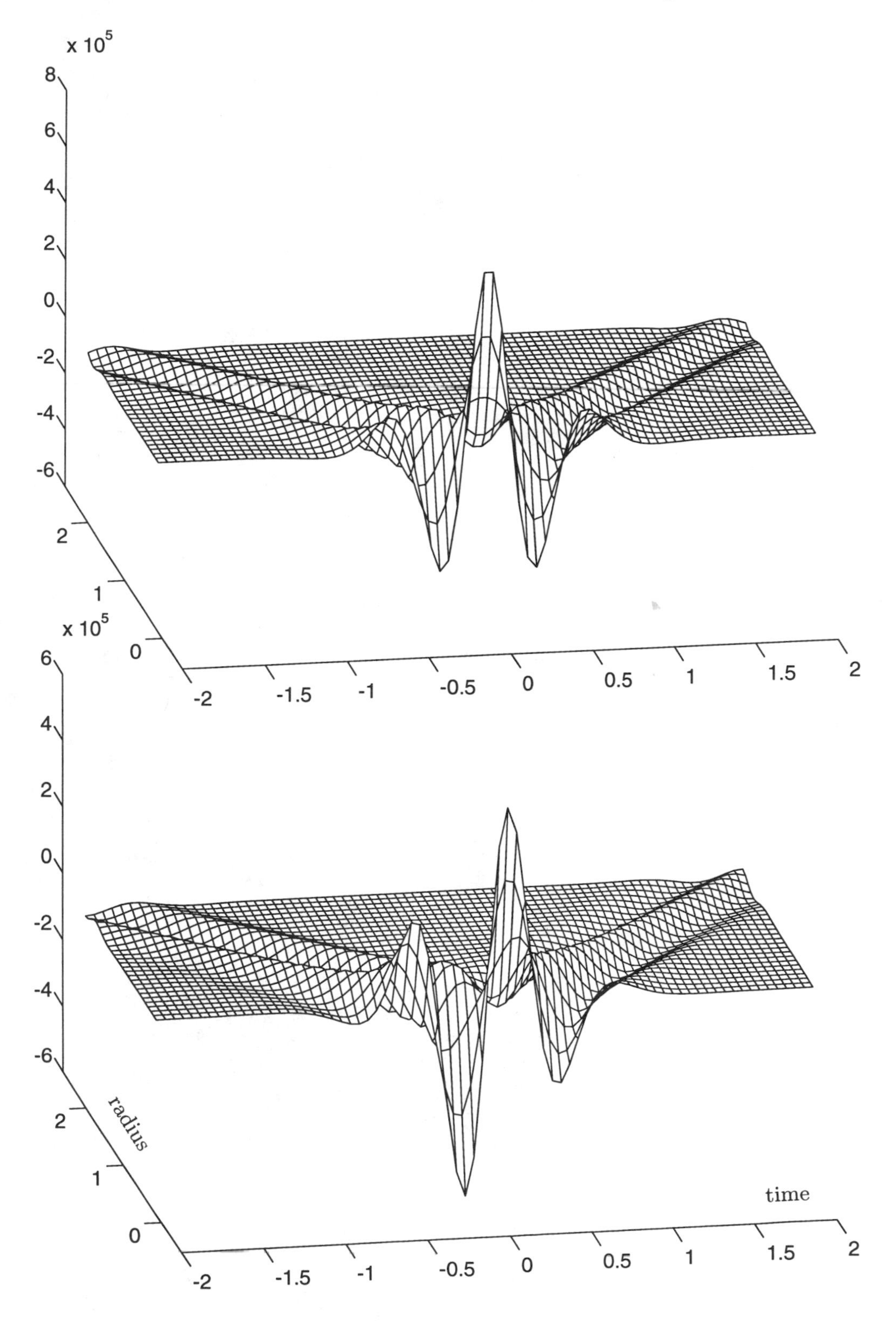

Figure 11.6. The real part (*top*) and imaginary part (*bottom*) of $\Psi(r, t)$ with $\alpha = 10$.

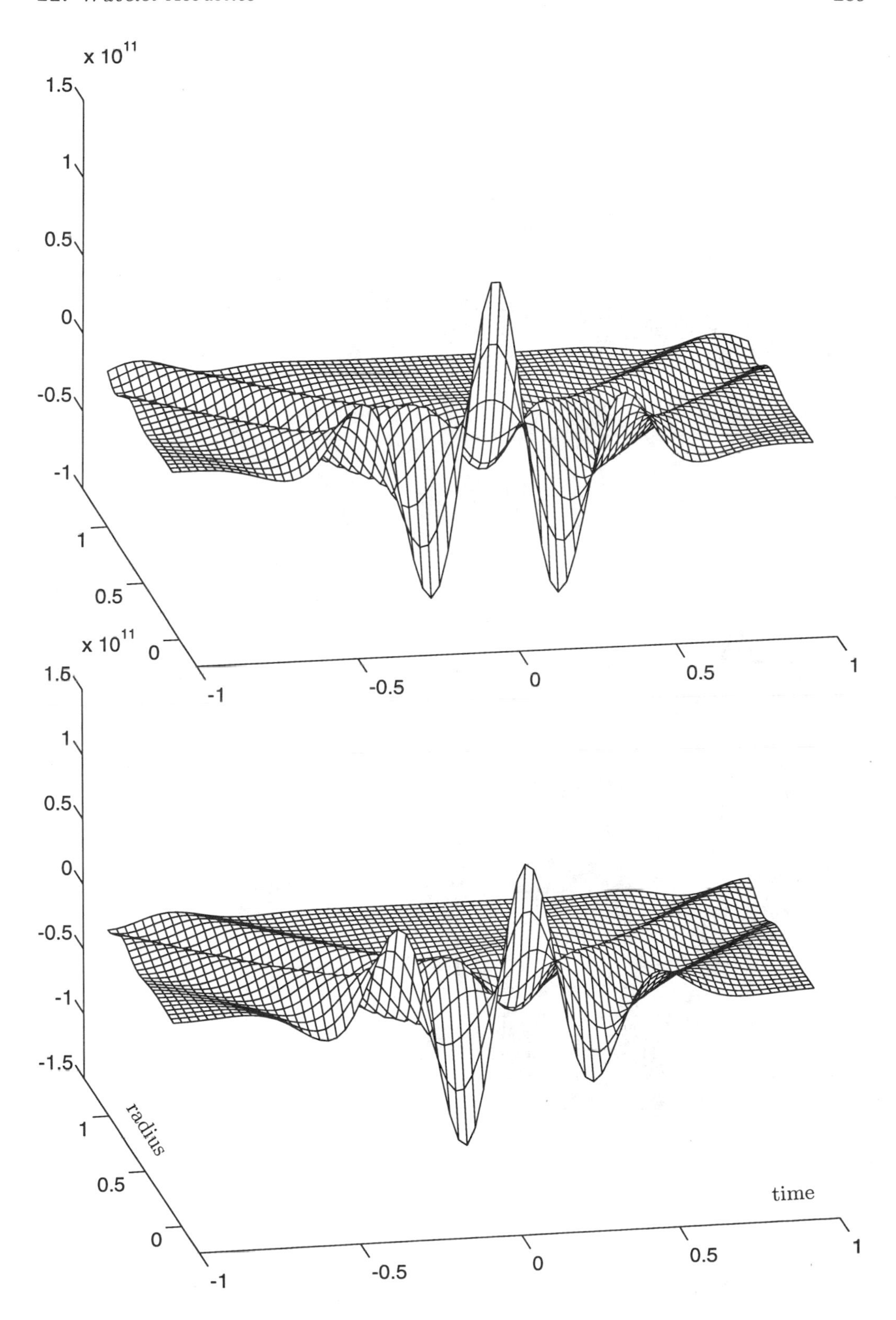

Figure 11.7. The real part (*top*) and imaginary part (*bottom*) of $\Psi(r,t)$ with $\alpha = 15$.

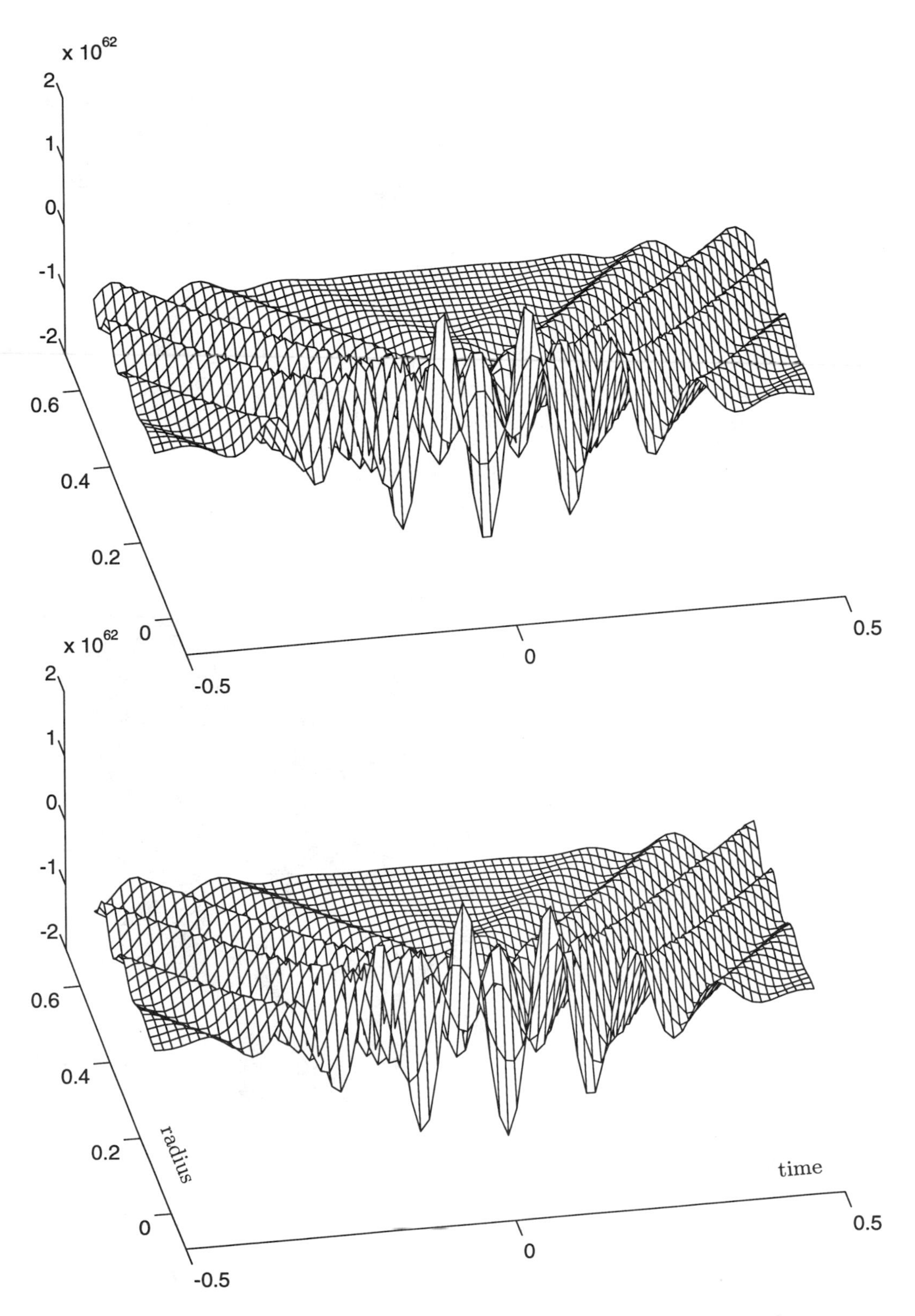

Figure 11.8. The real part (*top*) and imaginary part (*bottom*) of $\Psi(r,t)$ with $\alpha = 50$.

References

Ackiezer NI, *Theory of Approximation,* Ungar, New York, 1956.

Ackiezer NI and Glazman IM, *Theory of Linear Operators in Hilbert Space,* Ungar, New York, 1961.

Aslaksen EW and Klauder JR, Unitary representations of the affine group, *J. Math. Phys.* **9**(1968), 206–211.

Aslaksen EW and Klauder JR, Continuous representation theory using the affine group, *J. Math. Phys.* **10**(1969), 2267–2275.

Auslander L and Gertner I, Wideband ambiguity functions and the $a \cdot x + b$ group, in Auslander L, Grünbaum F A, Helton J W, Kailath T, Khargonekar P, and Mitter S, eds., *Signal Processing: Part I – Signal Processing Theory,* Springer-Verlag, New York, pp. 1–12, 1990.

Backus J, *The Acoustical Foundations of Music,* second edition, Norton, New York, 1977.

Bacry H, Grossmann A, and Zak J, Proof of the completeness of lattice states in the kq-representation, *Phys. Rev. B* **12**(1975), 1118–1120.

Balian R, Un principe d'incertitude fort en théorie du signal ou en mécanique quantique, *C. R. Acad. Sci. Paris,* **292**(1981), Série 2.

Baraniuk RG and Jones DL, New orthonormal bases and frames using chirp functions, *IEEE Transactions on Signal Processing* **41**(1993), 3543–3548.

Bargmann V and Wigner EP, Group theoretical discussion of relativistic wave equations, *Proc. Nat. Acad. Sci. USA* **34**(1948), 211–233.

Bargmann V, Butera P, Girardello L, and Klauder JR, On the completeness of coherent states, *Reps. Math. Phys.* **2**(1971), 221–228.

Barton DK, *Modern Radar System Analysis,* Artech House, Norwood, MA, USA, 1988.

Bastiaans MJ, Gabor's signal expansion and degrees of freedom of a signal, *Proc. IEEE* **68**(1980), 538–539.

Bastiaans MJ, A sampling theorem for the complex spectrogram and Gabor's expansion of a signal in Gaussian elementary signals, *Optical Engrg.* **20**(1981), 594–598.

Bastiaans MJ, Gabor's signal expansion and its relation to sampling of the sliding-window spectrum, in Marks II JR (1993), ed., *Advanced Topics in Shannon Sampling and Interpolation Theory,* Springer-Verlag, Berlin, 1993.

Bateman H, The transformation of the electrodynamical equations, *Proc. London Math. Soc.* **8**(1910), 223–264.

Battle G, A block spin construction of ondelettes. Part I: Lemarié functions, *Comm. Math. Phys.* **110**(1987), 601–615.

Battle G, Heisenberg proof of the Balian-Low theorem, *Lett. Math. Phys.* **15**(1988), 175–177.

Battle G, Wavelets: A renormalization group point of view, in Ruskai MB, Beylkin G, Coifman R, Daubechies I, Mallat S, Meyer Y, and Raphael L, eds., *Wavelets and their Applications,* Jones and Bartlett, Boston, 1992.

Benedetto JJ and Walnut DF, Gabor frames for L^2 and related spaces, in Benedetto JJ and Frazier MW, eds., *Wavelets: Mathematics and Applications,* CRC Press, Boca Raton, 1993.

Benedetto JJ and Frazier MW, eds., *Wavelets: Mathematics and Applications,* CRC Press, Boca Raton, 1993.

Bernfeld M, CHIRP Doppler radar, *Proc. IEEE* **72**(1984), 540–541.

Bernfeld M, On the alternatives for imaging rotational targets, in *Radar and Sonar, Part II,* Grünbaum FA, Bernfeld M and Bluhat RE, eds., Springer-Verlag, New York, 1992.

Beylkin G, Coifman R, and Rokhlin V, Fast wavelet transforms and numerical algorithms, *Comm. Pure Appl. Math.* **44**(1991), 141–183.

Białynicki-Birula I and Mycielski J, Uncertainty relations for information entropy, *Commun. Math. Phys.* **44**(1975), 129.

de Boor C, DeVore RA, and Ron A, Approximation from shift-invariant subspaces of $L^2(\mathbf{R}^d)$, *Trans. Amer. Math. Soc.* **341**(1994), 787–806.

Born M and Wolf E, *Principles of Optics,* Pergamon, Oxford, 1975.

Chui CK, *An Introduction to Wavelets,* Academic Press, New York, 1992a.

Chui CK, ed., *Wavelets: A Tutorial in Theory and Applications,* Academic Press, New York, 1992b.

Chui CK and Shi X, Inequalities of Littlewood-Paley type for frames and wavelets, *SIAM J. Math. Anal.* **24**(1993), 263–277.

Cohen A, Ondelettes, analyses multirésolutions et filtres miroir en quadrature, *Inst. H. Poincaré, Anal. non linéare* **7**(1990), 439–459.

Coifman R, Meyer Y, and Wickerhauser MV, Wavelet analysis and signal processing, in Ruskai MB, Beylkin G, Coifman R, Daubechies I, Mallat S, Meyer Y, and Raphael L, eds., *Wavelets and their Applications,* Jones and Bartlett, Boston, 1992.

Coifman R and Rochberg R, Representation theorems for holomorphic and harmonic functions in L^p, *Astérisque* **77**(1980), 11–66.

Cook CE and Bernfeld M, *Radar Signals,* Academic Press, New York; republished by Artech House, Norwood, MA, 1993 (1967).

Cunningham E, The principle of relativity in electrodynamics and an extension thereof, *Proc. London Math. Soc.* **8**(1910), 77–98.

Daubechies I, Grossmann A, and Meyer Y, Painless non-orthogonal expansions, *J. Math. Phys.* **27**(1986), 1271–1283.

Daubechies I, Orthonormal bases of compactly supported wavelets, *Comm. Pure Appl. Math.* **41**(1988), 909–996.

Daubechies I, The wavelet transform, time-frequency localization and signal analysis, *IEEE Trans. Inform. Theory* **36**(1990), 961–1005.

Daubechies I, *Ten Lectures on Wavelets,* SIAM, Philadelphia, 1992.

Deslauriers G and Dubuc S, Interpolation dyadique, in *Fractals, dimensions non enti'eres et applications,* G. Cherbit, ed., Masson, Paris, pp. 44–45, 1987.

Dirac PA M, *Quantum Mechanics,* Oxford University Press, Oxford, 1930.

Duffin RJ and Schaeffer AC, A class of nonharmonic Fourier series, *Trans. Amer. Math. Soc.* **72**(1952), 341–366.

Einstein A, Lorentz HA, Weyl H, *The Principle of Relativity,* Dover, 1923.

Feichtinger HG and Gröchenig K, A unified approach to atomic characterizations via integrable group representations, in *Proc. Conf. Lund, June 1986, Lecture Notes in Math.* 1302, (1986).

Feichtinger HG and Gröchenig K, Banach spaces related to integrable group representations and their atomic decompositions, I, *J. Funct. Anal.* **86**(1989), 307–340, 1989a.

Feichtinger HG and Gröchenig K, Banach spaces related to integrable group representations and their atomic decompositions, II, *Monatsch. f. Mathematik* **108**(1989), 129–148, 1989b.

Feichtinger HG and Gröchenig K, Gabor wavelets and the Heisenberg group: Gabor expansions and short-time Fourier transform from the group theoretical point of view, in Chui CK (1992), ed., *Wavelets: A Tutorial in Theory and Applications,* Academic Press, New York, 1992.

Feichtinger HG and Gröchenig K, Theory and practice of irregular sampling, in Benedetto JJ and Frazier MW (1993), eds., *Wavelets: Mathematics and Applications,* CRC Press, Boca Raton, 1993.

Folland GB, *Harmonic Analysis in Phase Space,* Princeton University Press, Princeton, NJ, 1989.

Gabor D, Theory of communication, *J. Inst. Electr. Eng.* **93**(1946) (III), 429–457.

Gel'fand IM and Shilov GE, *Generalized Functions,* Vol. I, Academic Press, New York, 1964.

Glimm J and Jaffe A (1981) *Quantum Physics: A Functional Point of View,* Springer-Verlag, New York.

Gnedenko BV and Kolmogorov AN, *Limit Distributions for Sums of Independent Random Variables,* Addison-Wesley, Reading, MA, 1954.

Gröchenig K, Describing functions: atomic decompositions versus frames, *Monatsch. f. Math.* **112**(1991), 1–41.

Gröchenig K, Acceleration of the frame algorithm, to appear in *IEEE Trans. Signal Proc,* 1993.

Gross L, Norm invariance of mass-zero equations under the conformal group, *J. Math. Phys.* **5**(1964), 687–695.

Grossmann A and Morlet J, Decomposition of Hardy functions into square-integrable wavelets of constant shape, *SIAM J. Math. Anal.* **15**(1984), 723–736.

Heil C and Walnut D, Continuous and discrete wavelet transforms, *SIAM Rev.* **31** (1989), 628–666.

Hill EL, On accelerated coordinate systems in classical and relativistic mechanics, *Phys. Rev.* **67**(1945), 358–363.

Hill EL, On the kinematics of uniformly accelerated motions and classical electromagnetic theory, *Phys. Rev.* **72**(1947), 143–149;

Hill EL, The definition of moving coordinate systems in relativistic theories, *Phys. Rev.* **84**(1951), 1165–1168.

Hille E, Reproducing kernels in analysis and probability, *Rocky Mountain J. Math.* **2**(1972), 319–368.

Jackson JD, *Classical Electrodynamics,* Wiley, New York, 1975.

Jacobsen HP and Vergne M, Wave and Dirac operators, and representations of the conformal group, *J. Funct. Anal.* **24**(1977), 52–106.

Janssen AJEM, Gabor representations of generalized functions, *J. Math. Anal. Appl.* **83**(1981), 377–394.

Kaiser G, *Phase-Space Approach to Relativistic Quantum Mechanics,* Thesis, University of Toronto Department of Mathematics, 1977a.

Kaiser G, Phase-space approach to relativistic quantum mechanics, Part I: Coherent-state representations of the Poincaré group, *J. Math. Phys.* **18**(1977), 952–959, 1977b.

Kaiser G, Phase-space approach to relativistic quantum mechanics, Part II: Geometrical aspects, *J. Math. Phys.* **19**(1978), 502–507, 1978a.

Kaiser G, Local Fourier analysis and synthesis, *University of Lowell preprint* (unpublished). Originally NSF Proposal #MCS-7822673, 1978b.

Kaiser G, Phase-space approach to relativistic quantum mechanics, Part III: Quantization, relativity, localization and gauge freedom, *J. Math. Phys.* **22**(1981), 705–714.

Kaiser G, A sampling theorem in the joint time-frequency domain, *University of Lowell preprint* (unpublished), 1984.

Kaiser G, Quantized fields in complex spacetime, *Ann. Phys.* **173**(1987), 338–354.

Kaiser G, *Quantum Physics, Relativity, and Complex Spacetime: Towards a New Synthesis,* North-Holland, Amsterdam, 1990a.

Kaiser G, Generalized wavelet transforms, Part I: The windowed X-ray transform, *Technical Reports Series #18,* Mathematics Department, University of Lowell. Part II: The multivariate analytic-signal transform, *Technical Reports Series #19,* Mathematics Department, University of Lowell, 1990b.

Kaiser G, An algebraic theory of wavelets, Part I: Complex structure and operational calculus, *SIAM J. Math. Anal.* **23**(1992), 222–243, 1992a.

Kaiser G, Wavelet electrodynamics, *Physics Letters A* **168**(1992), 28–34, 1992b.

Kaiser G, Space-time-scale analysis of electromagnetic waves, in *Proc. of IEEE-SP Internat. Symp. on Time-Frequency and Time-Scale Analysis,* Victoria, Canada, 1992c.

Kaiser G and Streater RF, Windowed Radon transforms, analytic signals and the wave equation, in Chui CK, ed., *Wavelets: A Tutorial in Theory and Applications,* Academic Press, New York, pp. 399–441, 1992.

Kaiser G, Wavelet electrodynamics, in Meyer Y and Roques S, eds., *Progress in Wavelet Analysis and Applications,* Editions Frontières, Paris, pp. 729–734 (1993).

Kaiser G, Deformations of Gabor frames, *J. Math. Phys.* **35**(1994), 1372–1376, 1994a.

Kaiser G, *Wavelet Electrodynamics: A Short Course,* Lecture notes for course given at the Tenth Annual ACES (Applied Computational Electromagnetics Society) Conference, March 1994, Monterey, CA, 1994b.

Kaiser G, Wavelet electrodynamics, Part II: Atomic composition of electromagnetic waves, *Applied and Computational Harmonic Analysis* **1**(1994), 246–260 (1994c).

Kaiser G, *Cumulants: New Path to Wavelets?* UMass Lowell preprint, work in progress, (1994d).

Kalnins EG and Miller W, A note on group contractions and radar ambiguity functions, in Grünbaum FA, Bernfeld M, and Bluhat RE, eds., *Radar and Sonar, Part II,* Springer-Verlag, New York, 1992.

Katznelson Y, *An Introduction to Harmonic Analysis,* Dover, New York, 1976.

Klauder JR and Surarshan ECG, *Fundamentals of Quantum Optics,* Benjamin, New York, 1968.

Lawton W, Necessary and sufficient conditions for constructing orthonormal wavelet bases, *J. Math. Phys.* **32**(1991), 57–61.

Lemarié PG, Une nouvelle base d'ondelettes de $L^2(\mathbf{R}^n)$, *J. de Math. Pure et Appl.* **67**(1988), 227–236.

Low F, Complete sets of wave packets, in *A Passion for Physics – Essays in Honor of Godfrey Chew,* World Scientific, Singapore, pp. 17–22, 1985.

Lyubarskii Yu I, Frames in the Bargmann space of entire functions, in *Entire and Subharmonic Functions, Advances in Soviet Mathematics* **11**(1989), 167–180.

Maas P, Wideband approximation and wavelet transform, in Grünbaum FA, Bernfeld M, and Bluhat RE, eds., *Radar and Sonar, Part II,* Springer-Verlag, New York, 1992.

Mallat S, Multiresolution approximation and wavelets, *Trans. Amer. Math. Soc.* **315** (1989), 69–88.

Mallat S and Zhong S, Wavelet transform scale maxima and multiscale edges, in Ruskai MB, Beylkin G, Coifman R, Daubechies I, Mallat S, Meyer Y, and Raphael L, eds., *Wavelets and their Applications,* Jones and Bartlett, Boston, 1992.

Messiah A, *Quantum Mechanics,* North-Holland, Amsterdam, 1961.

Meyer Y, *Wavelets and Operators,* Cambridge University Press, Cambridge, 1993a.

Meyer Y, *Wavelets: Algorithms and Applications,* SIAM, Philadelphia, 1993b.

Meyer Y and Roques S, eds., *Progress in Wavelet Analysis and Applications,* Editions Frontières, Paris, 1993.

Miller W, Topics in harmonic analysis with applications to radar and sonar, in Bluhat RE, Miller W, and Wilcox CH, eds., *Radar and Sonar, Part I,* Springer-Verlag, New York, 1991.

Morlet J, Sampling theory and wave propagation, in NATO ASI Series, *Vol. I, Issues in Acoustic Signal/Image Processing and Recognition,* Chen CH, ed., Springer-Verlag, Berlin, 1983.

Moses HE, Eigenfunctions of the curl operator, rotationally invariant Helmholtz theorem, and applications to electromagnetic theory and fluid mechanics, *SIAM J. Appl. Math.* **21**(1971), 114–144.

Naparst H, Radar signal choice and processing for a dense target environment, Ph. D. thesis, University of California, Berkeley, 1988.

Page L, A new relativity, *Phys. Rev.* **49**(1936), 254–268.

Papoulis A, *The Fourier Integral and its Applications,* McGraw-Hill, New York, 1962.

Papoulis A, *Signal Analysis,* McGraw-Hill, New York, 1977.

Papoulis A, *Probability, Random Variables, and Stochastic Processes,* McGraw-Hill, New York, 1984.

Rihaczek AW, *Principles of High-Resolution Radar,* McGraw-Hill, New York, 1968.

Roederer JG (1975), *Inroduction to the Physics and Psychophysics of Music,* Springer-Verlag, Berlin.

Rudin W, *Real and Complex Analysis,* McGraw-Hill, New York, 1966.

Rühl W, Distributions on Minkowski space and their connection with analytic representations of the conformal group, *Commun. Math. Phys.* **27**(1972), 53–86.

Ruskai MB, Beylkin G, Coifman R, Daubechies I, Mallat S, Meyer Y, and Raphael L, eds., *Wavelets and their Applications,* Jones and Bartlett, Boston, 1992.

Schatzberg A and Deveney AJ, Monostatic radar equation for backscattering from dynamic objects, *preprint,* AJ Deveney and Associates, 355 Boylston St., Boston, MA 02116; 1994.

Seip K and Wallstén R, Sampling and interpolation in the Bargmann-Fock space, *preprint,* Mittag-Leffler Institute, 1990.

Shannon CE, Communication in the presence of noise, *Proc. IRE,* January issue, 1949.

Smith MJ T and Banrwell TP, Exact reconstruction techniques for tree-structured subband coders, *IEEE Trans. Acoust. Signal Speech Process.* **34**(1986), 434–441.

Stein E, *Singular Integrals and Differentiability Properties of Functions,* Princeton University Press, Princeton, NJ, 1970.

Stein E and Weiss G (1971), *Fourier Analysis on Euclidean Spaces,* Princeton University Press, Princeton, NJ.

Strang G and Fix G, *A Fourier analysis of the finite-element method,* in *Constructive Aspects of Functional Analysis,* G. Geymonat, ed., C.I.M.E. II, Ciclo 1971, pp. 793–840, 1973.

Strang G, Wavelets and dilation equations, *SIAM Review* **31**(1989), 614–627.

Strang G, The optimal coefficients in Daubechies wavelets, *Physica D* **60**(1992), 239–244.

Streater RF and Wightman AS, *PCT, Spin and Statistics, and All That,* Benjamin, New York, 1964.

Strichartz RS, Construction of orthonormal wavelets, in Benedetto JJ and Frazier MW, eds., *Wavelets: Mathematics and Applications,* CRC Press, Boca Raton, 1993.

Swick DA, An ambiguity function independent of assumption about bandwidth and carrier frequency, *NRL Report 6471,* Washington, DC, 1966.

Swick DA, A review of wide-band ambiguity functions, *NRL Report 6994,* Washington, DC, 1969.

Tewfik AH and Hosur S, Recent progress in the application of wavelets in surveillance systems, *Proc. SPIE Conf. on Wavelet Applications,* 1994.

Vetterli M, Filter banks allowing perfect reconstruction, *Signal Process.* **10**(1986), 219–244.

Ward RS and Wells RO, *Twistor Geometry and Field Theory,* Cambridge University Press, Cambridge, 1990.

Woodward PM, *Probability and Information Theory, with Applications to Radar,* Pergamon Press, London, 1953.

Young RM, *An Introduction to Nonharmonic Fourier Series,* Academic Press, New York, 1980.

Zakai M, *Information and Control* **3**(1960), 101.

Zou H, Lu J, and Greenleaf F, Obtaining limited diffraction beams with the wavelet transform, *Proc. IEEE Ultrasonic Symposium,* Baltimore, MD, 1993a.

Zou H, Lu J, and Greenleaf F, A limited diffraction beam obtained by wavelet theory, *preprint,* Mayo Clinic and Foundation 1993b.

Index

The keywords cite the main references only.